FUZZY LOGIC

TRENDS IN LOGIC
Studia Logica Library

VOLUME 11

SCOPE OF THE SERIES

Trends in Logic is a bookseries covering essentially the same area as the journal *Studia Logica* – that is, contemporary formal logic and its applications and relations to other disciplines. These include artificial intelligence, informatics, cognitive science, philosophy of science, and the philosophy of language. However, this list is not exhaustive, moreover, the range of applications, comparisons and sources of inspiration is open and evolves over time.

The titles published in this series are listed at the end of this volume.

GIANGIACOMO GERLA
Department of Mathematics and Computer Sciences,
University of Salerno, Italy

FUZZY LOGIC

Mathematical Tools for Approximate Reasoning

Springer-Science+Business Media, B.V.

A C.I.P. Catalogue record for this book is available from the Library of Congress.

ISBN 978-90-481-5694-8 ISBN 978-94-015-9660-2 (eBook)
DOI 10.1007/978-94-015-9660-2

Printed on acid-free paper

To my wife Loredana,
to my daughters Brunella and Francesca

CONTENTS

Preface

Fuzzy logic in narrow sense is a promising new chapter of formal logic whose basic ideas were formulated by Lotfi Zadeh (see Zadeh [1975]a). The aim of this theory is to formalize the *"approximate reasoning"* we use in everyday life, the object of investigation being the human aptitude to manage vague properties (as, for example, *"beautiful"*, *"small"*, *"plausible"*, *"believable"*, etc.) that by their own nature can be satisfied to a degree different from 0 (false) and 1 (true). It is worth noting that the traditional deductive framework in many-valued logic is different from the one adopted in this book for fuzzy logic: in the former logics one always uses a *"crisp"* deduction apparatus, producing crisp sets of formulas, the formulas that are considered logically valid. By contrast, fuzzy logical deductive machinery is devised to produce a fuzzy set of formulas (the theorems) from a fuzzy set of formulas (the hypotheses). Approximate reasoning has generated a very interesting literature in recent years. However, in spite of several basic results, in our opinion, we are still far from a satisfactory setting of this very hard and mysterious subject.

The aim of this book is to furnish some theoretical devices and to sketch a general framework for fuzzy logic. This is also in accordance with the non-Fregean attitude of the book. Indeed, we aim to give some instruments to define rough mathematical models of the wonderful human capacity of reasoning with vague notions and not to propose a unique rigorous formalized (multivalued) logic able to settle such human activity. Consequently, in the book there is no definitive choice of the logical connectives and of their associated interpretations. Also, our constant usage of the set of truth values given by the unit real interval [0,1], is only due to the purpose of simplifying the treatment of the material and not to the conviction that different sets of truth-values are useless. In any case, the book is mainly an exposition of our ideas and research and it does not have any pretension of completeness. In particular, we do not expose the very important fuzzy logic related to Łukasiewicz truth-functional logic. Very good books and papers exist on this argument and we strongly suggest their direct reading (see, for example, Pavelka [1979]c, Cignoli, D'ottaviano and Mundici [2000], Häjek [1998], Novák, Perfilieva and Mockor [1999], Turunen [1999]).

Mainly, three tools are proposed and examined:

- the theory of fuzzy closure operators,
- an extension principle for closure operators,
- the theory of recursively enumerable fuzzy subsets.

Indeed, we embrace Tarski's viewpoint, according to which a monotone logic is a set (of formulas) together with a closure operator (the deduction operator). Consequently, in Chapters 1 and 2 the theory of closure operators in a lattice is exposed and applied to outline an abstract approach to fuzzy logic. In Chapters 3 and 6 an extension principle for classical closure operators is also proposed and largely used. This principle enables us to extend any crisp logic into a fuzzy logic. In Chapter 7 such an approach is generalized by showing that it is possible to associate any chain of crisp logics with a fuzzy logic. In such a way we obtain a very interesting class of fuzzy logics we call *"stratified"*. Necessity logic, graded consequence operator theory and similarity logic all belong to this class. Chapter 4 introduces the notion of fuzzy inferential apparatus in a *"Hilbert style"* by giving a fuzzy set of logical axioms and a suitable set of fuzzy inference rules. In particular, we prove that this approach is equivalent to the theory of continuous fuzzy closure systems (see Theorem 2.6). In Chapter 5 we extend the definition of approximate reasoning usually proposed in the literature by assuming that a deduction apparatus is a tool to calculate constraints on the possible truth values of the formulas. Truth-functional multi-valued logic is examined in Chapter 8 where a strong connection between axiomatizability and continuity of the logical connectives is established (see Theorems 4.5 and Theorem 5.5). In Chapter 9 we propose fuzzy logics probabilistic in nature which are strictly related to super-additive probabilities, upper-lower probabilities, lower envelopes and belief measures. In Chapter 10 we present a tentative approach to fuzzy control by translating any system of IF-THEN fuzzy rules into a system of fuzzy clauses, i.e., a fuzzy program. This enables us to unify the treatment based on the triangular norms and the treatment based on the implications. Finally, in Chapter 11 we extend to fuzzy sets the fundamental notions of decidability and recursive enumerability. In accordance, a definition of an effective Hilbert system is proposed and compared with a definition of enumeration fuzzy operator. This will enable us to get a constructive version of the just quoted Theorem 2.6 (see Theorem 9.2). Also, the notion of recursively enumerable fuzzy subset enables us to obtain several interesting limitative results for fuzzy logic.

Note that three types of fuzzy logics are considered. The first type arises from a *fuzzification of the metalogic.* Necessity logic, similarity logic and graded consequence theory, are typical examples. Indeed, these logics are obtained by admitting that the notions of hypothesis, identity between formulas, logical consequence, can be vague. In this case, the worlds we will describe are *"crisp"*; vagueness arises from the language and the deductive apparatus we use. The second type is related to truth-functional multivalued logic. In such a case *fuzzy worlds* are considered, i.e., worlds whose elements can have graded properties. Finally, we define logics, probabilistic in nature, which are related to *"belief measures"*. Perhaps, we can allocate these logics to the first class. Indeed, in this case vagueness concerns the metalogic notion of *"believable"*.

We expect the book to be read by people interested in artificial intelligence, fuzzy control, formal logic, philosophy. The book is almost completely self-contained but some familiarity with classical logic is required. Moreover, Chapter 11 assumes some acquaintance with the theory of recursive functions.

We are indebted to Loredana Biacino for the critical reading of the manuscript and for several ideas and suggestions. We wish to thank my collaborator Ferrante Formato for stimulating comments and cooperation and Professor Joe Saporito for checking the English style of the manuscript. Finally, we wish to thank the anonymous referee for spending many hours reading preliminary versions, and writing long and interesting reports.

Napoli, February 25, 2000.

CHAPTER 1

ABSTRACT LOGIC IN A LATTICE

1. INTRODUCTION

Let S be a nonempty set and denote by $\mathcal{P}(S)$ the class of all subsets of S. Then a *closure operator* in S is any map $J : \mathcal{P}(S) \to \mathcal{P}(S)$ such that, for $X, Y \in \mathcal{P}(S)$,

$$X \subseteq Y \Rightarrow J(X) \subseteq J(Y) \quad ; \quad X \subseteq J(X) \quad ; \quad J(J(X)) = J(X)$$

(see, e.g., Cohn [1965]). The theory of closure operators is a powerful and elegant tool for (monotone) logics. Indeed, given any deduction apparatus whose set of well-formed formulas is $\mathbb{F}$, we can consider the related deduction operator $\mathcal{D} : \mathcal{P}(\mathbb{F}) \to \mathcal{P}(\mathbb{F})$, i.e., the operator defined by setting, for any $X \in \mathcal{P}(\mathbb{F})$, $\mathcal{D}(X)$ equal to the set of formulas we can derive from X. Then $\mathcal{D}$ is a closure operator. This led several authors to propose a general approach in which an *abstract logic* is a pair $(\mathbb{F}, \mathcal{D})$ where $\mathbb{F}$ is the set of formulas in a given language and $\mathcal{D}$ a closure operator in $\mathbb{F}$ (see Tarski [1956], Brown and Suszko [1973], Wójcicki [1988]). The extension of such an approach to fuzzy logic is straightforward. It is enough to substitute the lattice of all subsets of $\mathbb{F}$ with the lattice of all fuzzy subsets of $\mathbb{F}$. This is in accordance with the definition of deduction operator given in Pavelka [1979].

In this chapter we are considering a very general notion of abstract logic through structures like $(L, \mathcal{D})$, where L is a complete lattice and $\mathcal{D} : L \to L$ is a closure operator in L, i.e., a map such that

$$x \leq y \Rightarrow \mathcal{D}(x) \subseteq \mathcal{D}(y) \quad ; \quad x \subseteq \mathcal{D}(x) \quad ; \quad \mathcal{D}(\mathcal{D}(x)) = \mathcal{D}(x).$$

We interpret an element $x \in L$ as a piece of information and $\mathcal{D}(x)$ as the whole information we can derive from x. By considering the case in which L is the lattice of all fuzzy subsets of $\mathbb{F}$, we obtain the notion of abstract fuzzy logic.

In the following, given a set X, we denote by $|X|$ the number of elements in X. The set of natural, positive natural, rational and real numbers are denoted by $\boldsymbol{N_0}$, $\boldsymbol{N}$, $\boldsymbol{Q}$ and $\boldsymbol{R}$ respectively. Also, if $a, b \in \boldsymbol{R}$, we set:

$$[a,b] = \{x \in \boldsymbol{R} : a \leq x \leq b\} \quad ; \quad [a,b) = \{x \in \boldsymbol{R} : a \leq x < b\},$$
$$(a,b] = \{x \in \boldsymbol{R} : a < x \leq b\} \quad ; \quad (a,b) = \{x \in \boldsymbol{R} : a < x < b\}.$$

2. LATTICES, BOOLEAN ALGEBRAS, TRIANGULAR NORMS

In this section we will give some basic notions on ordered structures which are useful for fuzzy logic (see, e.g., Birkhoff [1948]). An *ordered set* $(L, \leq)$ is a set L with a binary relation $\leq$ satisfying for all $x, y, z \in L$

(i) $x \leq x$ (reflexivity),

(ii) $x \leq y$ and $y \leq x \Rightarrow x = y$ (antisymmetry),

(iii) $x \le y$ and $y \le z \;\Rightarrow\; x \le z$ (transitivity).

We write $x < y$ to denote that $x \le y$ and $x \ne y$. If only reflexivity and transitivity are verified, we say that $\le$ is a *pre-order*. A *totally ordered set* (equivalently, a *chain*, a *linearly ordered set*) is an ordered set satisfying the additional condition

(iv) $x \le y$ or $y \le x$.

A nonempty class C of elements in L is *upward directed* if

$$x \in C \text{ and } y \in C \Rightarrow \exists z \in C, x \le z \text{ and } y \le z.$$

Dually, C is *downward directed*, if

$$x \in C \text{ and } y \in C \Rightarrow \exists z \in C, x \ge z \text{ and } y \ge z.$$

The totally ordered subsets of L are examples of upward and downward directed classes. In the sequel, we write "directed" to mean "upward directed".

Given a subset X of an ordered set L, we say that $u \in L$ is an *upper bound* of X if $x \le u$ for all $x \in X$. Trivially, every element of L is an upper bound of $\varnothing$. An element u is called the *least upper bound* of X, in symbols $u = Sup(X)$, or $u = SupX$, if

(i) u is an upper bound of X ; (ii) x upper bound of $X \Rightarrow u \le x$.

Sometimes $Sup(X)$ is also called the *join* of the elements in X. If X is upward directed, the element $z = Sup(X)$ is also called the *limit* of X and we write $z = limX$. We say that $m \in L$ is a *maximal element* in X if m is an element of X such that no element in X strictly dominates m, i.e.,

(i) $m \in X$; (ii) $x \in X$ and $m \le x \Rightarrow x = m$.

We say that m is the *greatest element of* X and we set $m = Max(X)$ if

(i) $m \in X$; (ii) $x \in X \Rightarrow x \le m$.

Trivially, the greatest element of X is also a maximal element in X. Likewise, we say that $l \in L$ is a *lower bound* of X if $l \le x$ for all $x \in X$ and that l is the *greatest lower bound of* X, in symbols $l = Inf(X)$ or $l = Inf\,X$, if

(i) l is a lower bound of X ; (ii) x lower bound of $X \Rightarrow x \le l$.

Sometimes $Inf(X)$ is also called the *meet* of the elements in X. We call a *minimal element* of X an element m such that

(i) $m \in X$; (ii) $x \in X$ and $x \le m \Rightarrow x = m$.

In other words, m is an element of X such that no element in X is strictly dominated by m. We say that m is the *least element of* X and we set $m = Min(X)$ if

(i) $m \in X$; (ii) $x \in X \Rightarrow m \le x$.

If X is totally ordered, then X contains at most one maximal element which is also the greatest element of X. Moreover, X contains at most one minimal element which is the least element of X. If the greatest (the least) element of L exists, we denote it by 1 (by 0). In this case $Inf(\varnothing) = 1$ and $Sup(\varnothing) = 0$. A minimal element in $L - \{0\}$ is called an *atom*, a maximal element in $L - \{1\}$ a *co-atom*.

Given two ordered sets L_1 and L_2, a map $h : L_1 \to L_2$ is *order-preserving* (*order-reversing*) if, for every $x, y \in L$,

$$x \le y \Rightarrow h(x) \le h(y) \quad (x \le y \Rightarrow h(x) \ge h(y)).$$

We say that h is an *embedding* if

$$x \le y \Leftrightarrow h(x) \le h(y),$$

and that h is an *isomorphism* if h is a surjective embedding. Trivially, an embedding is injective. An *automorphism* of L is an isomorphism from L onto itself.

Given a subset X of L, there exists at most one least upper bound (greatest lower bound, least element, greatest element) of X. If $X = \{x_1, x_2\}$, we write $x_1 \wedge x_2$ and $x_1 \vee x_2$ to denote $Inf(\{x_1, x_2\})$ and $Sup(\{x_1, x_2\})$, respectively. An ordered set L is a *lattice* provided that, for any x_1 and x_2 in L, both $x_1 \wedge x_2$ and $x_1 \vee x_2$ exist. If in L there is a greatest element 1 and a least element 0, then we say that the lattice is *bounded.* We represent a bounded lattice L as an algebraic structure $(L, \wedge, \vee, 0, 1)$. Since all the lattices in this book are bounded, we prefer to call them simply "lattices". A *sublattice* of L is a subset L' of L containing 0 and 1 and closed with respect to the operations $\wedge$ and $\vee$. Every totally ordered set with a least element and a greatest element is a bounded lattice. A lattice is *complete* if $Inf(X)$ and $Sup(X)$ exist for every subset X of L. Since $Inf(\varnothing) = 1$ and $Sup(\varnothing) = 0$, any complete lattice is bounded. A subset L' of a complete lattice L is called a *complete sublattice* of L if 0 and 1 belong to L' and the join and the meet in L of any family of elements in L' belong to L'. Note that the claim that L' is a complete sublattice of L is different from the claim that L' is a sublattice and L' is complete. As an example, let L be the whole class of subsets of a Euclidean space and L' the class of open subsets. Then L' is a sublattice of L, L' is complete but L' is not closed under infinite intersections.

Given two bounded lattices L_1 and L_2, a *homomorphism* is any map $h : L_1 \to L_2$ such that, for every x, y in L_1,

$$h(0) = 0 \;;\; h(1) = 1 \;;\; h(x \wedge y) = h(x) \wedge h(y) \;;\; h(x \vee y) = h(x) \vee h(y).$$

An *isomorphism* is a bijective homomorphism and an *automorphism* of a lattice L is an isomorphism from L onto itself. The map h is *join-preserving* if,

$$Sup(h(X)) = h(Sup(X)),$$

for every nonempty subset X of L, h is *meet-preserving* if,

$$Inf(h(X)) = h(Inf(X)).$$

It is immediate that both the join-preserving and the meet-preserving maps are order-preserving. Dually, we say that h is *join-reversing*, if

$$Inf(h(X)) = h(Sup(X)),$$

and *meet-reversing*, if

$$Sup(h(X)) = h(Inf(X)).$$

Both the join-reversing and the meet-reversing maps are order-reversing. A lattice L is *distributive* if, for every x, y, z in L,

$$x \wedge (y \vee z) = (x \wedge y) \vee (x \wedge z) \;,\; x \vee (y \wedge z) = (x \vee y) \wedge (x \vee z).$$

A distributive lattice $\boldsymbol{B}$ is called a *Boolean algebra* if an operation $- : \boldsymbol{B} \to \boldsymbol{B}$ is defined in such a way that, for any $x \in \boldsymbol{B}$,

$$x \vee (-x) = 1 \;;\; x \wedge (-x) = 0.$$

One proves that, for any x and y in $\boldsymbol{B}$,

$$-(-x) = x \;;\; -(x \vee y) = (-x) \wedge (-y),$$

$$-(x \wedge y) = (-x) \vee (-y) \;;\; -0 = 1 \;;\; -1 = 0.$$

The main example of a complete lattice we will consider in this book is the closed unit interval $[0,1]$ in which, for λ and λ' in $[0,1]$, we have

$$\lambda \vee \lambda' = max\{\lambda,\lambda'\} \quad \text{and} \quad \lambda \wedge \lambda' = min\{\lambda,\lambda'\}.$$

In [0,1] one considers also a *negation* $\sim : [0,1] \to [0,1]$ defined by setting $\sim(\lambda) = 1-\lambda$. Also, sometimes we consider a binary operation $\odot$ in [0,1] able to interpret the logical connective *"and"*. We assume that $\odot$ is a *continuous triangular norm*, i.e., a binary continuous operation in [0,1] such that :

- $(x \odot y) \odot z = x \odot (y \odot z)$ (associative),
- $x \odot y = y \odot x$ (commutative),
- $x \odot 1 = x$ (1 is a neutral element),
- $x \leq y \Rightarrow x \odot z \leq y \odot z$ (order-preserving),

where x, y and z are elements in [0,1] (see Schweizer and Sklar [1983]). Observe that, for any $x, y \in [0,1]$, $x \odot y \leq x \odot 1 = x$ and therefore $x \odot y \leq x \wedge y$. In particular, $x \odot x \leq x$. Examples are the minimum, ordinary product and *Łukasiewicz conjunction* defined by setting, for any x and y in [0,1],

$$x \odot y = (x + y - 1) \vee 0.$$

Any triangular norm is associated with a *triangular co-norm* $\oplus$ defined by setting

$$x \oplus y = \sim(\sim x \odot \sim y).$$

As an example, the co-norm associated with the minimum is the maximum. The co-norm associated with the product is defined by

$$x \oplus y = x + y - xy,$$

the co-norm related with Łukasiewicz conjunction, we call *Łukasiewicz sum*, by

$$x \oplus y = (x + y) \wedge 1.$$

3. CLOSURE OPERATORS AND CLOSURE SYSTEMS

The notions of closure operator and closure system can be defined in any complete lattice L (see Ward [1942]). In the following, we call an *operator* in L any map J from L into L and *class* or *system* in L any subset C of L.

Definition 3.1. Let L be a complete lattice. Then a *closure operator* in L is any operator $J : L \to L$ satisfying :

(i) $x \leq y \Rightarrow J(x) \leq J(y)$ (order-preserving),
(ii) $x \leq J(x)$ (inclusion),
(iii) $J(J(x)) = J(x)$ (idempotence).

If (iii) is skipped, J is called an *almost closure operator* in L, in brief *a-c-operator*.

Strictly related to the notion of closure operator, we introduce the concept of closure system.

Definition 3.2. A class C of elements of L is called a *closure system* if the meet of any family of elements of C is an element of C.

Since $Inf(\varnothing) = 1$ and $\varnothing$ is a subset of any class, every closure system contains 1. Given a closure system C and $x \in L$, the *element* $<x>$ *of* C *generated by* x is defined by setting

$$<x> = Inf\{x' \in C : x' \geq x\}. \tag{3.1}$$

The following proposition shows that any closure system is a complete lattice:

Proposition 3.3. *Let $C \subseteq L$ be a closure system. Then C is a complete lattice such that*

- *the least element of C is the meet of all the elements in C,*
- *the unity in C coincides with the unity 1 in L,*
- *the meets in C coincide with the meets in L,*
- *the join in C of a class X is $<Sup(X)>$.*

Such a lattice is not necessarily a sublattice of L.

Proof. We confine ourselves to observing that the joins in C are different from the joins in L. As an example, consider the class C of subalgebras of a given algebraic structure $\mathcal{A}$. Then C is a closure system and hence a complete lattice but, while the meet operator coincides with the intersection, the join of a family $(A_i)_{i\in I}$ of subalgebras coincides with the subalgebra generated by $\cup_{i\in I}A_i$. □

The class of closure operators and the class of closure systems define two closure systems. Indeed, denote by L^L the *direct power* of L with index set L, i.e., the class of operators in L equipped with the order relation $\leq$ defined by setting, for every $J, J' \in L^L$,

$$J \leq J' \Leftrightarrow J(x) \leq J'(x) \text{ for every } x \in L.$$

Then, it is easy to prove that L^L is a complete lattice whose operations are defined by setting, for every $x \in L$,

$$(J \vee J')(x) = J(x) \vee J'(x) \quad ; \quad (J \wedge J')(x) = J(x) \wedge J'(x).$$

Proposition 3.4. *Both the classes $CO(L)$ of closure operators and $AC(L)$ of almost closure operators are closure systems in L^L. The class $CS(L)$ of closure systems in L is a closure system in $\mathcal{P}(L)$.*

Proof. Let J be the meet of a family $(J_i)_{i\in I}$ of *a-c*-operators and $x \in L$. Then it is immediate that J satisfies (i) and (ii) of Definition 3.1. Assume that each J_i is idempotent. Then, for every $x \in L$ and $k \in I$,

$$J(J(x)) = J(Inf_{i\in I} J_i(x)) \leq J_k(Inf_{i\in I} J_i(x)) \leq J_k(J_k(x)) = J_k(x).$$

Hence

$$J(J(x)) \leq Inf_{k\in I} J_k(x) = J(x)$$

and therefore $J(J(x)) = J(x)$. This proves that $CO(L)$ is a closure system.

Let C be the intersection of a family $(C_i)_{i\in I}$ of closure systems and let $(x_j)_{j\in J}$ be any family of elements in C. We must prove that $x = Inf\{x_j : j \in J\}$ is an element of C. Indeed, since, for every $i \in I$, $(x_j)_{j\in J}$ is a family of elements in C_i, we have $x \in C_i$. Thus, $x \in \bigcap\{C_i : i \in I\} = C$. This proves that $CS(L)$ is a closure system. □

Let J be an operator in L and C a class of elements of L. Then we denote by $c(J)$ the closure operator generated by J and $c(C)$ the closure system generated by C.

4. A GALOIS CONNECTION BETWEEN OPERATORS AND CLASSES

In this section we will show that the closure systems and the closure operators are strictly related. To this purpose we define the map $Co : \mathcal{P}(L) \to L^L$ by assuming that, for each $C \subseteq L$, $Co(C) : L \to L$ is the operator defined by setting

$$Co(C)(x) = Inf\{y \in C : y \geq x\}, \tag{4.1}$$

for every $x \in L$. Moreover, we define the map $Cs : L^L \to \mathcal{P}(L)$ by setting, for any operator $J \in L^L$,

$$Cs(J) = \{x \in L : J(y) \leq x \text{ for every } y \leq x\}. \tag{4.2}$$

Proposition 4.1. *Given any $C \subseteq L$, the operator $Co(C) : L \to L$ is a closure operator. Given any $J \in L^L$, the class $Cs(J)$ is a closure system.*

Proof. The first part is trivial. Let $(x_i)_{i\in I}$ be a family of elements of $Cs(J)$. We claim that $Inf\{x_i : i \in I\} \in Cs(J)$. Indeed, suppose $y \leq Inf\{x_i : i \in I\}$, that is $y \leq x_i$ for every $i \in I$. Then $J(y) \leq x_i$ for every $i \in I$. Thus $J(y) \leq Inf\{x_i : i \in I\}$ and therefore $Inf\{x_i : i \in I\} \in Cs(J)$. This proves that $Cs(J)$ is a closure system. □

As claimed in Section 3, sometimes we write $<x>$ instead of $Co(C)(x)$. Given an operator J, we call *fixed point of J* any element $x \in L$ such that $J(x) = x$. In the case that J is an *a-c*-operator this is equivalent to saying that $J(x) \leq x$, i.e., *x is closed with respect to J.* Moreover, we have the following:

Proposition 4.2. *Let J be an a-c-operator. Then*

$$Cs(J) = \{x \in L : J(x) = x\}, \tag{4.3}$$

i.e., Cs(J) is the class of fixed points of J.

Proof. By (4.2), $J(x) \leq x$ for every $x \in Cs(J)$. Then, if J is an *a-c*-operator, $J(x) = x$. Conversely, if x is a fixed point of J and $y \leq x$, then $J(y) \leq J(x) = x$ and this proves that $x \in Cs(J)$. □

The proof of the next proposition is evident:

Proposition 4.3. *Let J and J' be operators and C, C' classes. Then,*

$$J \leq J' \Rightarrow Cs(J) \supseteq Cs(J') \quad ; \quad C \subseteq C' \Rightarrow Co(C) \geq Co(C').$$

The first implication says that, if J and J' are *a-c*-operators such that $J \leq J'$, then every fixed point for J' is a fixed point for J. The following theorem gives a way to obtain the closure operator $c(J)$ generated by J:

Theorem 4.4. *Let J be an operator. Then*

$$c(J) = Co(Cs(J)). \tag{4.4}$$

So, if J is an a-c-operator and $x \in L$, $c(J)(x)$ is the least fixed point of J greater than or equal to x.

Proof. Set $J' = Co(Cs(J))$, then J' is a closure operator. To prove that $J' \geq J$, it suffices to observe that for every $y \in L$, from $y \in Cs(J)$ and $x \leq y$, it follows that $J(x) \leq y$. Consequently,

$$J'(x) = Inf\{y \in Cs(J) : y \geq x\} \geq J(x).$$

Let H be a closure operator such that $H \geq J$ and suppose $y' \leq H(x)$. Then

$$J(y') \leq H(y') \leq H(H(x)) = H(x).$$

This proves that $H(x) \in Cs(J)$ and, since $H(x) \geq x$, that $H(x) \geq J'(x)$. □

Given any class C of elements of L, we can obtain the closure system $c(C)$ generated by C as follows:

$$c(C) = \{Inf(X) : X \subseteq C\}. \tag{4.5}$$

Moreover, we have the following theorem:

Theorem 4.5. *Given any class C of elements of L, we have*

$$c(C) = Cs(Co(C)). \tag{4.6}$$

Proof. Being every element of C a fixed point of $Co(C)$, $Cs(Co(C))$ is a closure system containing C. Let C' be a closure system containing C, and x an element of $Cs(Co(C))$. Then, since $x = Co(C)(x)$, x is a meet of elements of C and hence belongs to C. Thus $Cs(Co(C)) \subseteq C'$ and, therefore, $Cs(Co(C)) = c(C)$. □

Some interesting properties of the operators Co and Cs are listed in the following proposition:

Proposition 4.6. *Let J and J' be operators, and C and C' classes. Then*

$$Cs(J) = Cs(c(J)) \quad ; \quad Co(C) = Co(c(C)). \tag{4.7}$$

Also,

$$c(J) = c(J') \Leftrightarrow Cs(J) = Cs(J') \quad ; \quad c(C) = c(C) \Leftrightarrow Co(C) = Co(C'). \tag{4.8}$$

Moreover, if C and C' are closure systems, and J and J' closure operators, then

$$C \subseteq C' \Leftrightarrow Co(C) \geq Co(C') \quad ; \quad J \leq J' \Leftrightarrow Cs(J) \supseteq Cs(J') \tag{4.9}$$

$$Co(Cs(J)) = J \quad ; \quad Cs(Co(C)) = C. \tag{4.10}$$

We can summarize the main results in this section in terms of a Galois connection between the lattices L^L and $\mathcal{P}(L)$.

Definition 4.7. Let $(S, \leq)$ and $(T, \leq)$ be ordered sets, a *Galois connection* is a pair of order-reversing maps $h : S \to T$ and $k : T \to S$ such that

- $k \circ h : S \to S$ is a closure operator in S,
- $h \circ k : T \to T$ is a closure operator in T.

Notice that several equivalent definitions of a Galois connection can be achieved by duality (see Section 10) from Definition 4.7. For example, by substituting $(T, \leq)$ with its dual we can call any pair $h : S \rightarrow T$ and $k : T \rightarrow S$ of order-preserving maps a Galois connection if $k \circ h$ is a closure operator in S and $h \circ k$ an *interior operator* in T, i.e., $h \circ k$ is order-preserving, idempotent and $h(k(x)) \leq x$ for any $x \in T$.

Theorem 4.8. *The maps* $Cs : L^L \rightarrow \mathcal{P}(L)$ *and* $Co : \mathcal{P}(L) \rightarrow L^L$ *define a Galois connection in which*

- $Cs \circ Co$ *is the closure operator* $c : \mathcal{P}(L) \rightarrow \mathcal{P}(L)$ *which associates any* $C \subseteq L$ *with the closure system* $c(C)$ *generated by* C,
- $Co \circ Cs$ *is the closure operator in* $c : L^L \rightarrow L^L$ *which associates any* $J \in L^L$ *with the closure operator* $c(J)$ *generated by* J.

5. ABSTRACT LOGIC IN A LATTICE

The following is the main definition in this chapter:

Definition 5.1. Let $\mathcal{D}$ be a closure operator in a complete lattice L. Then we can say that the pair $\mathcal{S} = (L,\mathcal{D})$ is an *abstract deduction system* and that $\mathcal{D}$ is a *deduction operator* (Biacino and Gerla [1999]).

We call *pieces of information* the elements in L. Any classical logic $\mathcal{L}$ defines an abstract logic whose pieces of information are the set of formulas. Indeed, if $\mathbb{F}$ is the set of formulas of $\mathcal{L}$, we can set $L = \mathcal{P}(\mathbb{F})$ and $\mathcal{D}$ equal to the operator associating any $X \in \mathcal{P}(\mathbb{F})$ with the set $\mathcal{D}(X)$ of consequences of X.

A *theory* in an abstract deduction system $(L, \mathcal{D})$ is defined as a fixed point of $\mathcal{D}$, i.e., a piece of information τ closed under deductions. Proposition 4.2 says that the class $\mathcal{T} = Cs(\mathcal{D})$ of theories of a deduction system is a closure system and hence a complete lattice. If τ is a theory and $\mathcal{D}(x) = \tau$, then we can say that x is a *system of axioms* for τ. A piece of information $x \in L$ is *inconsistent* provided that $\mathcal{D}(x) = 1$. This extends the fact that in classical logic an inconsistent set of axioms generates the whole set $\mathbb{F}$ of formulas (i.e., the greatest element of $\mathcal{P}(\mathbb{F})$). In accordance, the piece of information 1 is called the *inconsistent theory* and a theory τ is *consistent* provided that $\tau \neq 1$. A *maximal theory* is a theory τ which is maximal in the class of consistent theories, i.e., no theory τ' exists such that $1 > \tau' > \tau$.

Definition 5.2. A class $\mathcal{M}$ of elements of L such that $1 \notin \mathcal{M}$ is called an *abstract semantics* and the elements in $\mathcal{M}$ are called *models*. If $x \in L$, $m \in \mathcal{M}$ and $x \leq m$, then we can say that m is a *model* of x and we can write $m \models x$. If $x, y \in L$ admit the same models, then we can say that x is *logically equivalent* to y.

In accordance with Proposition 4.1, any semantics $\mathcal{M}$ induces a closure operator $Co(\mathcal{M}) : L \to L$. We call this a *logical consequence operator* and we denote it by Lc. Then, Lc is defined by setting, given a piece of information x,

$$Lc(x) = Inf\{m \in \mathcal{M} : m \models x\}.$$

These definitions are in accordance with the classical definitions because we can identify the class of models in a classical logic $\mathcal{L}$ with the class $\mathcal{M}$ of complete theories of $\mathcal{L}$. In fact, each model m in $\mathcal{L}$ is associated with its theory, i.e., with the complete theory

$$T_m = \{\alpha \in \mathbb{F} : \alpha \text{ is true in } m\}.$$

Conversely, for every complete theory T a model m exists such that $T_m = T$. Moreover, it is easy to see that m is a model of a set X of formulas iff $X \subseteq T_m$ and that the set $Lc(X)$ of logical consequences of X is equal to the intersection of all the complete theories containing X.

If τ is a theory of Lc we can say also that τ is a theory of $\mathcal{M}$. Trivially,

$$x \text{ is logically equivalent to } y \Leftrightarrow Lc(x) = Lc(y).$$

Remark. We can also define Lc as follows: Consider the operators

$$mod : L \to \mathcal{P}(\mathcal{M}) \quad \text{and} \quad th : \mathcal{P}(\mathcal{M}) \to L$$

defined by setting, for every $x \in L$ and $C \in \mathcal{P}(\mathcal{M})$,

$$mod(x) = \{m \in \mathcal{M} : m \models x\} \;;\; th(C) = Sup\{x \in L : m \models x \;\forall m \in C\}.$$

Then, $mod(x)$ is the set of models of x and $th(C)$ the information shared by all the models in C. It is easy to verify that *mod* and *th* define a Galois connection such that $th \circ mod$ coincides with the closure operator Lc.

We define the *system of tautologies* as

$$Tau(\mathcal{M}) = Inf\{m : m \in \mathcal{M}\},$$

equivalently,

$$Tau(\mathcal{M}) = Lc(\emptyset).$$

If x is consistent with respect to Lc, then we prefer to say that x is *satisfiable*. Equivalently, x is satisfiable if a model of x exists. Also, x is *categorical* if just one model of x exists. We denote the class of satisfiable pieces of information by $Sat(\mathcal{M})$, i.e.,

$$Sat(\mathcal{M}) = \{x \in L : m \in \mathcal{M} \text{ exists such that } m \models x\}.$$

Definition 5.3. An *abstract logic* is a triplet $(L, \mathcal{D}, \mathcal{M})$ where L is a complete lattice, $\mathcal{D}$ is a closure operator and $\mathcal{M}$ an abstract semantics such that $\mathcal{D} = Co(\mathcal{M})$, i.e., the *"completeness theorem"* holds.

Given any abstract deduction system $(L, \mathcal{D})$ we can define an abstract logic in a trivial way by setting $\mathcal{M}$ equal to the class of consistent theories of $\mathcal{D}$. In this sense, any deduction system has an abstract semantics. Nevertheless, such a semantics is unsatisfactory because we look for semantics containing only theories that are complete pieces of information.

6. CONTINUITY FOR ABSTRACT LOGICS

In defining the notion of abstract logic it seems natural to require some additional properties expressing the effectiveness of the deduction processes. This will be done in Chapter 11 for fuzzy logic but it is very hard to say what *"effectiveness"* means in the general case. In this chapter we confine ourselves to extend the notion of compactness to abstract logic. We first recall the classical notion of compactness.

Definition 6.1. Let $J : \mathcal{P}(S) \to \mathcal{P}(S)$ be an operator in a set S. Then we say that J is *compact*, provided that, for every subset X of S,

$$x \in J(X) \Leftrightarrow \text{ a finite subset } X_f \text{ of } X \text{ exists such that } x \in J(X_f).$$

Due to the finiteness of any proof, the deduction operator of a crisp logic is compact. Now, being the notion of finite subset not defined on a generic lattice L, we cannot define a compactness property in a generic abstract logic. Then we propose the notion of continuity (see, for example, Lloyd [1987]) as a natural counterpart of the notion of compactness. To this aim, we give some definitions (see, for example, Lloyd [1987]).

If J is an order-preserving operator and C is directed, then the image $J(C) = \{J(x) : x \in C\}$ is also directed. Then we can give the following definition:

Definition 6.2. An order-preserving operator J is *continuous* if, for every directed class C,

$$J(lim\ C) = limJ(C). \tag{6.1}$$

A continuous closure operator is also called an *algebraic closure* operator.

The following proposition shows that the notion of continuity extends the notion of compactness.

Proposition 6.3. Assume that L is the lattice $\mathcal{P}(S)$ of all subsets of a given set S. Then J is continuous iff J is compact.

Proof. If J is continuous and $X \in \mathcal{P}(S)$, then, because $C = \{X_f : X_f \text{ is a finite subset of } X\}$ is directed, we have

$$J(X) = J(lim\ C) = lim\ J(C) = \cup\{J(X_f) : X_f \text{ is a finite subset of } X\}.$$

Conversely, let $J : \mathcal{P}(S) \to \mathcal{P}(S)$ be compact and observe that if C is a directed class of subsets of S then

$$X_f \subseteq lim\ C \text{ and } X_f \text{ finite} \Rightarrow \text{there exists } X \in C \text{ such that } X_f \subseteq X.$$

Consequently,

$$J(lim\ C) = \cup\ \{J(X_f)\ : X_f \subseteq lim\ C\ \}) = \cup\{J(X) : X \in C\} = lim\ J(C). \qquad \square$$

The connection between closure operators and closure systems suggests the following question:

is there a property for closure systems fitting the continuity property for closure operators well ?

The next definition enables us to give a positive answer.

Definition 6.4. A class C of elements of L is called *inductive* if the limit of every directed family of elements in C belongs to C. An inductive closure system is called *algebraic*.

Every finite subset of L is inductive and therefore every finite closure system is algebraic. The notion of an algebraic closure system is well related to the notion of an algebraic closure operator.

Theorem 6.5. *Given a nonempty class C,*

$$C \text{ is an algebraic closure system } \Leftrightarrow \ Co(C) \text{ is an algebraic closure operator.}$$

Given a closure operator J,

$$J \text{ is algebraic } \Leftrightarrow \ Cs(J) \text{ is an algebraic closure system.}$$

Proof. Suppose the C is an algebraic closure system and let $\mathcal{T}$ be any directed class. Then, since the set $\mathcal{H} = \{Co(C)(x) : x \in \mathcal{T}\}$ is a directed subclass of C, $Sup\{Co(C)(x) : x \in \mathcal{T}\}$ is an element of C and therefore a fixed point for $Co(C)$. Then,

$$\begin{aligned} Co(C)(Sup(\mathcal{T})) &\subseteq Co(C)(Sup\{Co(C)(x) : x \in \mathcal{T}\}) \\ &= Sup\{Co(C)(x) : x \in \mathcal{T}\}, \end{aligned}$$

and this proves that $Co(C)$ is algebraic.

Conversely, let $Co(C)$ be algebraic and let $\mathcal{T}$ be a directed subset of C. Then, as C is the class of fixed points of $Co(C)$,

$$Co(C)(Sup(\mathcal{T})) = Sup(\{Co(C)(x) : x \in \mathcal{T}\}) = Sup(\mathcal{T}).$$

This proves that $Sup(\mathcal{T})$ is a fixed point for $Co(C)$ and hence an element of C. In conclusion, C is an algebraic closure system.

In order to prove the second part of the proposition, recall that if J is a closure operator then $J = Co(Cs(J))$. □

Example. Examine the closure systems in the lattice $[0,1]$. Then a closure system is any subset C of $[0,1]$ closed with respect to the greatest lower bounds. Now, any nonempty subset of $[0,1]$ is directed and therefore the algebraic closure systems coincide with the subsets C both closed under least upper bounds of subsets and greatest lower bounds of nonempty subsets. Thus, the class of algebraic closure systems coincides with $\{X \subseteq [0,1] : X \text{ is closed and } 1 \in X\}$. For instance, set $C = [1/3, 2/3) \cup \{1\}$. Then C is a closure system which is not algebraic and the associated closure operator is defined by setting:

$Co(C)(x) = 1/3$ for every $x \in [0,1/3]$,

$Co(C)(x) = x$ for every $x \in [1/3, 2/3)$,

$Co(C)(x) = 1$ otherwise.

Moreover, we have that its topological closure $[1/3, 2/3] \cup \{1\}$ is an algebraic closure system. Notice that the continuity proposed in Definition 6.2 is different than the continuity with respect to the natural topology in [0,1]. In fact, an operator J satisfies (6.1) iff J is order-preserving and lower semicontinuous with respect to natural topology.

We conclude this section with the following basic definition:

Definition 6.6. An abstract deduction system (L, $\mathcal{D}$) (more generally, an abstract logic) is called *continuous* provided that $\mathcal{D}$ is continuous.

7. STEP-BY-STEP DEDUCTION SYSTEMS

Usually a deduction operator $\mathcal{D}$ is defined by giving a suitable set $\mathcal{A}$ of logical axioms and a suitable set of inference rules. In this case we can define the *immediate consequence operator* H by setting, for any set X of formulas, $J(X)$ equal to the set of formulas that can be obtained by one application of the inference rules to formulas in X and

$$H(X) = J(X) \cup \mathcal{A} \cup X.$$

In other words, $\alpha \in H(X)$ if either α is obtained by applying an inference rule to formulas in X, or α is a logical axiom or α is a hypothesis (a proper axiom). Also, we define H^n by induction on n, by setting

$$H^1 = H \text{ and } H^{n+1} = H \circ H^n.$$

Given a natural number n, $H^n(X)$ represents the set of formulas that can be achieved by an n-step inferential process from X. It is easy to prove that H is a compact almost closure operator and that $\mathcal{D}$ is the closure operator generated by H. Moreover,

$$\mathcal{D}(X) = \bigcup_{n\in N} H^n(X).$$

To extend such an approach to abstract logics, we must first examine how to obtain the closure operator generated by a continuous *a-c*-operator.

Proposition 7.1. *Let H be a continuous a-c-operator. Then the set $Cs(H)$ of fixed points of H is an algebraic closure system and the closure operator $c(H)$ generated by H, is an algebraic closure operator.*

Proof. Let $\mathcal{T}$ be a directed subclass of $Cs(H)$. Then, from the continuity of H it follows that:

$$H(Sup(\mathcal{T})) = Sup(\{H(x) : x \in \mathcal{T}\}) = Sup(\mathcal{T})$$

and, hence, $Sup(\mathcal{T}) \in Cs(H)$. This proves that $Cs(H)$ is algebraic. Thus, from the equality, $c(H) = Co(Cs(H))$, we can conclude that $c(H)$ is algebraic. □

Let H be a continuous *a-c*-operator. Then, the following simple and useful theorem enables us to calculate the closure operator $c(H)$ generated by H (see, for example, Lloyd [1987]).

Theorem 7.2. (Fixed-Point Theorem). *Let H be a continuous a-c-operator. Then*

$$c(H) = Sup_{n\in N} H^n. \tag{7.1}$$

In other words, for every $x \in L$, the least fixed point of H greater than or equal to x is given by $Sup_{n\in N}H^n(x)$.

Proof. We have to prove that, for every $x \in L$, $Sup_{n\in N}H^n(x)$ is the least fixed point of H greater than or equal to x. Now, the inequality $H(x) \geq x$ entails that $H^{n+1}(x) \geq H^n(x)$ for every n, and hence, that $(H^n(x))_{n\in N}$ is directed. By the continuity of H,

$$H(Sup_{n\in N}H^n(x)) = Sup_{n\in N}H^{n+1}(x) = Sup_{n\in N}H^n(x)$$

and $Sup_{n\in N}H^n(x)$ is a fixed point for H greater than or equal to x. Let y be any fixed point such that $y \geq x$. Then, for every $n \in N$, $y = H^n(y) \geq H^n(x)$ and therefore, $y \geq Sup_{n\in N}H^n(x)$. This proves that $Sup_{n\in N}H^n(x) = c(H)(x)$. □

In accordance with the above considerations, we propose the following definition:

Definition 7.3. An *abstract step-by-step* deduction system is an triplet like (L, J, a) where

- L is a complete lattice,
- J is a continuous operator in L,
- a is an element of L (the system of *logical axioms*).

Let (L, J, a) be a step-by-step deduction system and define H by setting

$$H(x) = J(x) \vee x \vee a, \tag{7.2}$$

for every $x \in L$. Then, H is a continuous a-c-operator we call *the immediate consequence operator.*

Definition 7.4. Let (L, J, a) be a step-by-step-deduction system and denote by $\mathcal{D}$ the closure operator generated by the immediate consequence operator H. Then the abstract deduction system $(L, \mathcal{D})$ is called the *deduction system associated with* (L, J, a).

The proof of the following theorem is trivial:

Theorem 7.5. *Let (L, J, a) be a step-by-step-deduction system and $(L, \mathcal{D})$ the associated deduction system. Then $\mathcal{D}$ is continuous and*

$$\mathcal{D}(x) = Sup_{n\in N}H^n(x). \tag{7.3}$$

Moreover, τ is a theory of $(L, \mathcal{D})$ iff $\tau \geq J(\tau)$ and $\tau \geq a$.

8. LOGICAL COMPACTNESS

We call *logically compact*, in brief *l-compact*, a closure operator $\mathcal{D}$ if the related class of consistent pieces of information is inductive (see Biacino and Gerla [1999]). We call *logically compact* a semantics $\mathcal{M}$ such that the related logical consequence operator Lc is logically compact, i.e., the class $Sat(\mathcal{M})$ of satisfiable pieces of information is inductive. An abstract logic is *logically compact* if its deduction operator (and hence its semantics) is logically compact.

The notion of logical compactness enables us to obtain an extension of Lindenbaum Theorem (see, for example Chang and Keisler [1973], pag. 26).

Proposition 8.1. *If $\mathcal{D}$ is logically compact every consistent piece of information is contained in a maximal theory.*

Proof. This is a direct consequence of Zorn's Lemma. □

Note that logical compactness is different from continuity. As we will show in the next chapters, while the deduction operator of a fuzzy logic in Hilbert's style is always continuous, there are very interesting examples of Hilbert systems whose deduction operator is not logically compact. The following theorem characterizes the deduction operators that are both continuous and logically compact.

Theorem 8.2. *Let $\mathcal{D}$ be a deduction operator. Then the following are equivalent:*
(a) *$\mathcal{D}$ is continuous and logically compact.*
(b) *the class of consistent theories is inductive.*

Proof. (a) ⇒ (b). Let $\mathcal{T}$ be a directed family of consistent theories and denote its limit by τ. Then, by the continuity of $\mathcal{D}$,

$$\mathcal{D}(\tau) = \mathcal{D}(lim\, \mathcal{T}) = lim\, \mathcal{D}(\mathcal{T}) = lim\, \mathcal{T} = \tau.$$

This proves that τ is a theory. Since $\mathcal{D}$ is logically compact, τ is consistent.

(b) ⇒ (a). In order to prove that $\mathcal{D}$ is continuous we must prove that the class of theories is inductive. Let $\mathcal{T}$ be any directed family of theories and denote the limit of $\mathcal{T}$ by τ. Then, if all the elements in τ are consistent, τ is a consistent theory. If the inconsistent theory 1 belongs to τ, then $\tau = 1$. So, in any case, τ is a theory. Let $(x_i)_{i\in I}$ be a directed family of consistent pieces of information and denote its limit by x. Then $(\mathcal{D}(x_i))_{i\in I}$ is a directed family of consistent theories and, hence, its limit τ is a consistent theory. Since $x \leq \tau$, this proves that x is consistent. Thus, $\mathcal{D}$ is logically compact. □

The interest of logically compact semantics is expressed in the following proposition:

Proposition 8.3. *If the semantics $\mathcal{M}$ is logically compact, then every satisfiable piece of information admits a maximal model. Equivalently, every element of $\mathcal{M}$ is contained in a maximal element in $\mathcal{M}$.*

Proof. Assume that x is satisfiable, then, because $Sat(\mathcal{M})$ is inductive, the class $C = \{y \in Sat(\mathcal{M}) : y \geq x\}$ is inductive. By Zorn's Lemma, a maximal element z of C exists. Since z is satisfiable, $m \in \mathcal{M}$ exists such that $m \geq z$. Since $m \in C$ and z is maximal of z we can conclude that $z = m$. This proves both that z belongs to $\mathcal{M}$ and that z is a maximal element in $\mathcal{M}$. □

9. PRODUCT OF TWO ABSTRACT DEDUCTION SYSTEMS

Given two abstract deduction systems $(L,\mathcal{D})$ and $(L,\mathcal{D}')$, it is natural to search for a new deduction apparatus able to use both the inferential instruments of $(L,\mathcal{D})$ and $(L,\mathcal{D}')$. This suggests considering the operators $\mathcal{D} \circ \mathcal{D}'$, $\mathcal{D} \circ \mathcal{D}'$ and $\mathcal{D} \vee \mathcal{D}'$. Now, the composition (and the join) of two closure operators is, in general, an almost closure operator and not a closure operator. Consequently, we have to refer to the closure operators generated by these operators which coincide as the following theorem shows:

Theorem 9.1. *Let J and J' be a-c-operators. Then*

$$Cs(J \circ J') = Cs(J) \cap Cs(J') = Cs(J \vee J') = Cs(J' \circ J). \quad (9.1)$$

Consequently,

$$c(J \circ J') = c(J' \circ J) = c(J \vee J'), \quad (9.2)$$

i.e., $J \circ J'$, $J' \circ J$ and $J \vee J'$ generate the same closure operator.

Proof. Let x be a fixed point of $J \circ J'$. Then $J'(x) \leq J(J'(x)) = x$ and therefore x is a fixed point of J'. Moreover, the equalities $x = J(J'(x)) = J(x)$ show that x is also a fixed point of J. Conversely, it is apparent that if x is fixed for both J and J', then x is fixed for $J \circ J'$ and this proves the first equality. The remaining part of the proposition follows from (9.1) and Theorem 4.4. □

Definition 9.2. Let $(L, \mathcal{D})$ and $(L, \mathcal{D}')$ be two deduction systems. Then we call *product* of $(L, \mathcal{D})$ and $(L, \mathcal{D}')$ the deduction system

$$(L, c(\mathcal{D}\circ\mathcal{D}')) = (L, c(\mathcal{D}'\circ\mathcal{D})) = (L, c(\mathcal{D}\vee\mathcal{D}')) .$$

In Chapter 2 we will use such a definition to define similarity-based fuzzy logics. From the first equality in (9.1) it follows that a piece of information x is a theory of the product $(L, c(\mathcal{D} \circ \mathcal{D}'))$ iff x is a theory of both the deduction system $(L, \mathcal{D})$ and $(L, \mathcal{D}')$. The proof of the following theorem is evident:

Theorem 9.3. *The product of deduction systems is a commutative and associative operation. Moreover, the product of two continuous (logically compact) deduction systems is a continuous (logically compact) deduction system.*

Sometimes it is possible that the composition of two closure operators is a closure operator. The following theorem gives some information to this regard:

Theorem 9.4. *Let J and J' be closure operators. Then the following are equivalent:*

(i) *$J \circ J'$ is a closure operator.*

(ii) $J \circ J' \geq J' \circ J$.
(iii) $J(J'(x))$ *is a fixed point of* J' *for every* $x \in L$.

Proof. (i) $\Rightarrow$ (ii). Let $x \in L$. Then, from $J'(x) \geq x$ it follows that $J(J'(x)) \geq J(x)$ and therefore $J'(J(J'(x))) \geq J'(J(x))$. Thus, since by hypothesis $J \circ J'$ is a closure operator, by the inclusion property for J,

$$J(J'(x)) = J(J'(J(J'(x)))) \geq J'(J(J'(x))) \geq J'(J(x)).$$

(ii) $\Rightarrow$ (iii). Observe that

$$J'(J(J'(x))) \leq J(J'(J'(x))) = J(J'(x)).$$

(iii) $\Rightarrow$ (i). Observe that

$$J(J'(J(J'(x)))) = J(J(J'(x))) = J(J'(x)).$$

□

10. DUALITY PRINCIPLE FOR ORDERED SETS

Let $(L, \leq)$ be an ordered set and denote the inverse of $\leq$ by $\leq^d$, i.e.,

$$x \leq^d y \;\Leftrightarrow\; y \leq x.$$

Then the pair $(L, \leq^d)$ is an ordered set we call the *dual* of $(L, \leq)$. Let $\mathcal{P}$ be any definition or proposition in a language suitable for ordered sets. Then, we denote by $\mathcal{P}^d$ the definition or proposition obtained by interchanging $\leq$ and $\geq$ and we say that $\mathcal{P}^d$ is *the dual of* $\mathcal{P}$. As an example, the dual of the definition of least element is the definition of greatest element and conversely. Also, let $\mathcal{P}$ be the statement which claims that there is at most one least element. Then $\mathcal{P}^d$ is the statement which claims that there is at most one greatest element. Trivially, a statement $\mathcal{P}$ is true in $(L, \leq)$ if and only if $\mathcal{P}^d$ is true in $(L, \leq^d)$. Consequently, we have the following duality principle for ordered sets:

If $\mathcal{P}$ *holds for every ordered set, then* $\mathcal{P}^d$ *holds for every ordered set, too.*

For instance, assume that we have proved that in any ordered set there is at most one least element. Then we are authorized to claim that in any ordered set there is at most one greatest element. On the other hand, if a proposition was proved only for a particular ordered set L, its dual is not necessarily true for L. As an example, consider the ordered set $L = \{1/n : n \in N\}$. Then while "there is no least element" holds for L, the dual proposition "there is no greatest element" is false because 1 is the greatest element.

Define a *dual isomorphism* from L_1 onto L_2 as any isomorphism from L_1 onto $L_1{}^d$, i.e., any surjective map $h : L_1 \to L_2$ such that, for $x, y \in L_1$

$$x \leq y \;\Leftrightarrow\; h(x) \geq h(y).$$

A *dual automorphism* of L is a dual isomorphism from L onto itself. Let L be an ordered set admitting a dual automorphism $h : L \to L$. Then L satisfies a duality principle as

if a property $\mathcal{P}$ *is satisfied in* L, *then its dual* $\mathcal{P}^d$ *is satisfied in* L.

As an example, consider the ordered set $\boldsymbol{R}$ of real numbers and set $h(x) = -x$. Then h is a dual automorphism and to any proven property for $\boldsymbol{R}$ (for instance, there is no least element) corresponds a dual property for $\boldsymbol{R}$ (there is no greatest element).

Also, a duality principle applies to (complete) lattices. Indeed, the dual L^d of a (complete) lattice L is a (complete) lattice, so we have

if $\mathcal{P}$ holds for any (complete) lattice, then $\mathcal{P}^d$ holds for any (complete)lattice.

Trivially, the meet $\wedge^d$ in L^d is the join in L, the join $\vee^d$ is the meet in L, the greatest element 1^d in L^d is the least element 0 in L and the least element 0^d is the greatest element 1 in L. As an example, the fact that equation $x \vee 1 = 1$ holds in any lattice entails that its dual $x \wedge 0 = 0$ holds in any lattice, too. Given two lattices L_1 and L_2, define a *dual homomorphism* (*isomorphism*) from L_1 to L_2, as any homomorphism (isomorphism) from L_1 to $L_2{}^d$, i.e., any map $h : L_1 \to L_2$, such that

$$h(0) = 1 \ , \ h(1) = 0 \ , \ h(x \vee y) = h(x) \wedge h(y) \ , \ h(x \wedge y) = h(x) \vee h(y).$$

If a dual isomorphism $h : L \to L$ exists, then L satisfies a duality principle as

if a property $\mathcal{P}$ is satisfied in L, then its dual $\mathcal{P}^d$ is also satisfied in L.

For instance, the above principle holds for the class $\mathcal{P}(S)$ of all subsets of a given set S because the equality $h(X) = -X$ defines a dual automorphism.

Due to the symmetry of the system of axioms characterizing a Boolean algebra, duality principle can be extended to Boolean algebras. Notice that the complement operator $- : \boldsymbol{B} \to \boldsymbol{B}$ is a dual automorphism of any Boolean algebra $\boldsymbol{B}$.

Duality principle applies also to the concepts of closure operator and closure system. Namely, we call *interior operator* in L, any closure operator in L^d, i.e., any operator $I : L \to L$ such that

$$\text{(i)}^d \ \ x \le y \ \Rightarrow I(x) \le I(y) \qquad , \qquad \text{(ii)}^d \ \ x \ge I(x) \qquad , \qquad \text{(iii)}^d \ \ I(I(x)) = I(x).$$

In the case in which only (i)^d and (ii)^d are satisfied, I is called an *almost interior operator*. Likewise, we define an *open system or interior system* in L as any class C of elements of L closed under arbitrary joins. Equivalently, an open system is any closure system in L^d. It is evident how translate the results in this chapter in the dual results. For instance, we have the following proposition:

Proposition 10.1. *Any interior system C is a complete lattice such that*
- *the greatest element of C is the join of all the elements in C,*
- *the zero in C coincides with the zero 0 in L,*
- *the joins in C coincide with the joins in L,*
- *the meet in C of a class X is* $Sup\{x \in C : x \le Inf(X)\}$.

The dual of the operator $Co : \mathcal{P}(L) \to L^L$ is the operator $Io : \mathcal{P}(L) \to L^L$ defined by setting, for any $C \in \mathcal{P}(L)$ and $x \in L$,

$$Io(C)(x) = Sup\{y \in C : \ y \le x\}. \tag{10.1}$$

The dual of the operator $Cs : L^L \to \mathcal{P}(L)$ is the operator $Is : L^L \to \mathcal{P}(L)$ defined by setting, for any $J \in L^L$,

$$Is(J) = \{x \in L : J(y) \ge x \text{ for every } y \ge x\}. \tag{10.2}$$

By duality, we obtain:

Proposition 10.2. *For any $C \in \mathcal{P}(L)$, $Io(C)$ is an interior operator. For any $J \in L^L$, $Is(J)$ is an interior system.*

So, any class C is able to generate both a closure operator $Co(C)$ and an interior operator $Io(C)$, and any operator J is able to generate both a closure system $Cs(J)$ and an interior system $Is(J)$.

CHAPTER 2

ABSTRACT FUZZY LOGIC

1. FUZZY SUBSETS FOR VAGUENESS

Let $\mathcal{P}$ be a *"well defined"* property in a set S, i.e., a property such that given any element x in S, either x satisfies $\mathcal{P}$ or not. Then, the axiom of separation in classical set theory enables us to assert that the elements of S satisfying $\mathcal{P}$ define a subset of S we denote by $\{x \in S : x \text{ satisfies } \mathcal{P}\}$. As an example, if S is the set of natural numbers and $\mathcal{P}$ is the property *"odd"*, then the set $\{x \in S : x \text{ is an odd number}\}$ is defined. Now, there are properties that are *"vague"* and therefore not well defined. This since they can be satisfied with several different degrees. For instance, *"short"*, *"big"*, *"near"*, are vague properties. Then, the question arises whether these properties isolate some type of subset and therefore whether we can give a precise meaning to intuitive notions as *"the set of short men"*, *"the set of big numbers"*, *"the set of shops near to the station"*, and so on.

An answer can be obtained by starting from the notion of characteristic function of a subset. Recall that every subset X of S can be represented by its *"characteristic function"* $c_X : S \to \{0,1\}$ defined by setting $c_X(x) = 1$ if $x \in X$, and $c_X(x) = 0$ if $x \notin X$. Hear, 1 represents the truth-values *"true"* and 0 the truth-value *"false"*. The idea is that, by admitting truth-values different from 0 and 1, we can represent the extension of a vague predicate by a generalized characteristic function, namely a function assuming values in the interval [0,1] or, more generally, in a complete lattice U.

Definition 1.1. Let U be a complete lattice and S a set. Then a *fuzzy subset* or *U-subset* of S is any map $s : S \to U$ from S into U. We denote by $\mathcal{F}(S)$ or by U^S the class of fuzzy subsets of S (see Zadeh [1965]).

For every $x \in S$, we say that $s(x)$ is the *degree of membership* of x to s. For instance, if S is the set of cakes, we can represent the fuzzy subset of big cakes by a suitable map $s : S \to [0,1]$. Then, for any $x \in S$, the number $s(x)$ is the truth-value of the sentence *"x is big"*. From this point of view:

- $s(x) = 1$ means that x is big,
- $s(x) = 0$ means that x is not big,
- $s(x) - 0.5$ means that x is half-big,

and so on.

Definition 1.1 is basic for the whole theory of fuzzy subsets and in this book we refer constantly to it. Nevertheless, vagueness cannot be captured by such a simple notion. Indeed, in everyday linguistic activity not only several notions are vague but also the interpretations of these vague notions are, in turn, vague. For instance, it is reasonable that the value $s(x)$ depends on the diameter of the cake x, i.e., $s(x) = n(h(x))$ where $h(x)$ is the diameter of x and $n : \boldsymbol{R} \to [0,1]$ is a suitable function able to *"normalize"* the information contained by $h(x)$. Then, since there are an infinity of reasonable normalization functions, infinite fuzzy subsets exist that are adequate to represent the intuitive notion of *"big"*. This is a basic feature of vagueness because the interpretation of any vague concept is *"elastic"*, in a sense. This interpretation depends on the context and it is subjective in a relevant way.

Nevertheless, we can transmit information by a vague language !

For example, imagine that John says by telephone to Mary that a big cake is on his table. Then, some kind of information was transmitted since Mary is able to give a meaning to words as "big", "cake", "table". Now, by using fuzzy formalisms, we can interpret the vague notion "big" by a fuzzy subset, obviously. But we cannot assume that the fuzzy subset (unconsciously) used by John to interpret the word "big" is equal to the fuzzy subset used by Mary. The same argument holds for the words "cake" and "table". As a matter of fact, as observed in Hájek [1999]b, when two persons speak it is not clear whether they speak about the same things or not. This leads to the question:

how do we communicate ?

We agree with Hájek's idea that conversation is a cooperative game where in each person the meaning of the words may change (see also Section 1 in Chapter 10).

We can ask ourselves whether vagueness can be avoided by defining an ideal language where only precise concepts are admitted. Certainly, classical mathematical logic seems to furnish such a language but, unfortunately, it has achieved success only in speaking and reasoning about mathematical structures. As a matter of fact, we have learned to tolerate a range of vagueness and inaccuracy both in our communication and in our thinking only because this is necessary. Any attempt to do otherwise would generally obstruct rather than facilitate these activities. Imagine that a rigid discipline conditioned John to speak only in a completely precise way. Then, the only way John can communicate that a big cake is on the table is to communicate the exact diameter of the cake. Probably this is not possible. In actuality, nobody constantly carries a ruler in his pocket to avoid the vagueness of natural language. Very probably, this is not useful (see also Section 1 in Chapter 10). Then, in our opinion:

the aim of a logician cannot be to eliminate vagueness by an *ideal* language but to try to explain how and why vagueness is so useful.

For further information on fuzzy set theory and its philosophical implications readers may refer, e.g., to Goguen [1968/69], Hájek [1998] and Novák, Perfilieva, Mockor [1999].

In Sections 2, 3 and 4 we will give some basic notions on fuzzy subsets. In these sections there is no foundational ambition since fuzzy set theory (in particular, fuzzy logic) is considered only as a chapter of classical mathematics. Interesting attempts to give a general foundation to fuzzy set theory are related to category theory (see, e.g., Höhle [1991] and Höhle [1996]).

2. BASIC NOTIONS

In the following we assume that U coincides with the interval [0,1]. Nevertheless, almost all definitions and results can be extended to any complete lattice. Given a set S we say that a fuzzy subset s of S is *normal* if it attains the value 1 at some element in S. A special notation will be useful to denote the fuzzy subsets that are constant maps. Namely, for every $\lambda \in U$, we denote by s^{λ} the fuzzy subset of S constantly equal to λ. In such a way s^1 is the characteristic function of S and s^0 the characteristic function of $\varnothing$. The fuzzy subset $s^{0.5}$ is called the *completely undetermined* fuzzy subset.

The basic notions on classical sets are extended to the fuzzy sets as follows:

Inclusion. Observe that, if X and Y are subsets of S, then

$$X \subseteq Y \iff c_X(x) \leq c_Y(x) \quad \text{for every } x \in S.$$

So, we define the *inclusion* by setting, for every pair s and s' of fuzzy subsets of S

$$s \subseteq s' \iff s(x) \leq s'(x) \quad \text{for every } x \in S.$$

Union. In accordance with the equality:

$$c_{X \cup Y}(x) = c_X(x) \vee c_Y(x)$$

for every $x \in S$, we define the *union* $s \cup s'$ of two fuzzy subsets s and s' by setting

$$(s \cup s')(x) = s(x) \vee s'(x)$$

for every $x \in S$.

Intersection. Analogously, observe that

$$c_{X \cap Y}(x) = c_X(x) \wedge c_Y(x).$$

Then we can define the *intersection* $s \cap s'$ by setting

$$(s \cap s')(x) = s(x) \wedge s'(x).$$

for every $x \in S$. More generally, given a family $(s_i)_{i \in I}$ of fuzzy subsets of S, the *union* $\bigcup_{i \in I} s_i$ and the *intersection* $\bigcap_{i \in I} s_i$ are defined by

$$(\bigcup_{i \in I} s_i)(x) = Sup\{s_i(x) : i \in I\} \;\; ; \;\; (\bigcap_{i \in I} s_i)(x) = Inf\{s_i(x) : i \in I\}.$$

Complement. Consider the involution $\sim : U \rightarrow U$ defined by setting $\sim(x) = 1-x$, for every $x \in S$. By observing that

$$c_{-X}(x) = \sim c_X(x),$$

we define the *complement* $-s$ of s by setting

$$(-s)(x) = \sim s(x).$$

for every $x \in S$.

Observe that a *"normalization"* in the same lattice is necessary if we will define the union or the intersection of two fuzzy subsets, or, equivalently, evaluate the conjunction or the disjunction of two vague predicates. As an example, we cannot evaluate a composed sentence as *"x is high and weighty"* if height is measured by centimeters in the interval [100,130] and weight by grams in the interval [20000,150000]. Also, different notions of intersection and union can be defined by substituting the operations $\wedge$ and $\vee$ with new operations (for example, with a triangular norm and co-norm, respectively). The same holds for any definition in this section. This is in accordance with the non-Fregean nature of fuzzy logic where, in our opinion, the basic notions cannot be compressed into a unique definitive interpretation.

The usual language for classical sets is used for fuzzy sets, too. For example, if $s \subseteq s'$, we say that s is *contained* in s' or that s is a *part* of s'. The following theorem summarizes the main properties of the class of fuzzy subsets:

Theorem 2.1. *The structure* $(\mathcal{F}(S), \cup, \cap, -)$ *is a complete lattice with an involution, namely the direct power, with index set S, of the structure* $(U, \vee, \wedge, \sim)$. *Such a structure extends the Boolean algebra* $(\mathcal{P}(S), \cup, \cap, -)$. *Indeed, define the map* $H : \mathcal{P}(S) \rightarrow \mathcal{F}(S)$ *by setting* $H(X) = c_X$ *for every* $X \in \mathcal{P}(S)$. *Then H is a "complete embedding" of* $(\mathcal{P}(S), \cup, \cap, -, \varnothing, S)$ *into* $(\mathcal{F}(S), \cup, \cap, -, s^0, s^1)$, *i.e., H is injective, and*

$$H(\bigcup_{i\in I} X_i) = \bigcup_{i\in I} H(X_i) \;\; ; \;\; H(\bigcap_{i\in I} X_i) = \bigcap_{i\in I} H(X_i) \;\; ;$$
$$H(-X) = -H(X) \;\; ; \;\; H(\varnothing) = s^0 \;\; ; \;\; H(S) = s^1.$$

We call *crisp* a fuzzy subset s such that $s(x) \in \{0,1\}$ for every $x \in S$. We can identify the classical subsets of S with the crisp fuzzy subsets of S via the characteristic functions. Moreover, given a subset X of S and x in S, sometimes we write $X(x)$ to denote the element $c_X(x)$. The next theorem emphasizes that when dealing with fuzzy set theory, classical logic does not work well:

Theorem 2.2. *The structure* $(\mathcal{F}(S), \cup, \cap, -)$ *is not a Boolean algebra.*

Proof. Observe that $s^{0.5} = -s^{0.5}$. Hence,

$$s^{0.5} \cup -s^{0.5} = s^{0.5} \neq s^1 \text{ and } s^{0.5} \cap - s^{0.5} = s^{0.5} \neq s^0. \qquad \square$$

The main notions in classical mathematics can be extended to the fuzzy framework. As an example, let X and Y be two sets, then a *fuzzy relation* from X to Y is any fuzzy subset $f : X \times Y \rightarrow U$ of $X \times Y$. A fuzzy relation f is also called a (*non deterministic*) *fuzzy function* or *fuzzy map* and it is denoted by $f : X \sim> Y$. Given $x \in X$ and $y \in Y$, $f(x,y)$ represents the truth degree of the claim *"y is the image of x"*. The *domain* of a fuzzy function f is the fuzzy subset $Dom(f)$ of X defined by setting, for every $x \in X$,

$$Dom(f)(x) = Sup\{f(x,y) : y \in Y\}.$$

The *codomain* of f is the fuzzy subset of Y defined by setting, for every $y \in Y$,

$$Cod(f)(y) = Sup\{f(x,y) : x \in X\}.$$

Let $\odot$ be a binary operation in U interpreting the logical connective *"and"*. Then, given two fuzzy subsets $a : X \rightarrow U$ and $b : Y \rightarrow U$, we define the *Cartesian* $\odot$*-product*, in brief *Cartesian product*, as the fuzzy relation $a \times b : X \times Y \rightarrow U$ defined by setting $(a \times b)(x,y) = a(x) \odot b(y)$ for any $x \in X$ and $y \in Y$.

Remark. (Principle of duality for the lattice of all fuzzy subsets). The complement $- : \mathcal{F}(S) \rightarrow \mathcal{F}(S)$ is a dual automorphism in the lattice of fuzzy subsets. Then, the duality principle stated in Chapter 1 is satisfied. If a property $\mathcal{P}$ holds in $\mathcal{F}(S)$, then its dual $\mathcal{P}^d$ is also satisfied in $\mathcal{F}(S)$. So, for example, from the infinite distributive property of the intersection with respect to the union we can derive the infinite distributive property of the union with respect to the intersection. Also, since no atom exists in $\mathcal{F}(S)$ we can claim that there is no co-atom, and so on.

Definition 2.3. Given a set S, we call an *operator in S* any operator in the lattice $\mathcal{P}(S)$, and we call a *fuzzy operator in S* any operator in the lattice $\mathcal{F}(S)$. We call *class* or *system* in S any subset of $\mathcal{P}(S)$, and *fuzzy class* or *fuzzy system* in S any subset of $\mathcal{F}(S)$.

If J is a fuzzy operator and X a subset of S, then sometimes we write $J(X)$ to denote $J(c_X)$. Moreover, it is natural to associate any fuzzy operator J with its *dual* J_d defined by setting

$$J_d(s) = -J(-s)$$

for every fuzzy subset s. A similar definition holds for the crisp operators. Also, if C is a fuzzy class, then we set

$$C_d = \{-s : s \in C\}.$$

Proposition 2.4. *Let J be a fuzzy operator and C a fuzzy class. Then*

$$(J_d)_d = J \quad , \quad (C_d)_d = C$$

$$J \text{ fuzzy closure operator} \Leftrightarrow J_d \text{ fuzzy interior operator}$$

$$J \text{ fuzzy interior operator} \Leftrightarrow J_d \text{ fuzzy closure operator}$$

$$C \text{ fuzzy closure system} \Leftrightarrow C_d \text{ fuzzy interior system}$$

$$C \text{ fuzzy interior system} \Leftrightarrow C_d \text{ fuzzy closure system.}$$

Proof. Let J be a fuzzy closure operator. To prove that J_d is a fuzzy interior operator, let s be any fuzzy subset. Then, since $J(-s) \supseteq -s$, $J_d(s) = -J(-s) \subseteq s$. Let $s_1 \subseteq s_2$ be two fuzzy subsets. Then $-s_1 \supseteq -s_2$ and therefore $J(-s_1) \supseteq J(-s_2)$. Hence,

$$J_d(s_1) = -J(-s_1) \subseteq -J(-s_2) = J_d(s_2).$$

Finally,

$$J_d(J_d(s)) = -J(-J_d(s)) = -J(-(-J(-s))) = -J(J(-s)) = -J(-s) = J_d(s).$$

The rest is evident.

The following theorem shows how the operators Is and Io defined in Section 10 of Chapter 1 are related to the operators Cs and Co, respectively.

Theorem 2.5. *Let J be a fuzzy operator and C a fuzzy class. Then*

(a) $$Is(J) = (Cs(J_d))_d,$$

(b) $$Cs(J) = (Is(J_d))_d,$$

(c) $$Io(C) = Co(C_d)_d,$$

(d) $$Co(C) = Io(C_d)_d.$$

Proof. In order to prove (a) observe that

$s \in Is(J) \Leftrightarrow J(s') \supseteq s$ for every $s' \supseteq s \Leftrightarrow -J(s') \subseteq -s$ for every $-s' \subseteq -s$

$\Leftrightarrow -J(-s') \subseteq -s$ for every s' such that $s' \subseteq -s$

$\Leftrightarrow J_d(s'') \subseteq -s$ for every $s'' \subseteq -s \Leftrightarrow -s \in Cs(J_d) \Leftrightarrow s \in Cs(J_d)_d$.

In order to prove (b), let I be any fuzzy operator and set $J = I_d$ in (a). Then, since $(I_d)_d = I$, we can conclude that $Is(I_d) = Cs(I)_d$.

In a similar way one demonstrates the remaining part of the theorem. □

3. CLOSED AND OPEN CUTS

Given a fuzzy subset s of S and $\lambda \in U$, the subset

$$C(s,\lambda) = \{x \in S : s(x) \geq \lambda\}$$

is called the *closed λ-cut* of s and

$$O(s,\lambda) = \{x \in S : s(x) > \lambda\}$$

is called the *open λ-cut* of s. The concept of cut of a fuzzy subset is basic as it enables us to establish a connection between fuzzy and classical theory (see the extension principle in Chapter 3). If s is the characteristic function of a subset X, then $C(s,\lambda) = X$ for every $\lambda \neq 0$ and $O(s,\lambda) = X$ for every $\lambda \neq 1$. In the propositions below we summarize some basic properties of the closed cuts.

Proposition 3.1. *Let s and s' be fuzzy subsets. Then, for every $\lambda \in U$,*

(a) $C(s,0) = S$; (b) $\lambda \leq \lambda' \Rightarrow C(s,\lambda) \supseteq C(s,\lambda')$

(c) $s \subseteq s' \Rightarrow C(s,\lambda) \subseteq C(s',\lambda)$; (d) $C(s,\lambda) = \bigcap_{\mu<\lambda} O(s,\mu)$

(e) $C(s \cup s',\lambda) = C(s,\lambda) \cup C(s',\lambda)$; (f) $C(s \cap s',\lambda) = C(s,\lambda) \cap C(s',\lambda)$.

Proof. We confine ourselves to demonstrate (d). Indeed:

$x \in C(s,\lambda) \Leftrightarrow s(x) \geq \lambda \Leftrightarrow s(x) > \mu$ for every $\mu < \lambda \Leftrightarrow x \in \bigcap_{\mu<\lambda} O(s,\mu)$. □

Proposition 3.2. *The family of closed cuts of a fuzzy subset is join-reversing, i.e., for every family $(\lambda_i)_{i\in I}$ of elements of U*

$$C(s, Sup_{i\in I}\, \lambda_i) = \bigcap_{i\in I} C(s,\lambda_i). \tag{3.1}$$

Moreover, given any family $(s_i)_{i\in I}$ of fuzzy subsets

$$C(\bigcap_{i\in I} s_i,\lambda) = \bigcap_{i\in I} C(s_i,\lambda). \tag{3.2}$$

Proof. In order to prove (3.1) observe that

$x \in C(s, Sup_{i\in I}\, \lambda_i) \Leftrightarrow s(x) \geq \lambda_i$ for every $i \in I \Leftrightarrow x \in \bigcap_{i\in I} C(s,\lambda_i)$.

In order to prove (3.2) observe that

$x \in C(\bigcap_{i\in I} s_i, \lambda) \Leftrightarrow s_i(x) \geq \lambda$ for every $i \in I \Leftrightarrow x \in \bigcap_{i\in I} C(s_i,\lambda)$. □

Likewise, Proposition 3.3 and 3.4 refer to the open cuts.

Proposition 3.3. *Let s and s' be fuzzy subsets. Then, for every $\lambda \in U$,*

(a) $O(s,1) = \emptyset$; (b) $\lambda \leq \lambda' \Rightarrow O(s, \lambda) \supseteq O(s, \lambda')$

(c) $s \subseteq s' \Rightarrow O(s, \lambda) \subseteq O(s', \lambda)$; (d) $O(s, \lambda) = \bigcup_{\mu>\lambda} C(s, \mu)$

(e) $O(s \cup s', \lambda) = O(s, \lambda) \cup O(s', \lambda)$; (f) $O(s \cap s', \lambda) = O(s, \lambda) \cap O(s', \lambda)$.

Proposition 3.4. *The family of open cuts of a fuzzy subset s is meet-reversing, i.e., for every family $(\lambda_i)_{i\in I}$ of elements of U*

$$O(s, Inf_{i\in I}\, \lambda_i) = \bigcup_{i\in I} O(s, \lambda_i). \tag{3.3}$$

Given a family $(s_i)_{i\in I}$ *of fuzzy subsets*

$$O(\cup_{i\in I} s_i, \lambda) = \cup_{i\in I} O(s_i, \lambda). \tag{3.4}$$

The open cut $O(s,0) = \{x \in S : s(x) \neq 0\}$ is also called the *support* of s and is denoted by $Supp(s)$. If the support of s is finite, then we say that s is *finite*. If it contains just one element, we say that s is a *fuzzy singleton*. Trivially, any fuzzy subset is union of fuzzy singletons and a fuzzy subset is finite iff is union of a finite number of fuzzy singletons.

The next theorem shows that a fuzzy subset is characterized both by its closed cuts and its open cuts. For every $\lambda \in U$, we denote by $\lambda \wedge X$ the fuzzy subset $s^\lambda \cap c_X$. This means that we set $(\lambda \wedge X)(x) = \lambda$ if $x \in X$ and $(\lambda \wedge X)(x) = 0$ if $x \notin X$. Dually, we set $\lambda \vee X = s^\lambda \cup c_X$, that is $(\lambda \vee X)(x) = 1$ if $x \in X$ and $(\lambda \vee X)(x) = \lambda$ in the case $x \notin X$.

Theorem 3.5. *For every fuzzy subset s*

$$s = \cup_{\lambda\in U}(\lambda \wedge C(s, \lambda)), \tag{3.5}$$
$$s = \cup_{\lambda\in U}(\lambda \wedge O(s, \lambda)), \tag{3.6}$$

and, dually,

$$s = \cap_{\lambda\in U}(\lambda \vee O(s, \lambda)), \tag{3.7}$$
$$s = \cap_{\lambda\in U}(\lambda \vee C(s, \lambda)). \tag{3.8}$$

Proof. It is sufficient to observe that it is possible to rewrite (3.5), (3.6), (3.7) and (3.8) in the following equalities:

$$s(x) = Sup\{\lambda \in U : x \in C(s, \lambda)\}, \tag{3.9}$$
$$s(x) = Sup\{\lambda \in U : x \in O(s, \lambda)\}, \tag{3.10}$$
$$s(x) = Inf\{\lambda \in U : x \notin O(s, \lambda)\}, \tag{3.11}$$
$$s(x) = Inf\{\lambda \in U : x \notin C(s, \lambda)\}, \tag{3.12}$$

where $x \in S$. □

Remark. (Duality principle for cuts). We can arrive at Formulas (3.3), (3.4), (3.7), (3.8) and Proposition 3.3 in a simple way from (3.1), (3.2), (3.5), (3.6) and Proposition 3.1, respectively. We have to exchange only the closed cuts with the open cuts and 0, S, $\cap$, $\cup$, $\leq$, $\subseteq$, $\cap$, $\cup$ with 1, $\emptyset$, $\cup$, $\cap$, $\geq$, $\supseteq$, $\cup$, $\cap$, respectively. In fact, a duality principle holds inherited by the analogous principle for the Boolean algebra $\mathcal{P}(S)$ and for the lattice U (the complement $- : \mathcal{P}(S) \to \mathcal{P}(S)^d$ and the negation $\sim : U \to U^d$ are dual automorphisms). Then we are allowed to claim that every result in this book holds true with reference to the structures $\mathcal{P}(S)^d$ and U^d instead of $\mathcal{P}(S)$ and U. If $\mathcal{P}$ is either a proposition or a definition, then we denote its dual by $\mathcal{P}^d$, i.e., the proposition or definition obtained by substituting any subset of S with its complement and 1, 0, S, $\cap$, $\cup$, $\leq$, $\subseteq$, $\wedge$, $\vee$, $\cap$, $\cup$ respectively with 0, 1, $\emptyset$, $\cup$, $\cap$, $\geq$, $\supseteq$, $\vee$, $\wedge$, $\cup$, $\cap$. For instance, the dual of the notion "closed cut" coincides with the notion "open cut". Indeed, given a fuzzy subset s and $\lambda \in U$, if we substitute $\{x \in S : s(x) \geq \lambda\}$ with the complement of $\{x \in S : s(x) \leq \lambda\}$, we get the open cut $\{x \in S : s(x) > \lambda\}$.

4. FUZZY SUBSETS AND CONTINUOUS CHAINS

This section refers to the identification of fuzzy subsets with continuous chains as was proposed in Negoita and Ralescu [1975]a. At first, we consider the class $\mathcal{P}(S)^U$ whose elements are the families of subsets of S with index set U. We consider such a class as an algebraic structure, namely the direct power of the complete lattice $\mathcal{P}(S)$ with index set U. Then $\mathcal{P}(S)^U$ is a complete lattice whose order relation is defined by setting, for every $(A_\lambda)_{\lambda\in U}$ and $(B_\lambda)_{\lambda\in U}$ in $\mathcal{P}(S)^U$,

$$(A_\lambda)_{\lambda\in U} \le (B_\lambda)_{\lambda\in U} \Leftrightarrow A_\lambda \subseteq B_\lambda \text{ for every } \lambda \in U.$$

Further, the join and the meet of these two families are $(A_\lambda \cup B_\lambda)_{\lambda\in U}$, and $(A_\lambda \cap B_\lambda)_{\lambda\in U}$, respectively. One similarly defines the infinite join and meet. We are interested in a particular class of elements of $\mathcal{P}(S)^U$.

Definition 4.1. We call a *chain* in S any order-reversing family $(C_\lambda)_{\lambda\in U}$ of subsets of S such that $C_0 = S$ and we denote by $Ch(S)$ the set of chains in S. We say that $(C_\lambda)_{\lambda\in U}$ is *continuous* provided that it is join-reversing or, equivalently,

$$C_\lambda = \bigcap_{\mu<\lambda} C_\mu, \tag{4.1}$$

for every $\lambda \in U$. We denote by $CCh(S)$ the set of continuous chains in S.

The family of closed cuts of a given fuzzy set is an example of continuous chain.

Proposition 4.2. *The class $Ch(S)$ of chains in S is a complete sub-lattice of $\mathcal{P}(S)^U$. Hence, $Ch(S)$ is a closure system in $\mathcal{P}(S)^U$ and, given any family $(C_\lambda)_{\lambda\in U}$ of subsets of S, the chain $(C'_\lambda)_{\lambda\in U}$ generated by $(C_\lambda)_{\lambda\in U}$ exists. Furthermore, $(C'_\lambda)_{\lambda\in U}$ can be obtained by setting $C'_0 = S$ and, if $\lambda \neq 0$,*

$$C'_\lambda = \bigcup_{\mu\ge\lambda} C_\mu. \tag{4.2}$$

Proof. The first part of the proposition is trivial. To prove (4.2), observe that $(C'_\lambda)_{\lambda\in U}$ is a chain containing $(C_\lambda)_{\lambda\in U}$. Let $(A_\lambda)_{\lambda\in U}$ be a chain containing $(C_\lambda)_{\lambda\in U}$. Then, given $\lambda \in U$, for every $\mu \ge \lambda$, $A_\lambda \supseteq A_\mu \supseteq C_\mu$ and therefore $A_\lambda \supseteq C'_\lambda$. □

Proposition 4.3. *The class $CCh(S)$ of continuous chains is a closure system in $Ch(S)$ (hence in $\mathcal{P}(S)^U$). Let $(C_\lambda)_{\lambda\in U}$ be a chain and define $(C''_\lambda)_{\lambda\in U}$ by setting, for every $\lambda \in U$,*

$$C''_\lambda = \bigcap_{\mu<\lambda} C_\mu. \tag{4.3}$$

Then $(C''_\lambda)_{\lambda\in U}$ is the continuous chain generated by $(C_\lambda)_{\lambda\in U}$.

Proof. Let I be a set and, for every $i \in I$, let $(C^i_\lambda)_{\lambda\in U}$ be a continuous chain. Then the intersection of such a family of continuous chains is the chain $(C_\lambda)_{\lambda\in U}$ defined by setting $C_\lambda = \bigcap_{i\in I} C^i_\lambda$. Since

$$\bigcap_{\mu<\lambda} C_\mu = \bigcap_{\mu<\lambda} (\bigcap_{i\in I} C^i_\mu) = \bigcap_{i\in I}(\bigcap_{\mu<\lambda} C^i_\mu) = \bigcap_{i\in I} C^i_\lambda = C_\lambda,$$

we can claim that $(C_\lambda)_{\lambda\in U}$ is a continuous chain. Suppose $(C_\lambda)_{\lambda\in U}$ is a chain. Then, since $C_\mu \supseteq C_\lambda$ for every $\lambda > \mu$, we have $C''_\lambda \supseteq C_\lambda$. To prove that $(C''_\lambda)_{\lambda\in U}$ is a continuous chain, observe that

$$\bigcap_{\mu<\lambda} C''_\mu = \bigcap_{\mu<\lambda} (\bigcap_{\nu<\mu} C_\nu) = \bigcap_{\nu<\lambda} C_\nu = C''_\lambda.$$

Finally, let $(A_\lambda)_{\lambda\in U}$ be a continuous chain containing $(C_\lambda)_{\lambda\in U}$, then

$$A_\lambda = \bigcap_{\mu<\lambda} A_\mu \supseteq \bigcap_{\mu<\lambda} C_\mu = C''_\lambda. \quad \square$$

The proof following proposition is trivial:

Proposition 4.4. *Let $(C_\lambda)_{\lambda\in U}$ be any family of subsets of S and set*

$$C^*_\lambda = \bigcap_{x<\lambda}(\bigcup_{y\geq x} C_y). \tag{4.4}$$

Then $(C^*_\lambda)_{\lambda\in U}$ is the continuous chain generated by $(C_\lambda)_{\lambda\in U}$. Moreover,

$$C^*_\lambda = \bigcap_{x<\lambda}(\bigcup_{y>x} C_y). \tag{4.5}$$

The following basic definition enables us to associate any family of subsets of S with a fuzzy subset of S.

Definition 4.5. Let $(C_\lambda)_{\lambda\in U}$ be any family of subsets of S and set

$$s = \bigcup_{\lambda\in U} \lambda \wedge C_\lambda. \tag{4.6}$$

Then s is said to be *the fuzzy subset associated with* $(C_\lambda)_{\lambda\in U}$.

Observe that (4.6) is equivalent to setting, for any $x \in S$,

$$s(x) = Sup\{\lambda \in U : x \in C_\lambda\}. \tag{4.7}$$

Indeed,

$$\begin{aligned} s(x) &= Sup\{\lambda \wedge C_\lambda(x) : \lambda \in U\} \\ &= (Sup\{\lambda \wedge C_\lambda(x) : \lambda \in U, C_\lambda(x) = 0\}) \vee (Sup\{\lambda \wedge C_\lambda(x) : \lambda \in U, C_\lambda(x) = 1\}) \\ &= 0 \vee (Sup\{\lambda \in U : \lambda \in U, C_\lambda(x) = 1\}) = Sup\{\lambda \in U : x \in C_\lambda\}. \end{aligned}$$

Proposition 4.6. *Let $(C_\lambda)_{\lambda\in U}$ be any family of subsets of S and s the associated fuzzy subset. Then*

$$O(s,\mu) = \bigcup_{\lambda>\mu} C_\lambda \subseteq C'_\mu \subseteq C^*_\mu = C(s,\mu). \tag{4.8}$$

Consequently, both the chain $(C'_\lambda)_{\lambda\in U}$ and the continuous chain $(C^*_\lambda)_{\lambda\in U}$ generated by $(C_\lambda)_{\lambda\in U}$ define the same fuzzy subset s.

Proof. Observe that

$$x \in O(s,\mu) \Leftrightarrow s(x) > \mu \Leftrightarrow \exists\lambda > \mu \text{ such that } x \in C_\lambda \Leftrightarrow x \in \bigcup_{\lambda>\mu} C_\lambda,$$

and that

$$x \in C(s,\mu) \Leftrightarrow \forall\nu < \mu\, \exists\lambda > \nu \text{ s.t. } x \in C_\lambda \Leftrightarrow x \in \bigcap_{\nu<\mu}(\bigcup_{\lambda\geq\nu} C_\lambda). \quad \square$$

Proposition 4.7. *Let $(C_\lambda)_{\lambda\in U}$ be any chain of subsets of S and s the associated fuzzy subset. Then we also have*

$$s = \bigcap_{\lambda\in U} \lambda \vee C_\lambda. \tag{4.9}$$

Moreover,

$$O(s,\mu) = \bigcup_{\lambda>\mu} C_\lambda \subseteq C_\mu \subseteq \bigcap_{\lambda<\mu} C_\lambda = C(s,\mu). \tag{4.10}$$

Proof. Equality (4.9) is equivalent to

$$s(x) = Inf\{\lambda \in U : x \notin C_\lambda\}. \tag{4.11}$$

Indeed,

$$Inf\{\lambda \vee C_\lambda(x) : \lambda \in U\}$$
$$= (Inf\{\lambda \vee C_\lambda(x) : \lambda \in U, C_\lambda(x) = 1\}) \wedge (Inf\{\lambda \vee C_\lambda(x) : \lambda \in U, C_\lambda(x) = 0\})$$
$$= 1 \wedge (Inf\{\lambda : \lambda \in U, C_\lambda(x) = 0\}) = Inf\{\lambda \in U : x \notin C_\lambda\}.$$

On the other hand, since $(C_\lambda)_{\lambda\in U}$ is order-reversing, $\{\lambda \in U : x \in C_\lambda\}$ is an interval containing 0. Then,

$$s(x) = Sup\{\lambda \in U : x \in C_\lambda\} = Inf\{\lambda \in U : x \notin C_\lambda\}.$$

The rest of the proposition is evident. □

Observe that if $(C_\lambda)_{\lambda\in U}$ is a chain, the subsets C_1 and C_0 have no influence in defining the fuzzy subset s. In fact, the fuzzy subset $0 \wedge C_0 = \varnothing$ gives no contribution to the union in (4.6). The fuzzy subset $1 \vee C_1 = S$ gives no contribution to the intersection in (4.9). As a direct consequence of Proposition 4.6, we also have:

Proposition 4.8. *Let $(C_\lambda)_{\lambda\in U}$ be a continuous chain and s the associated fuzzy subset. Then, for every $\mu \in U$, $C(s,\mu) = C_\mu$.*

The following theorem shows that we can identify the lattices $\mathcal{F}(S)$ of all fuzzy subsets with the lattice $CCh(S)$ of all continuous chains.

Theorem 4.9. *Let $H : \mathcal{F}(S) \to CCh(S)$ be the map defined by setting, for every $s \in \mathcal{F}(S)$,*

$$H(s) = (C(s,\lambda))_{\lambda\in U}. \tag{4.12}$$

Then H is a lattice isomorphism between $\mathcal{F}(S)$ and $CCh(S)$. Moreover, its inverse $H^{-1} : CCh(S) \to \mathcal{F}(S)$ associates every continuous chain $(C_\lambda)_{\lambda\in U}$ with the fuzzy subset s defined by (4.6).

Proof. By Proposition 3.2 $(C(s,\lambda))_{\lambda\in U}$ is a continuous chain and this proves that $H(s) \in CCh(S)$. To prove that H is surjective, let $(C_\lambda)_{\lambda\in U}$ be any continuous chain and define s by (4.6). Then, since $C_\mu = C(s,\mu)$, $H(s)$ coincides with $(C_\lambda)_{\lambda\in U}$. □

Remark. The dual of the notion of chain is the notion of *co-chain.* Then a co-chain is an order-reversing family $(C_\lambda)_{\lambda\in U}$ of subsets of S such that $C_1 = \varnothing$. This is due to the fact that, when we apply the duality principle, both the order relations $\leq$ and $\subseteq$ are to be reversed and S and 0 are to be substituted by $\varnothing$ and 1, respectively. Likewise, a *continuous co-chain* is defined as any meet-reversing co-chain, i.e., a co-chain $(C_\lambda)_{\lambda\in U}$ such that $C_\lambda = \cup_{\mu>\lambda} C_\mu$ for every $\mu \in U$. The continuous co-chains are also known in the literature under the name of *flou subsets* (see Menger [1951]). A typical example of continuous co-chain is the family of open cuts of a fuzzy subset.

5. ABSTRACT FUZZY LOGIC

Let $\mathcal{F}$ be a set whose elements we call *formulas*. We call an *abstract crisp deduction system in $\mathcal{F}$* any abstract deduction system in the lattice $\mathcal{P}(\mathcal{F})$ of all subsets of $\mathbb{F}$. The notions of *abstract crisp semantics* and *abstract crisp logic in* $\mathbb{F}$ are defined in a similar way. Then, a crisp semantics is any class $\mathcal{M}$ of subsets of formulas such that $\mathbb{F} \notin \mathcal{M}$. Given a set X of formulas and $M \in \mathcal{M}$, we say that M is a (crisp) *model of* X, in brief $M \models X$, provided that $X \subseteq M$. In particular, we say that a formula α is *true* in M and then we write $M \models \alpha$, if $\alpha \in M$. Moreover, we write $X \models \alpha$ to denote that α is true in any model of X. We denote by Lc the logical consequence operator $Co(\mathcal{M})$ associated with $\mathcal{M}$, i.e.,

$$Lc(X) = \bigcap\{M \in \mathcal{M} : M \supseteq X\}.$$

Equivalently,

$$Lc(X) = \{\alpha \in \mathbb{F} : X \models \alpha\}.$$

An *abstract crisp logic* is a triplet $(\mathbb{F}, \mathcal{D}, \mathcal{M})$ such that $\mathcal{D} = Lc$. A justification of the above definitions was given in Section 5 of Chapter 1. The *system of tautologies* is the set

$$Tau(\mathcal{M}) = \bigcap\{M : M \in \mathcal{M}\},$$

and its elements are called *tautologies*. Then a tautology is a formula which is true in any model. We can also define the *set of contradictions* by

$$Contr(\mathcal{M}) = \bigcap\{-M : M \in \mathcal{M}\},$$

i.e., a contradiction is a formula which is false in any model.

Definition 5.1. We call an *abstract fuzzy deduction system* (*abstract fuzzy semantics*, *abstract fuzzy logic*) any abstract deduction system (abstract semantics, abstract logic) in the lattice of all fuzzy subsets of $\mathbb{F}$.

Then, a fuzzy semantics is a class $\mathcal{M}$ of fuzzy subsets of formulas such that the map s^1 constantly equal to 1 is not in $\mathcal{M}$. The meaning of such a condition is obvious: no world in which every formula is true exists. The elements in $\mathcal{M}$ are named *fuzzy models*. Examples of fuzzy semantics are obtained by setting $\mathcal{M}$ equal to the class of the truth-functional valuations of the formulas in a multivalued logic. Other examples are furnished by the probability logics considered in Chapter 9.

Sometimes we call an *initial valuation* (or a *piece of fuzzy information* or a *fuzzy set of proper axioms*) any piece of information v in the lattice of all fuzzy subsets of $\mathbb{F}$, i.e., any subset v of formulas. We interpret v as incomplete information about an unknown world m. Namely, as m is a model of v provided that $m \supseteq v$, the information carried on by v is that, given any formula α,

"the actual truth value of α is at least $v(\alpha)$".

From this point of view, an initial valuation v is not a fuzzy subset since the value $v(\alpha)$ is not a truth degree but a constraint on the possible truth degree of α (see Chapter 5). This is in accordance with classical logic where the available information is expressed by a set T of formulas (the assumptions) arising from a partial knowledge of a world m and the information carried on by T is that

"at least the formulas in T are true in m".

The logical consequence operator Lc is defined by setting, given an initial valuation v,

$$Lc(v) = \bigcap\{m \in \mathcal{M} : m \supseteq v\}.$$

Like the initial valuation v, the meaning of $Lc(v)(\alpha)$ is still

"*α is true at least to degree $Lc(v)(\alpha)$*",

further, we also have that

"*$Lc(v)(\alpha)$ is the best possible lower constraint we can draw from v.*"

Observe that, while $Lc(v)(\alpha) = 1$ entails that α is true in any model of v, $Lc(v)(\alpha) = 0$ does not mean that α is false but that the available information v does not say anything to support α. On the other hand, this also occurs in classical logic. Indeed, assume that T is a set of sentences expressing our partial knowledge about a world m and let α be a formula which is not a logical consequence of T. In such a case we cannot say that α is false in m but only that the information contained in T is not sufficient to prove α. In other words $Lc(v)(\alpha)$ only carries positive information about α. The negative information about α is represented by the number $Lc(v)(\neg\alpha)$ (provided that the language under consideration is equipped with a negation $\neg$).

Moreover, recall that, in accordance with the definition given in Chapter 1, the *fuzzy subset of tautologies* is defined by

$$Tau(\mathcal{M}) = \bigcap\{m : m \in \mathcal{M}\}.$$

The meaning of such a fuzzy subset is that, given any formula α, α is true in any model m at least to degree $Tau(\mathcal{M})(\alpha)$. We define a *tautology* as any formula α such that $Tau(\mathcal{M})(\alpha) = 1$, i.e., $m(\alpha) = 1$ for every $m \in \mathcal{M}$. We can also define the *fuzzy subset of contradictions* $Contr(\mathcal{M})$ as the intersection of all the complements of the elements in $\mathcal{M}$, i.e.,

$$Contr(\mathcal{M}) = \bigcap\{-m : m \in \mathcal{M}\}.$$

obviously,

$$Tau(\mathcal{M})(\alpha) + Contr(\mathcal{M})(\alpha) \leq 1$$

and

$$Contr(\mathcal{M})(\alpha) = 1 - Sup\{m(\alpha) : m \in \mathcal{M}\}.$$

The meaning is that, given any formula α, α is true in any model m at most to degree $\sim Contr(\mathcal{M})(\alpha)$. Thus, the interval $[Tau(\mathcal{M})(\alpha), \sim Contr(\mathcal{M})(\alpha)]$ represents the *a-prior* information about a formula α. Such information arises from the *"logical structure"* of the formula α and from the particular conformation of the given semantics. We define a *contradiction* as any formula α such that $Contr(\mathcal{M})(\alpha) = 1$, i.e., α is a contradiction if $m(\alpha) = 0$ for every $m \in \mathcal{M}$.

6. COMPACTNESS AND CONTINUITY

In this section we will examine continuous abstract fuzzy logics and therefore we will examine the notion of continuity for fuzzy operators. As observed in Chapter 1, given a set S, the notion of finite subset enables us to express the continuity of the operators in the lattice $\mathcal{P}(S)$. Indeed, if J is any order-preserving operator in S, then J is continuous iff J is compact, i.e., for any $X \in \mathcal{P}(S)$

$$J(X) = \bigcup\{J(X_f) : X_f \text{ is a finite part of } X\}. \quad (6.1)$$

Also, this is equivalent to saying that for every $X \subseteq S$ and $x \in S$,

$$x \in J(X) \Leftrightarrow \text{a finite subset } X_f \text{ exists such that } x \in J(X_f). \quad (6.2)$$

Now, since the concept of finite fuzzy subset is also defined, we can try for similar characterizations of the continuity in the lattice of all fuzzy subsets of S. Then we consider the following definitions:

Definition 6.1. A fuzzy operator J is called *compact* if, for every fuzzy set s

$$J(s) = \cup\{J(s_f) : s_f \text{ finite and } s_f \subseteq s\}. \tag{6.3}$$

J is called *p-compact* (see Pavelka [1979]a) provided that J is order-preserving and, for every fuzzy subset s and $x \in S$, a finite fuzzy subset s_f of s exists such that

$$J(s)(x) = J(s_f)(x). \tag{6.4}$$

Observe that in Turunen [1999] an operator is called *continuous* provided that

$$J(s) = \cup\{J(s \cap X_f) : X_f \text{ finite}\}.$$

It is easy to prove that for order-preserving operators such a notion coincides with compactness. The following proposition shows that compactness, p-compactness and continuity are three different notions and that they are related in accordance with the following diagram:

$$\begin{array}{ccc} & & \text{continuous} \\ & & \Downarrow \\ p\text{-compact} & \Rightarrow & \text{compact.} \end{array}$$

Proposition 6.2. *Let J be a fuzzy operator. Then the following implications hold*

(i) *J p-compact $\Rightarrow$ J compact* ; (ii) *J continuous $\Rightarrow$ J compact.*

Instead,

(iii) *J continuous $\not\Rightarrow$ J p-compact* ; (iv) *J p-compact $\not\Rightarrow$ J continuous,*

(v) *J compact $\not\Rightarrow$ J p-compact* ; (vi) *J compact $\not\Rightarrow$ J continuous.*

Proof. (i) is trivial. In order to prove (ii) suppose J is continuous, let s be a fuzzy subset and denote by $FS(s)$ the class of finite fuzzy subsets of s. Then, since $FS(s)$ is directed,

$$J(s) = J(lim\ FS(s)) = lim\ J(FS(s)) = \cup\{J(s_f) : s_f \in FS(s)\}$$

and this demonstrates that J is compact. In order to falsify (iii) and (v), consider the operator J defined by setting $J(s) = s^{\lambda}$, where $\lambda = Sup\{s(x) : x \in S\}$ and s^{λ} is the fuzzy subset constantly equal to λ. Then J is a continuous (and therefore compact) closure operator which is not p-compact. Indeed, it is sufficient to set $S = [0,1]$ and to consider the fuzzy subset s of S defined by setting $s(x) = x$ if $x < 1$ and $s(x) = 0$ if $x = 1$. Then it is clear that, $J(s)(x) = s^1(x) = 1$ and, for every finite fuzzy subset s_f of s, $J(s_f)(x) \neq 1$.

In order to disprove (iv) and (vi), consider the operator J defined by $J(s)(x) = s(x)$ if $s(x) < 0.5$ and, otherwise, $J(s)(x) = 1$. Then J is a p-compact (and therefore compact) closure operator which is not continuous. In fact, it is obvious that J is p-compact. Let

$$\lambda(1) < \lambda(2) < \ldots < \lambda(n) < \ldots$$

be an injective increasing sequence of elements of U such that $Sup_{n\in N}\lambda(n) = 0.5$. Then $(s^{\lambda(n)})_{n\in N}$ is a directed family but, while $J(lim\ s^{\lambda(n)}) = J(s^{0.5}) = s^1$, it is

$$lim\ J(s^{\lambda(n)}) = lim\ s^{\lambda(n)} = s^{0.5}. \qquad \square$$

A very interesting characterization of the continuous closure operators, in terms of finite fuzzy subsets, was established in Murali [1991]. The basic idea is to define a new relation $\ll$ by setting, for any $s_1, s_2 \in \mathcal{F}(S)$,

$$s_1 \ll s_2 \Leftrightarrow s_1(x) < s_2(x) \text{ for every } x \in Supp(s_1).$$

A basic property of this relation is that, given any directed class of fuzzy subsets $\mathcal{T}$,

$$s_f \text{ finite fuzzy subset and } s_f \ll \cup\{s' : s' \in \mathcal{T}\} \Rightarrow \exists s' \in \mathcal{T} \text{ such that } s_f \subseteq s'.$$

Indeed, let $Supp(s_f) = \{x_1,...,x_n\}$. Then for $i = 1,...,n$, the inequality $s_f(x_i) < Sup\{s'(x_i) : s' \in \mathcal{T}\}$ entails that $s_i \in \mathcal{T}$ exists such that $s_f(x_i) < s_i(x_i)$. So, if s' is any element of $\mathcal{T}$ containing $s_1,...,s_n$, we have that $s_f \subseteq s'$.

Theorem 6.3. *A fuzzy operator J is continuous iff, for every fuzzy subset s,*

$$J(s) = \cup\{J(s_f) : s_f \text{ is finite and } s_f \ll s\}. \tag{6.5}$$

Proof. Assume (6.5). Then it is evident that J is order-preserving. Let $\mathcal{T}$ be any directed class of fuzzy subsets. Then we have to prove that

$$J(\cup\{s' : s' \in \mathcal{T}\}) = \cup\{J(s') : s' \in \mathcal{T}\}.$$

Set $s = \cup\{s' : s' \in \mathcal{T}\}$, then,

$$J(s) = \cup\{J(s_f) : s_f \text{ is finite and } s_f \ll s\} \subseteq \cup\{J(s') : s' \in \mathcal{T}\}. \tag{6.6}$$

The converse inclusion is trivial. Thus, J is continuous.

Conversely, assume that J is continuous. Then, since $\{s_f : s_f \text{ is finite and } s_f \ll s\}$ is a directed family whose union is s, (6.5) is self-evident. $\square$

7. LOGICAL COMPACTNESS

The relation $\ll$ enables us also to characterize the logical compactness for fuzzy operators.

Theorem 7.1. *A fuzzy closure operator J is logically compact iff, for every initial valuation v,*

$$v \text{ is consistent} \Leftrightarrow \text{every finite } v_f \ll v \text{ is consistent.} \tag{7.1}$$

Proof. Assume that J is logically compact. Then, trivially, v consistent implies that every finite fuzzy subset v_f such that $v_f \ll v$ is consistent. In order to prove the converse implication, assume that every finite fuzzy subset v_f such that $v_f \ll v$ is consistent. Then, since v is the inductive limit of the class $\{v_f \in \mathcal{F}(S) : v_f \ll v, v_f \text{ finite}\}$, v is consistent. This demonstrates (7.1). Conversely, assume (7.1) and let $\mathcal{H}$ be an inductive class of consistent fuzzy subsets. We have to prove that $v = \cup\mathcal{H}$ is consistent. Now, for every finite fuzzy set v_f such that $v_f \ll v$ an element $s \in \mathcal{H}$ exists such that $v_f \subseteq s$. Since s is consistent, v_f is consistent too, and by (7.1) v is consistent. $\square$

We can rewrite Theorem 7.1 in terms of fuzzy semantics as follows:

Theorem 7.2. *A fuzzy semantics is logically compact iff, for any initial valuation v,*

$$v \text{ is satisfiable } \Leftrightarrow \text{ every finite } v_f \ll v \text{ is satisfiable.} \quad (7.2)$$

Theorem 7.2 suggests to call *compact* a fuzzy semantics $\mathcal{M}$ such that, for every fuzzy set of formulas v,

$$v \text{ satisfiable } \Leftrightarrow \text{ every finite fuzzy subset of } v \text{ is satisfiable.}$$

Proposition 7.3. *Let $\mathcal{M}$ be fuzzy semantics. Then*

$$\mathcal{M} \text{ is logically compact } \Rightarrow \mathcal{M} \text{ is compact.}$$

The converse is not true.

Proof. The first part of the proposition is trivial. The fuzzy class

$$\mathcal{M} = \{s \in \mathcal{F}(F) : s(\alpha) < 1 \text{ for every } \alpha \in F\}$$

is an example of compact semantics which is not logically compact. Indeed, s^1 is the limit of the class of satisfiable fuzzy subsets s^λ, $\lambda \neq 1$. □

The following theorem shows that if the fuzzy subset of contradictions is nonempty, then the semantics of a fuzzy logic is logically compact.

Theorem 7.4. *Let $\mathcal{M}$ be the fuzzy semantics of a continuous fuzzy logic. Then,*

$$Contr(\mathcal{M}) \neq \varnothing \Rightarrow \mathcal{M} \text{ is logically compact.}$$

Proof. Let $(v_i)_{i\in I}$ be a directed family of consistent valuations and let v be its limit. Moreover, for every $i \in I$, denote by τ_i the theory generated by v_i. Then $(\tau_i)_{i\in I}$ is directed and, by the continuity of the logical consequence operator, its limit τ is a theory. Let α be a formula such that $Contr(\mathcal{M})(\alpha) = 1-Sup\{m(\alpha) : m \in \mathcal{M}\} \neq 0$. Then $Sup\{m(\alpha) : m \in \mathcal{M}\} = b \neq 1$ and for every $i \in I$, a model m_i of v_i exists such that $\tau_i(\alpha) \leq m_i(\alpha) \leq b$. Consequently, $\tau(\alpha) \leq b$ and this demonstrates that τ is consistent. Since $v \subseteq \tau$, we can conclude that v is consistent. □

8. ULTRAPRODUCT OF A FAMILY OF FUZZY MODELS

A compactness criterion for a fuzzy semantics is obtained by using the notion of ultraproduct of a family of fuzzy subsets proposed in Gerla [1994]b. Given any pre-ordered set $(L,\leq)$, we say that a nonempty subset $\mathcal{F}$ of L is *upper* if

$$x \in \mathcal{F} \text{ and } y \geq x \Rightarrow x \in \mathcal{F}.$$

If $\mathcal{F} \neq L$ we can say that $\mathcal{F}$ is *proper*. An upper subset $\mathcal{F}$ of a lattice L is a *filter* provided that

$$x \in \mathcal{F} \text{ and } y \in \mathcal{F} \Rightarrow x \wedge y \in \mathcal{F}.$$

$\mathcal{F}$ is called *principal* if an element $c \in L$ exists such that $\mathcal{F} = \{x \in L : x \geq c\}$. A proper filter is called *prime* if

$$x \vee y \in \mathcal{F} \Rightarrow \text{either } x \in \mathcal{F} \text{ or } y \in \mathcal{F}.$$

A class C of elements of L is contained in a proper filter if and only if it satisfies the *finite intersection property*, i.e., the intersection of a finite number of elements in C is nonempty. Every proper filter is contained in a prime filter. Then, any class C satisfying the finite intersection property is contained in a prime filter. If L is a Boolean algebra then the following are equivalent:

i) $\mathcal{F}$ is a prime filter

ii) for any $x \in L$, either $x \in \mathcal{F}$ or $-x \in \mathcal{F}$

iii) $\mathcal{F}$ is an *ultrafilter* (i.e., a filter maximal in the class of proper filters).

Let I be a set, $(\lambda_i)_{i\in I}$ a family of real numbers, and $\mathcal{F}$ a filter in the Boolean algebra $\mathcal{P}(I)$. Then we can write $lim_{\mathcal{F}}\lambda_i = \lambda$ provided that

$$\forall\varepsilon > 0\ \exists X \in \mathcal{F}\ \forall i \in X\ |\lambda-\lambda_i| < \varepsilon.$$

Equivalently, $lim_{\mathcal{F}}\lambda_i = \lambda$ if

$$\{i \in I:\ \lambda_i \in (a,b)\} \in \mathcal{F} \text{ for every interval } (a,b) \text{ containing } \lambda.$$

This notion of convergence satisfies the same properties of the classical convergence but, in addition, if $\mathcal{F}$ is prime, then for any bounded family $(\lambda_i)_{i\in I}$ of real numbers the limit $lim_{\mathcal{F}}\lambda_i$ always exists. Assume that I is the set N of natural numbers, that $\mathcal{F}$ is a prime filter on N and that $\mathcal{F}$ is not principal. Then, as $\mathcal{F}$ contains the filter of co-finite subsets of N, the usual convergence entails the convergence with respect to $\mathcal{F}$,

$$lim_{n\to\infty}\lambda_n = \lambda \Rightarrow lim_{\mathcal{F}}\lambda_n = \lambda$$

(see, e.g., Chang and Keisler [1966]).

Given a family $(s_i)_{i\in I}$ of fuzzy subsets of a set S and a prime filter $\mathcal{F}$ on I, we call an *ultraproduct modulo* $\mathcal{F}$ of $(s_i)_{i\in I}$ the fuzzy subset $s = lim_{\mathcal{F}}s_i$ defined by setting, for any x in S,

$$s(x) = lim_{\mathcal{F}}s_i(x).$$

Theorem 8.1. *Let $\mathcal{M}$ be a fuzzy semantics closed under ultraproducts and v an initial valuation. Then:*

(i) *The ultraproduct of a family of models of v is a model of v.*

(ii) *$\mathcal{M}$ is logically compact.*

(iii) *For every formula α there is a model m of v such that $Lc(v)(\alpha) = m(\alpha)$.*

Proof. (i) is trivial. To prove (ii) we refer to Theorem 7.1. Let v be an initial valuation such that v_f is satisfiable for every v_f finite such that $v_f \ll v$. At first we prove that every finite fuzzy subset s of v is satisfiable. Indeed, it is easy to find an increasing sequence v_n of finite fuzzy subsets such that $s(x) = lim_{n\to\infty} v_n(x)$ for every x and $v_n \ll s$. Since we also have that $v_n \ll v$, by hypothesis a sequence of models m_n exists such that $m_n \supseteq v_n$. Let $\mathcal{U}$ be a non-principal prime filter and let m be the ultraproduct of $(m_n)_{n\in N}$ modulo $\mathcal{U}$. Then

$$m(\alpha) = lim_{\mathcal{U}}\, m_n(\alpha) \geq lim_{\mathcal{U}}v_n(\alpha) = lim_{n\to\infty} v_n(\alpha) = s(\alpha),$$

and this proves that m is a model of s.

Denote by I the class of finite subsets of F and let $i \in I$. Then, since the restriction of v to i is satisfiable, there is an element m_i of $\mathcal{M}$ such that $m_i(x) \geq v(x)$ for every $x \in i$. We find a model m of v as a suitable ultraproduct m of the obtained family $(m_i)_{i\in I}$. To this purpose, we must find an ultrafilter $\mathcal{U}$ such that for every $x \in F$, the set $B(x) = \{i \in I :\ m_i(x) \geq v(x)\}$ belongs to $\mathcal{U}$. In turn, this is possible provided that the class $\{B(x) : x \in F\}$ of subsets of I satisfies the finite intersection property. Now, let $x_1,...,x_n$ be formulas and let $i = \{x_1,...,x_n\}$. Then $m_i(x_j) \geq v(x_j)$ for $j = 1,...,n$ and, therefore, i belongs to $B(x_1) \cap ... \cap B(x_n)$. Hence, such an intersection is nonempty. This completes the proof of ii).

To prove (iii), let v be an initial valuation and α a formula. Then, since $Lc(v)(\alpha) = Inf\{m(\alpha) : m \in \mathcal{M}, m \supseteq v\}$, a sequence $(m_n)_{n\in N}$ of models of v exists such that $m_n(\alpha)$ is decreasing and $Lc(v)(\alpha) = lim_{n\to\infty}\ m_n(\alpha)$. Let $\mathcal{U}$ be a non-principal ultrafilter on N and m the ultraproduct of $(m_n)_{n\in N}$ modulo $\mathcal{U}$. Then, m is a model of v such that

$$m(\alpha) = lim_{\mathcal{U}}\ m_n(\alpha) = lim_{\ n\to\infty}\ m_n(\alpha) = Lc(v)(\alpha).$$

This completes the proof of iii). □

Note. If $\mathcal{F}$ is numerable we can demonstrate logical compactness in a simple way. Indeed assume that every finite fuzzy subset v_f such that $v_f \ll v$ is satisfiable and denote by α_1, α_2, . . . the sequence of all the formulas in $\mathcal{F}$. Moreover, for every $h \in N$, consider the fuzzy subset v_h defined by setting

$$v_h(x) = \begin{cases} 0 \vee (v(\alpha_i) - 1/h) & \text{if } x = \alpha_i \text{ with } i \leq h, \\ 0 & \text{otherwise.} \end{cases}$$

Since v_h is finite and $v_h \ll v$, an element $m_h \in \mathcal{M}$ exists such that

$$m_h(\alpha_1) \geq v(\alpha_1) - 1/h, \ . \ . \ . \ , m_h(\alpha_h) \geq v(\alpha_h) - 1/h.$$

Let $\mathcal{U}$ be a non-principal prime filter and m the ultraproduct of $(m_n)_{n\in N}$ modulo $\mathcal{U}$. We claim that m is a model of v. Indeed, given any formula α_j, we have that $m_h(\alpha_j) \geq v(\alpha_j) - 1/h$ for every $h \geq j$ and therefore, $m(\alpha_j) = lim_{\mathcal{U}} m_n(\alpha_j) \geq v(\alpha_j)$.

Theorem 8.1 suggests a general method to arrive at logically compact fuzzy semantics (in the next chapters the interest of such a method will be apparent). In the following a *closed k-ary relation* is a closed subset $\mathcal{R}$ of $\boldsymbol{R}^k$. As usual, if $x_1,...,x_k$ are real numbers, we write $\mathcal{R}(x_1,...,x_k)$ to denote that $(x_1,...,x_k) \in \mathcal{R}$.

Proposition 8.2. *Denote by $\mathcal{M}$ the class of all fuzzy subsets m of $\mathcal{F}$ satisfying a set of conditions like*

$$\mathcal{R}(m(p_0(x_1,...,x_h)),...,m(p_k(x_1,...,x_h))), \tag{8.1}$$

where

- *$p_0,..., p_k$ are partial operations on $\mathcal{F}$ defined on a domain $D \subseteq \mathcal{F}^h$,*
- *$\mathcal{R} \subseteq U^{k+1}$ is a closed relation.*

Then $\mathcal{M}$ is closed under ultraproducts. Hence, if $s^1 \notin \mathcal{M}$, $\mathcal{M}$ is a logically compact fuzzy semantics.

Proof. Let $\mathcal{U}$ an ultrafilter on a set I. It is easy to prove that, due to the closure of $\mathcal{R}$, if $(\lambda_i^0)_{i\in I},...,(\lambda_i^k)_{i\in I}$ are families of real numbers such that $\mathcal{R}(\lambda_i^0,...,\lambda_i^k)$ for every $i \in I$, then $\mathcal{R}(lim_{\mathcal{U}}\lambda_i^0,...,lim_{\mathcal{U}}\lambda_i^k)$. Now, let $(m_i)_{i\in I}$ be a family of elements of $\mathcal{M}$, and m the ultraproduct of $(m_i)_{i\in I}$ modulo $\mathcal{U}$. Then, since for every $i \in I$

$$\mathcal{R}(m_i(p_0(x_1,...,x_h)),...,m_i(p_k(x_1,...,x_h))),$$

we have

$$\mathcal{R}(lim_{\mathcal{U}}\, m_i(p_0(x_1,...,x_h)),...,lim_{\mathcal{U}} m_i(p_k(x_1,...,x_h))).$$

Thus, $m \in \mathcal{M}$. □

If $f(x_0,...,x_k)$ and $g(x_0,...,x_k)$ are polynomial functions in $\boldsymbol{R}$, then the equality $f(x_0,...,x_k) = g(x_0,...,x_k)$ and the inequality $f(x_0,...,x_k) \geq g(x_0,...,x_k)$ define a closed relation. Then, interesting examples of closed relations can be achieved by algebraic systems of equalities and inequalities. In Chapter 8, we will apply Proposition 8.2 to continuous truth-functional logics. In Chapter 9 we will apply it to probability logics.

9. FUZZY LOGIC IS NOT MONOTONE

Given a fuzzy closure operator J and a fuzzy subset of formulas v, the *degree of inconsistency* of v with respect to J is the number $Inc(v)$ defined by

$$Inc(v) = Inf\{J(v)(\alpha) : \alpha \in \mathcal{F}\}. \tag{9.1}$$

The *degree of consistency* is defined as the number $Cons(v) = \sim Inc(v)$. It is evident that $Inc(v) = 1$ iff v is inconsistent. We can say that v is *totally consistent* if $Cons(v) = 1$. The *degree of inconsistency* of J is the number

$$Inc(J) = Sup\{Inc(v) : v \text{ consistent}\}. \tag{9.2}$$

The *degree of consistency* of J is the number $Cons(J) = \sim Inc(J)$. If $\mathcal{M}$ is a fuzzy semantics, then $Inc(v)$, $Cons(v)$, $Inc(\mathcal{M})$ and $Cons(\mathcal{M})$ are defined with reference to the related logical consequence operator. Consequently,

$$Inc(\mathcal{M}) = Sup_{m\in\mathcal{M}} Inf_{\alpha\in\mathcal{F}}\, m(\alpha). \tag{9.3}$$

As an example, consider a simple propositional calculus in which $\wedge$, $\vee$, $\neg$ are assumed as logical connectives and where $\mathcal{F}$ denotes the resulting set of formulas. Then, if we denote again by $\wedge$ and $\vee$ the minimum and the maximum in [0,1] and we set $\neg x = 1\text{-}x$, then the class $\mathcal{M}$ of fuzzy subsets m of $\mathcal{F}$ such that

$$m(\alpha\wedge\beta) = m(\alpha)\wedge m(\beta)\ ,\ m(\alpha\vee\beta) = m(\alpha)\vee m(\beta)\ ,\ m(\neg\alpha) = \sim m(\alpha)$$

define a fuzzy semantics (see Section 3 in Chapter 5). In this case the completely undetermined fuzzy subset $v^{0.5}$ is an element of $\mathcal{M}$ such that $Inc(v^{0.5}) = 0.5$. Also, given a propositional variable p, for every $m \in \mathcal{M}$, $Inc(m) \leq m(p \wedge \neg p) = m(p) \wedge (\sim m(p)) \leq 0.5$. Thus, $Inc(\mathcal{M}) = 0.5$. This is in accordance with the fact that there is no contradiction for such a semantics. There are also very interesting examples of fuzzy semantics such that $Inc(\mathcal{M}) = 1$ (see Chapters 3 and 6).

Proposition 9.1. *For every fuzzy semantics $\mathcal{M}$,*

$$Cons(\mathcal{M}) = 0 \Rightarrow Contr(\mathcal{M}) = \emptyset.$$

Proof. Let $\alpha \in \mathcal{F}$, then

$$Cons(\mathcal{M}) = 1 - Sup_{m\in\mathcal{M}} Inf_{x\in\mathcal{F}}\, m(x) \geq 1 - Sup_{m\in\mathcal{M}}\, m(\alpha) = Contr(\mathcal{M})(\alpha).$$ □

The degree of inconsistency of an abstract fuzzy logic is necessary to evaluate the actual degree of validity of a formula α. Indeed, even if we are able to prove α to a degree $\mathcal{D}(v)(\alpha)$ different from zero, in the case $\mathcal{D}(v)(\alpha) = Inc(v)$ no reason exists for the validity of α. So, we may affirm that v confirms α only in the case that $\mathcal{D}(v)(\alpha) > Inc(v)$. Accordingly, a fuzzy logic such that $Inc(\mathcal{D}) \neq 0$ is not monotone in spite of the monotonicity of $\mathcal{D}$. Indeed, the degree of validity of a formula α is represented by $\mathcal{D}(v)(\alpha)$-$Inc(v)$ rather than by $\mathcal{D}(v)(\alpha)$. On the other hand, since both $\mathcal{D}(v)(\alpha)$ and $Inc(v)$ are increasing functions of v, their difference is not an increasing function of v, in general. In fact, assume that the available information is represented by a fuzzy subset v and that this information confirms α, that is $\mathcal{D}(v)(\alpha) > Inc(v)$. Moreover, imagine new information may be available, so that v can be extended to a larger fuzzy subset v'. Now, if the added information is partially inconsistent with the previous one, then the degree of inconsistency increases, and it is possible that $\mathcal{D}(v')(\alpha) = Inc(v')$. So, while the primitive information v confirms α, the new information shows that no reason exists to believe in the validity of α. Owing to the above consideration, perhaps it would be better to substitute the operator $\mathcal{D}$ with a *"normalized"* operator.

Definition 9.2. Let $(\mathbb{F},\mathcal{D})$ be a fuzzy deduction system and define the *normalized deduction operator* $\mathcal{D}_n$ by setting, for any $v \in \mathcal{F}(\mathbb{F})$ and $\alpha \in \mathbb{F}$,

$$\mathcal{D}_n(v)(\alpha) = \frac{\mathcal{D}(v) - Inc(v)}{1 - Inc(v)} \tag{9.4}$$

if $Inc(v) \neq 1$ and by setting $\mathcal{D}_n(v)$ equal to the totally inconsistent theory if $Inc(v) = 1$.

The operator $\mathcal{D}_n$ is not monotone. Indeed, in the above example, $\mathcal{D}_n(v)(\alpha) \neq 0$, $\mathcal{D}_n(v')(\alpha) = 0$ while $v' \supseteq v$. The following theorem shows that non-monotone reasoning is strictly related to the possibility of managing partial inconsistency.

Theorem 9.3. *Let $(\mathbb{F},\mathcal{D})$ be a fuzzy deduction system and $(\mathbb{F},\mathcal{D}_n)$ its associated normalized system. Then $(\mathbb{F},\mathcal{D}_n)$ is a monotone deduction apparatus if and only if $Inc(\mathcal{D}) = 0$. In this case $(\mathbb{F},\mathcal{D}_n)$ coincides with $(\mathbb{F},\mathcal{D})$.*

We can now naturally ask whether fuzzy logic encompasses non-monotonic reasoning in its generality. Interesting examples in this direction can be found in Chapter 6. It is interesting to observe that $\mathcal{D}_n(v)(\alpha) = 1$ entails that $\mathcal{D}_n(v')(\alpha) = 1$ for every extension v' of v. This means that if we are sure of the validity of α, then no new information is able to change our primitive opinion.

In the following, in spite of the fact that $\mathcal{D}_n$ is more meaningful than $\mathcal{D}$, we prefer to refer to the operator $\mathcal{D}$. Indeed, in this way we can apply the very simple and powerful theory of closure operators without losing anything since, given $\mathcal{D}$, we can calculate $\mathcal{D}_n$.

10. ABSTRACT SIMILARITY LOGIC

In Ying [1994] a logic was proposed in which an approximate reasoning can be obtained by allowing the antecedent clauses of a rule to match its premises only approximately. As an example, consider an inference like

1.	*x is a thriller* $\Rightarrow$ *x good for me*	+
2.	*b is a detective story*	+
3.	*"detective story"* <u>is similar to</u> *"thriller"*	=

b is good for me.

Now the similarity is a fuzzy notion and, as it is usual in fuzzy logic, the truth degree of the conclusion *"b is good for me"* depends on the degree of similarity between *"detective story"* and *"thriller"*, obviously. Note that such an approach is rather different from the approach proposed in Ruspini [1991] (see also Dubois, Esteva, Garcia, Godo and Prade [1996]). Indeed, in the latter the similarity is defined on the set of worlds and not on the language. In this section we attempt to formalize in an abstract way such a type of approximate reasoning.

Given a continuous triangular norm $\odot$ and a fuzzy relation $R : \mathbb{F} \times \mathbb{F} \to U$, we say that R is a $\odot$-*similarity*, in brief *similarity*, if, for every $x, y, z \in \mathbb{F}$,

(a) $R(x,x) = 1$ (reflexivity),

(b) $R(x,y) \geq R(x,z) \odot R(z,y)$ (transitivity),

(c) $R(x,y) = R(y,x)$ (symmetry).

The crisp similarities coincide with the equivalence relations. The first definition of similarity was given in Zadeh [1971] ($\odot$ equal to minimum), in Menger [1951] ($\odot$ equal to the usual product) and Ruspini [1997] ($\odot$ equal to Łukasiewicz conjunction). Moreover, the notion of similarity is strictly related to the foundation of fuzzy sets theory based on category theory (see, e.g., Höhle [1991]).

Proposition 10.1. *Let R be a $\odot$-similarity and define SIM : $\mathcal{F}(\mathbb{F}) \to \mathcal{F}(\mathbb{F})$ by setting, for any $s \in \mathcal{F}(\mathbb{F})$ and $x \in \mathbb{F}$,*

$$SIM(s)(x) = Sup\{R(x',x) \odot s(x') : x' \in \mathbb{F}\}. \tag{10.1}$$

Then SIM is a continuous closure operator we call the fuzzy closure operator associated with R.

Proof. From $R(x,x) = 1$ it follows that $SIM(s)(x) \geq s(x)$. It is obvious that SIM is order preserving. In order to prove that $SIM(SIM(s))(x) \leq SIM(s)(x)$ observe that by the transitivity

$$\begin{aligned} SIM(SIM(s))(x) &= Sup_{x'\in\mathbb{F}}R(x',x) \odot SIM(s)(x') \\ &= Sup_{x'\in\mathbb{F}}R(x',x) \odot (Sup_{z\in\mathbb{F}}R(z,x') \odot s(z)) \\ &= Sup_{x'\in\mathbb{F}}Sup_{z\in\mathbb{F}}R(x',x) \odot R(z,x') \odot s(z) \\ &= Sup_{x'\in\mathbb{F}}Sup_{z\in\mathbb{F}}R(z,x') \odot R(x',x) \odot s(z) \\ &\leq Sup_{z\in\mathbb{F}}R(z,x) \odot s(z) = SIM(s)(x). \end{aligned}$$

One immediately sees that SIM is continuous. □

We can interpret $SIM(s)$ as the fuzzy subset of formulas similar to any formula in s. We say that a fuzzy subset s is *extensional* or that s is *closed* with respect to R if s is a fixed point of SIM. In the crisp case, s is extensional if it is union of equivalence classes. $SIM(s)$ is also called the *extensional hull* of s. The proof of the next proposition is trivial:

Proposition 10.2. *A fuzzy subset s is extensional with respect to R if and only if*

$$s(x) \geq s(x') \odot R(x',x) \tag{10.2}$$

for every x, x' in $\mathbb{F}$.

In particular, since $SIM(s)$ is extensional,

$$SIM(s)(x) \geq SIM(s)(x') \odot R(x',x). \tag{10.3}$$

We can interpret the pair ($\mathbb{F}$, SIM) as a deduction system and therefore, the extensional fuzzy subsets as theories of such a system. This enables us to give the following definition:

Definition 10.3. Let R be a similarity on the set $\mathbb{F}$ and ($\mathbb{F}$, $\mathcal{D}$) a continuous fuzzy deduction system. Then we call an *abstract similarity deduction system* the product ($\mathbb{F}$, $\mathcal{D}_R$) of ($\mathbb{F}$, $\mathcal{D}$) with ($\mathbb{F}$, SIM).

Such a definition means that for a similarity-based reasoning we have to start from the operator

$$K = \mathcal{D} \circ SIM. \tag{10.4}$$

Given an initial valuation v, $\mathcal{D}(SIM(v))$ is the fuzzy subset of formulas that can be deduced by $\mathcal{D}$ from formulas that are either in v or are similar to formulas in v. Now, it is possible that such a fuzzy subset is not closed with respect to the similarity relation. Then we have to consider the formulas similar to formulas in $\mathcal{D}(SIM(v))$, i.e., the fuzzy subset $SIM(\mathcal{D}(SIM(v)))$ and, again, the formulas we can deduce from $SIM(\mathcal{D}(SIM(v)))$. By extending this procedure, we get $\mathcal{D}_R(v)$ by a limit process

$$v \subseteq SIM(v) \subseteq \mathcal{D}(SIM(v)) \subseteq SIM(\mathcal{D}(SIM(v))) \subseteq \ldots \longrightarrow \mathcal{D}_R(v),$$

i.e.,

$$\mathcal{D}_R(v) = \cup_{n \in N} K^n(v). \tag{10.5}$$

Note that, by Theorem 9.1 in Chapter 1, the abstract similarity system ($\mathbb{F}$, $\mathcal{D}_R$) can be achieved also by starting either from $\mathcal{D} \vee SIM$ or from $SIM \circ \mathcal{D}$ instead of the operator $\mathcal{D} \circ SIM$.

From the results in Section 9 in Chapter 1 we have the following theorem:

Theorem 10.4. *Let ($\mathbb{F}$, $\mathcal{D}_R$) be an abstract similarity deduction system. Then the theories of ($\mathbb{F}$, $\mathcal{D}_R$) are the theories of $\mathcal{D}$ that are extensional with respect to R. In accordance, given an initial valuation v, $\mathcal{D}_R(v)$ is the least fuzzy subset of formulas containing v closed with respect to R and $\mathcal{D}$.*

11. ANY FUZZY LOGIC IS EQUIVALENT TO A CRISP LOGIC

In accordance with Stone's representation theorem for distributive lattices (see Stone [1937]), $\mathcal{F}(\mathbb{F})$ can be represented as a lattice of subsets of a given set. In this section we find a representation of such a lattice working well for closure operators and, hence, for abstract fuzzy logics. Namely, we prove that every abstract fuzzy logic is, in a sense, equivalent to a crisp logic.

Definition 11.1. We define a *signed formula* as any pair (α,λ) where $\alpha \in \mathbb{F}$ and $\lambda \in U$. The set of signed formulas is denoted by $\mathbb{F}_s$, i.e., $\mathbb{F}_s = \mathbb{F} \times U$.

Such a definition, for $\lambda \neq 0$, corresponds to the concept of fuzzy point (see Kerre [1980], Pu Pao-Mine and Liu Ying-Ming [1980], Gerla [1983] and Gerla [1986]). The intended meaning of a signed formula (α,λ) is that *"the truth value of α is at least to degree λ"*. In order to establish a connection between the notion of signed formula and the notion of fuzzy subset, we consider the map $K : \mathcal{P}(\mathbb{F}_s) \to \mathcal{F}(\mathbb{F})$ defined by associating any set X of signed formulas with the fuzzy subset of formulas $K(X)$ defined by setting

$$K(X)(x) = Sup\{\lambda \in U : (x,\lambda) \in X\} \tag{11.1}$$

for $x \in \mathbb{F}$. Moreover, we define the map $E : \mathcal{F}(\mathbb{F}) \to \mathcal{P}(\mathbb{F}_s)$ associating any fuzzy subset v with the set $E(v)$ of signed formulas defined by setting:

$$E(v) = \{(\alpha,\lambda) \in \mathbb{F}_s : v(\alpha) \geq \lambda\}. \tag{11.2}$$

If we say that (α,λ) *satisfies* v if $v(\alpha) \geq \lambda$, then $E(v)$ is the set of signed formulas satisfying v. Observe that $K(\emptyset)$ coincides with the fuzzy subset s^0 constantly equal to 0, i.e., with the empty set. Instead, $E(s^0) = \{(\alpha, 0) \in \mathbb{F}_s : \alpha \in \mathbb{F}\} \neq \emptyset$.

Proposition 11.2. *The function $E : \mathcal{F}(\mathbb{F}) \to \mathcal{P}(\mathbb{F}_s)$ is injective and preserves finite unions and finite and infinite intersections.*

Proof. It is obvious that E is injective. Also, for every $s_1, s_2 \in \mathcal{F}(\mathbb{F})$ have

$$\begin{aligned}(x,\lambda) \in E(s_1 \cup s_2) &\Leftrightarrow s_1(x) \vee s_2(x) \geq \lambda \Leftrightarrow \text{ either } s_1(x) \geq \lambda \text{ or } s_2(x) \geq \lambda\\ &\Leftrightarrow \text{ either } (x,\lambda) \in E(s_1) \text{ or } (x,\lambda) \in E(s_2)\\ &\Leftrightarrow (x,\lambda) \in E(s_1) \cup E(s_2).\end{aligned}$$

Let $(s_i)_{i\in I}$ be any family of fuzzy subsets of $\mathbb{F}$. Then

$$\begin{aligned}(x,\lambda) \in E(\textstyle\bigcap_{i\in I} s_i) &\Leftrightarrow Inf\{s_i(x) : i \in I\} \geq \lambda \Leftrightarrow \forall\, i \in I, s_i(x) \geq \lambda\\ &\Leftrightarrow \forall\, i \in I, (x,\lambda) \in E(s_i) \Leftrightarrow (x,\lambda) \in \textstyle\bigcap_{i\in I} E(s_i).\end{aligned}$$

E does not preserve infinite unions and is not surjective. To characterize the codomain of E, consider the following definition:

Definition 11.3. We call an *hypograph* any subset X of $\mathbb{F}_s$ such that, for every $x \in \mathbb{F}$,

(i) $(x,\lambda) \in X$ and $\mu \leq \lambda \Rightarrow (x,\mu) \in X$,
(ii) $\mu = Sup\{\lambda \in U : (x,\lambda) \in X)\} \Rightarrow (x,\mu) \in X$,
(iii) $(x,0) \in X$.

We denote by $G(\mathbb{F})$ the class of hypographs.

Lemma 11.4. *$G(\mathbb{F})$ is the codomain of E. Moreover $G(\mathbb{F})$ is a (non-algebraic) closure system in $\mathbb{F}_s$ and therefore a complete lattice.*

Proof. It is obvious that $G(\mathbb{F})$ is the codomain of E. Let $(X_i)_{i\in I}$ be a family of elements of $G(\mathbb{F})$ and set $X = \bigcap_{i\in I} X_i$. Then it is evident that X satisfies (i) and (iii).

To prove (ii), let x be an element of $\mathbb{F}$ such that $\{\lambda \in U : (x,\lambda) \in X)\}$ is nonempty. Then, for every $i \in I$, $\{\lambda \in U : (x,\lambda) \in X_i)\}$ is nonempty and, by (ii), it admits a greatest element μ_i. We set $\mu = Inf_{i\in I}\ \mu_i$. Then, since X_i satisfies (i), we have that $(x,\mu) \in X_i$ for every $i \in I$ and, hence, that $(x,\mu) \in X$. On the other hand, if $(x,\lambda) \in X$, then, since $(x,\lambda) \in X_i$, it is also $\lambda \leq \mu_i$ for every $i \in I$ and therefore $\lambda \leq \mu$. Thus μ is the greatest element of $\{\lambda \in U : (x,\lambda) \in X)\}$ and therefore X satisfies (ii). This demonstrates that X is an element of $G(\mathbb{F})$.

In order to prove that $G(\mathbb{F})$ is not algebraic, set $X_\lambda = \{(x,\mu) : \mu \leq \lambda \text{ and } x \in \mathbb{F}\}$. Then $(X_\lambda)_{\lambda\neq 1}$ is a directed family of elements of $G(\mathbb{F})$ whose union $\{(x,\lambda) : x \in \mathbb{F}, \lambda \neq 1\}$ does not satisfy (ii) and, hence, is not an element of $G(\mathbb{F})$.

Theorem 11.5. *The pair $K : \mathcal{P}(\mathbb{F}_s) \to \mathcal{F}(\mathbb{F})$, $E : \mathcal{F}(\mathbb{F}) \to \mathcal{P}(\mathbb{F}_s)$ defines a Galois connection in which*

- *$E \circ K$ is the closure operator generated by $G(\mathbb{F})$,*
- *$K \circ E$ is the identity map.*

Proof. Given $X \subseteq \mathbb{F}_s$, we must prove that $E(K(X))$ is the element of $G(\mathbb{F})$ generated by X. Set $s = K(X)$. Then, it is evident that $E(s)$ is an element of $G(\mathbb{F})$ containing X. Let Y be any element of $G(\mathbb{F})$ containing X and suppose $(x,\mu) \in E(s)$, i.e., $s(x) \geq \mu$. Then, in the case $\mu = 0$ it is obvious that $(x,\mu) \in Y$. In the case $\mu \neq 0$, it is $s(x) \neq 0$ and therefore $\{\lambda \in U : (x,\lambda) \in X\}$ is nonempty. Since $Y \supseteq X$ and Y satisfies (ii), the pair $(x, s(x))$ belongs to Y. By (i) this entails that $(x,\mu) \in Y$. Thus $Y \supseteq E(s)$. This demonstrates that $E(s)$ is the element of $G(\mathbb{F})$ generated by X. It is immediate that $K \circ E$ is the identity map.

Now, we are able to demonstrate the following representation theorem:

Theorem 11.6. *The map $E : \mathcal{F}(\mathbb{F}) \to G(\mathbb{F})$ is an isomorphism between $\mathcal{F}(\mathbb{F})$ and $G(\mathbb{F})$. Its inverse E^{-1} coincides with K.*

Proof. It is obvious.

A representation theorem for fuzzy closure operators holds. Given a fuzzy operator J in $\mathbb{F}$, we denote by J_s the operator in $\mathbb{F}_s$ defined by setting, for any $X \subseteq \mathbb{F}_s$,

$$J_s(X) = E(J(K(X))), \tag{11.3}$$

that is,

$$\begin{array}{ccc} \mathcal{P}(\mathbb{F}_s) & \xrightarrow{\quad K \quad} & \mathcal{F}(\mathbb{F}) \\ J_s \downarrow & & \downarrow J \\ \mathcal{P}(\mathbb{F}_s) & \xleftarrow{\quad E \quad} & \mathcal{F}(\mathbb{F}) \end{array}$$

Theorem 11.7. *Given a fuzzy operator J in $\mathbb{F}$,*

$$J \text{ is a fuzzy closure operator in } \mathbb{F} \iff J_s \text{ closure operator in } G(\mathbb{F})$$
$$\iff J_s \text{ closure operator in } \mathbb{F}_s.$$

Proof. The fact that E is an isomorphism between $\mathcal{F}(\mathbb{F})$ and $G(\mathbb{F})$ proves the first equivalence. Assume that J is a fuzzy closure operator. It is evident that J_s satisfies the inclusion property and that it is order-preserving. Moreover, since $K \circ E$ is the identity map, given $X \subseteq \mathbb{F}_s$, we have that

$$J_s(J_s(X)) = E(J(K(E(J(K(X)))))) = E(J(J(K(X)))) = E(J(K(X))) = J_s(X).$$

this proves that J_s is a closure operator in $\mathbb{F}_s$.

Conversely, let J_s be a closure operator in $\mathbb{F}_s$. Then, since $J_s(X) \in G(\mathbb{F})$ for any X, J_s is a closure operator in $G(\mathbb{F})$ and therefore a fuzzy closure operator in $\mathbb{F}$. □

Obviously, if J is the identity map, then J_s coincides with the closure operator associated with $G(\mathbb{F})$ and therefore, J_s is not the identity map in $\mathcal{P}(\mathbb{F}_s)$.

Theorem 11.8. *The map associating to any fuzzy closure operator J in $\mathbb{F}$ the closure operator J_s in $G(\mathbb{F})$ is an isomorphism between the lattice $CO(\mathcal{F}(\mathbb{F}))$ of all fuzzy closure operators in $\mathbb{F}$ and the lattice $CO(G(\mathbb{F}))$ of all closure operators in $G(\mathbb{F})$.*

To give a representation theorem for fuzzy closure systems, given any class C of fuzzy subsets, we set

$$C_s = E(C) = \{E(s) : s \in C\}. \tag{11.4}$$

Theorem 11.9. *Let C be a class of fuzzy subsets of $\mathbb{F}$. Then*

$$C \text{ fuzzy closure system in } \mathbb{F} \iff C_s \text{ is a closure system in } G(\mathbb{F})$$
$$\iff C_s \text{ is a closure system in } \mathbb{F}_s.$$

Proof. The first equivalence is obvious because E preserves the intersections. As $G(\mathbb{F})$ is a closure system in $\mathbb{F}_s$, the second equivalence follows. □

Such a theorem enables us to prove the following representation theorem for fuzzy closure systems:

Theorem 11.10. *The map associating any fuzzy closure system in $\mathbb{F}$ with the closure system C_s in $G(\mathbb{F})$ is an isomorphism between the lattice $CS(\mathcal{F}(\mathbb{F}))$ of all fuzzy closure systems in $\mathbb{F}$ and the lattice $CS(G(\mathbb{F}))$ of all closure systems in $G(\mathbb{F})$.*

It is worthwhile to note that, in accordance with the fact that E does not preserve unions, the implication

$$C \text{ algebraic} \Rightarrow C_s \text{ algebraic}$$

is false. For instance, the whole class $\mathcal{F}(\mathbb{F})$ of fuzzy subsets is an algebraic fuzzy closure system but $E(\mathcal{F}(\mathbb{F})) = G(\mathbb{F})$ is not algebraic.

The definitions given by (11.3) and (11.4) are related in accordance with the following diagram:

$$\begin{array}{ccc} J & \longrightarrow & Cs(J) \\ \downarrow & & \downarrow \\ J_s & \longleftarrow & Cs(J)_s \end{array}$$

Proposition 11.11. *Let J be a fuzzy closure operator. Then,*

$$J_s = Co(Cs(J)_s), \tag{11.5}$$

or, equivalently,

$$J_s(X) = \bigcap\{E(s) : J(s) = s,\ E(s) \supseteq X\} \tag{11.6}$$

for every $X \subseteq \mathbb{F}_s$.

Given a fuzzy semantics $\mathcal{M}$ we can consider the associated crisp semantics $\mathcal{M}_s$. This means that we identify a model $m \in \mathcal{M}$ with the set of signed formulas satisfied by m. We denote by Lc_s the related logical consequence operator, and therefore, for every set X of signed formulas,

$$Lc_s(X) = \bigcap\{E(m) : E(m) \supseteq X \text{ and } m \in \mathcal{M}\}.$$

An equivalent more expressive way to introduce Lc_s is the following: We say that a model $m \in \mathcal{M}$ *satisfies* the signed formula (α,λ) if $m(\alpha) \geq \lambda$ or, equivalently, $(\alpha,\lambda) \in E(m)$. In this case we write $m \models (\alpha,\lambda)$. Given a set X of signed formulas, we write $m \models X$ and we say that *m is a model of X* if any signed formula in X is satisfied by m, i.e., $X \subseteq E(m)$. We write $X \models (\alpha, \lambda)$ to denote that (α,λ) is satisfied by any model of X. Then we have the following proposition:

Proposition 11.12. *Let* $Lc_s : \mathcal{P}(\mathbb{F}_s) \to \mathcal{P}(\mathbb{F}_s)$ *be the logical consequence operator, then*

$$Lc_s(X) = \{(\alpha,\lambda) \in \mathbb{F}_s : X \models (\alpha, \lambda)\}.$$

Furthermore, it is interesting to get Lc_s by a Galois connection. Indeed, define the mappings

$$mod : \mathcal{P}(\mathbb{F}_s) \to \mathcal{P}(\mathcal{M}) \quad \text{and} \quad th : \mathcal{P}(\mathcal{M}) \to \mathcal{P}(\mathbb{F}_s)$$

by setting, for any $X \in \mathcal{P}(\mathbb{F}_s)$

$$mod(X) = \{m \in \mathcal{M} : m \models X\}$$

and, for any $C \subseteq \mathcal{M}$,

$$th(C) = \{(\alpha, \lambda) \in \mathbb{F}_s : m \models (\alpha, \lambda)\ \forall m \in C\}.$$

Then, it is easy to verify that the pair *mod* and *th* defines a Galois connection and that

$$Lc_s = th \circ mod.$$

Now we are able to demonstrate the following representation theorem for fuzzy logic. It shows that we can reduce any deduction in a fuzzy logic to a deduction in a crisp logic in accordance with the following diagram:

$$\begin{array}{ccc} \mathcal{F}(\mathbb{F}) & \xrightarrow{\ E\ } & G(\mathbb{F}) \\ \mathcal{D}\downarrow & & \downarrow\mathcal{D}_s \\ \mathcal{F}(\mathbb{F}) & \xleftarrow{\ E^{-1}\ } & G(\mathbb{F}) \end{array}$$

Theorem 11.13. *Let* ($\mathcal{F}$, $\mathcal{D}$, $\mathcal{M}$) *be an abstract fuzzy logic. Then* ($\mathcal{F}_s$, $\mathcal{D}_s$, $\mathcal{M}_s$) *is an abstract crisp logic equivalent to* ($\mathcal{F}$, $\mathcal{D}$, $\mathcal{M}$), *i.e.,*

$$\mathcal{D}(v) = E^{-1}(\mathcal{D}_s(E(v))). \tag{11.7}$$

Moreover,

$$\tau \text{ is a (fuzzy) theory of } (\mathcal{F}, \mathcal{D}) \iff E(\tau) \text{ is a (crisp) theory of } (\mathcal{F}_s, \mathcal{D}_s).$$

Proof. We must prove that $\mathcal{M}_s$ is a semantics for ($\mathcal{F}_s$, $\mathcal{D}_s$), that is, that $\mathcal{D}_s = Lc_s$. Observe that by hypothesis every theory τ in ($\mathcal{F}$, $\mathcal{D}$) is the intersection of a family $(m_i)_{i\in I}$ of models in $\mathcal{M}$. Hence, since E preserves the intersections,

$$E(\tau) = E(\textstyle\bigcap_{i\in I} m_i) = \bigcap_{i\in I} E(m_i).$$

Then,

$$\begin{aligned}\mathcal{D}_s(X) &= \bigcap\{E(v) : \mathcal{D}(v) = v,\ E(v) \supseteq X\}\\ &= \bigcap\{E(m) : m \in \mathcal{M},\ E(m) \supseteq X\} = Lc_s(X).\end{aligned}$$

The remaining part of the proposition is evident. □

About the meaning of such a reduction theorem, see the remark at the end of Section 7 in Chapter 4.

CHAPTER 3

EXTENDING AN ABSTRACT CRISP LOGIC

1. AN EXTENSION PRINCIPLE FOR CLOSURE OPERATORS

Several notions in crisp mathematics can be translated into the corresponding notions in fuzzy mathematics in a uniform way by Zadeh Extension Principle (see Zadeh [1975]b). So, it is natural to ask the following question:

Given a crisp logic, does there exist a canonical way to extend it in a fuzzy logic ?

This question leads to the problem of finding an extension principle for crisp operators. Now, given any set S, a simple way to extend an operator $J : \mathcal{P}(S) \to \mathcal{P}(S)$ is to consider the fuzzy operator $J' : \mathcal{F}(S) \to \mathcal{F}(S)$ defined by setting, for every fuzzy subset s,

$$J'(s) = J(Supp(s)).$$

Indeed, J' is an extension of J and J' is a closure operator everywhere J is a closure operator. In logical terms, if $(\mathcal{F}, \mathcal{D})$ is a crisp deduction system, then $(\mathcal{F}, \mathcal{D}')$ is the fuzzy deduction system achieved by applying the crisp deduction apparatus $(\mathcal{F}, \mathcal{D})$ to those formulas that belong to s with a degree different from zero. This is rather unsatisfactory because it doesn't consider all the information contained in s. A more interesting way to extend a crisp operator is suggested by the identification of a fuzzy subset with a continuous chain was proposed in Chapter 2 (see Ramik [1983], Biacino and Gerla [1996], Gerla and Scarpati [1997]). Indeed, given $s \in \mathcal{F}(S)$, it is natural to proceed in the following way:

- identify s with the continuous chain $(C(s,\lambda))_{\lambda\in U}$,
- apply J to each element of this chain by obtaining the chain $(J(C(s,\lambda)))_{\lambda\in U}$,
- consider the continuous chain $(J(C(s,\lambda))'')_{\lambda\in U}$ generated by such a chain,
- assume as image of s the fuzzy subset $J^*(s)$ corresponding to such a continuous chain.

Such a procedure can be pictured by the following diagram :

$$\begin{array}{ccccc} s & \longrightarrow & (C(s,\lambda))_{\lambda\in U} & \longrightarrow & (J(C(s,\lambda)))_{\lambda\in U} \\ \downarrow & & & & \downarrow \\ J^*(s) & & \longleftarrow & & (J(C(s,\lambda))'')_{\lambda\in U}. \end{array}$$

As observed in Chapter 2, all this can be obtained in a more direct way by the following definition:

Definition 1.1. For every operator J we define the *canonical extension* J^* of J as the fuzzy operator defined by setting, for every $s \in \mathcal{F}(S)$

$$J^*(s) = \bigcup_{\lambda\in U} \lambda \wedge J(C(s,\lambda)) \qquad (1.1)$$

or, equivalently,

$$J^*(s)(x) = Sup\{\lambda \in U : x \in J(C(s,\lambda))\}. \quad (1.2)$$

Proposition 1.2. *Let $J : \mathcal{P}(S) \to \mathcal{P}(S)$ be an operator. Then $J^* : \mathcal{F}(S) \to \mathcal{F}(S)$ is an extension of J.*

Proof. Let c_X be the characteristic function of a subset X of S. We claim that

$$J^*(c_X) = c_{J(X)} \quad (1.3)$$

Indeed, for every $\lambda \neq 0$, $C(c_X,\lambda) = X$ and hence

$x \in J(X) \Rightarrow x \in J(C(c_X,\lambda))$ for every $\lambda \neq 0 \Rightarrow J^*(c_X)(x) = 1$,
$x \notin J(X) \Rightarrow x \notin J(C(c_X,\lambda))$ for every $\lambda \neq 0 \Rightarrow J^*(c_X)(x) = 0$.

This demonstrates that the restriction of J^* to the crisp subsets coincides with J. □

The connection between the cuts of the image of s by J^* and the image of the cuts of s by J is examined in the following lemma:

Lemma 1.3. *Let J be an order-preserving operator. Then, for every fuzzy subset s and $\mu \in U$,*

$$O(J^*(s),\mu) = \bigcup_{\lambda>\mu} J(C(s,\lambda)) \subseteq J(O(s,\mu)) \subseteq J(C(s,\mu)) \subseteq \bigcap_{\lambda<\mu} J(C(s,\lambda)) = C(J^*(s),\mu).$$

Proof. Observe that for every $\lambda > \mu$ we have $C(s,\lambda) \subseteq O(s,\mu)$. Then, since J is order-preserving, $J(C(s,\lambda)) \subseteq J(O(s,\mu))$. Consequently, $J(O(s,\mu)) \supseteq \bigcup_{\lambda>\mu} J(C(s,\lambda))$ and the proposition is a consequence of Proposition 4.7 in Chapter 2. □

Proposition 1.4. *Let J be an order-preserving operator. Then*

$$J^*(s) = \bigcap_{\lambda\in U} \lambda \vee J(C(s,\lambda)), \quad (1.4)$$
$$J^*(s) = \bigcup_{\lambda\in U} \lambda \wedge J(O(s,\lambda)), \quad (1.5)$$
$$J^*(s) = \bigcap_{\lambda\in U} \lambda \vee J(O(s,\lambda)), \quad (1.6)$$

or, equivalently,

$$J^*(s)(x) = Inf\{\lambda \in U : x \notin J(C(s,\lambda))\}, \quad (1.7)$$
$$J^*(s)(x) = Sup\{\lambda \in U : x \in J(O(s,\lambda))\}, \quad (1.8)$$
$$J^*(s)(x) = Inf\{\lambda \in U : x \notin J(O(s,\lambda))\}. \quad (1.9)$$

Proof. Equality (1.4) and the equivalence between (1.5) and (1.6) follow from Proposition 4.7 in Chapter 2. In order to prove (1.5) observe that, since J is order preserving, $J(O(s,\lambda)) \subseteq J(C(s,\lambda))$ and, hence, for every fuzzy subset s and $x \in S$

$$Sup\{\lambda \in U : x \in J(O(s,\lambda))\} \leq Sup\{\lambda \in U : x \in J(C(s,\lambda))\} = J^*(s)(x).$$

Moreover, by Lemma 1.3, $O(J^*(s),\lambda) \subseteq J(O(s,\lambda))$ and

$$J^*(s)(x) = Sup\{\lambda \in U : x \in O(J^*(s),\lambda)\} \leq Sup\{\lambda \in U : x \in J(O(s,\lambda))\}. \quad □$$

Theorem 1.5. *Let J be an operator. Then*

J^* almost closure operator $\Leftrightarrow$ J almost closure operator.
J^* closure operator $\Leftrightarrow$ J closure operator.

Proof. Assume that J^* is an *a-c*-operator (a closure operator). Then, since J is the restriction of J^* to the crisp subsets, J is an *a-c*-operator (a closure operator). Let J be an *a-c*-operator. Then it is evident that J^* is order-preserving. In order to prove that $J^*(s) \supseteq s$, observe that, because of the inclusion $C(s,\lambda) \subseteq J(C(s,\lambda))$,

$$s(x) = Sup\{\lambda \in U : x \in C(s,\lambda)\} \leq Sup\{\lambda \in U : x \in J(C(s,\lambda))\}.$$

Assume that J is a closure operator. In order to prove that $J^*(J^*(s)) = J^*(s)$, we first prove that every cut $C(J^*(s),\lambda)$ is a fixed point for J. Indeed, observe that the intersection of a class of fixed points for J is a fixed point for J and that $C(J^*(s),\lambda) = \bigcap_{\mu<\lambda} J(C(s,\mu))$. Thus

$$J^*(J^*(s))(x) = Sup\{\lambda \in U : x \in J(C(J^*(s),\lambda))\}$$
$$= Sup\{\lambda \in U : x \in C(J^*(s),\lambda)\} = J^*(s)(x).$$ □

Recall that in Section 2 of Chapter 2 the dual J_d of an operator J was defined. The next proposition shows that the following diagram commutes:

$$\begin{array}{ccc} J & \longrightarrow & J_d \\ \downarrow & & \downarrow \\ J^* = ((J_d)^*)_d & \longleftarrow & (J_d)^*. \end{array}$$

Proposition 1.6. *Given any operator J,*

$$J^* = ((J_d)^*)_d. \tag{1.10}$$

As a consequence,

J^* almost interior operator $\Leftrightarrow$ J almost interior operator.
J^* interior operator $\Leftrightarrow$ J interior operator.

Proof. By (1.9), for any $x \in S$,

$$J^*(s)(x) = Inf\{\lambda \in U : x \notin J(O(s,\lambda))\} = 1 - Sup\{1-\lambda \in U : x \notin J(O(s,\lambda))\}$$
$$= 1 - Sup\{1-\lambda \in U : x \in J_d(-O(s,\lambda))\}$$
$$= 1 - Sup\{\lambda \in U : x \in J_d(-O(s,1-\lambda))\}$$
$$= 1 - Sup\{\lambda \in U : x \in J_d(C(-s,\lambda))\}$$
$$= 1 - (J_d)^*(-s)(x).$$

Implications

$$J^* \text{ (almost) interior operator} \Rightarrow J \text{ (almost) interior operator}$$

are evident. Conversely, assume that J is an (almost) interior operator. Then, since J_d is a (an almost) closure operator, $(J_d)^*$ is a (an almost) fuzzy closure operator. Thus, $J^* = (J_d)^*{}_d$ is an (almost) interior operator. □

In Proposition 1.7 below we show that the extension principle defines an embedding from the lattice $AC(\mathcal{P}(S))$ of all (crisp) *a-c*-operators in S into the lattice $AC(\mathcal{F}(S))$ of all fuzzy *a-c*-operators in S.

Proposition 1.7. *Assume that J_1, J_2 are two ac-operators. Then*

$$(J_1 \vee J_2)^* = J_1^* \vee J_2^* \tag{1.11}$$

and

$$(J_1 \wedge J_2)^* = J_1^* \wedge J_2^*. \tag{1.12}$$

Consequently, the extension principle defines an embedding from the lattice of all (crisp) a-c-operators (closure operators) in S into the lattice of all fuzzy a-c-operators (closure operators) in S.

Proof. In order to prove (1.11), let $s \in \mathcal{F}(S)$ and $x \in S$. Then

$$\begin{aligned}(J_1 \vee J_2)^*(s)(x) &= Sup\{\lambda \in U : x \in (J_1 \vee J_2)(C(s,\lambda))\}\\ &= Sup\{\lambda \in U : x \in J_1(C(s,\lambda)) \cup J_2(C(s,\lambda))\}\\ &= (Sup\{\lambda \in U : x \in J_1(C(s,\lambda))\}) \vee (Sup\{\lambda \in U : x \in J_2(C(s,\lambda))\})\\ &= J_1^*(s)(x) \vee J_2^*(s)(x).\end{aligned}$$

In order to prove (1.12), i.e.,

$$(J_1 \wedge J_2)^*(s)(x) = J_1^*(s)(x) \wedge J_2^*(s)(x),$$

assume, for example, that $J_1^*(s)(x) \leq J_2^*(s)(x)$. Then since $\{\lambda \in U : x \in J_1(C(s,\lambda))\}$ and $\{\lambda \in U : x \in J_2(C(s,\lambda))\}\}$ are intervals of type $[0, \beta)$ or $[0, \beta]$, we also have that

$$\{\lambda \in U : x \in J_1(C(s,\lambda))\} \subseteq \{\lambda \in U : x \in J_2(C(s,\lambda))\}$$

and, accordingly,

$$\begin{aligned}(J_1 \wedge J_2)^*(s)(x) &= Sup\{\lambda \in U : x \in J_1(C(s,\lambda)) \cap J_2(C(s,\lambda))\}\\ &= Sup\{\lambda \in U : x \in J_1(C(s,\lambda))\}\\ &= J_1^*(s)(x) = J_1^*(s)(x) \wedge J_2^*(s)(x).\end{aligned}$$

□

Remark. As observed in Ramik [1983] and in Gerla and Scarpati [1998], it is possible to extend Definition 1.1 to obtain an extension principle for several variables operators. Indeed, let $S_1,...,S_n$ and S be sets and consider a several variables operator $J : \mathcal{P}(S_1) \times ... \times \mathcal{P}(S_n) \to \mathcal{P}(S)$. Then we can define a fuzzy operator $J^* : \mathcal{F}(S_1) \times ... \times \mathcal{F}(S_n) \to \mathcal{F}(S)$ by setting, for every $s_1 \in \mathcal{F}(S_1), ... , s_n \in \mathcal{F}(S_n)$ and $x \in S$,

$$J^*(s_1,...,s_n)(x) = Sup\{\lambda \in U : x \in J(C(s_1,\lambda),...,C(s_n,\lambda))\}. \tag{1.13}$$

This enables us to restate Zadeh Extension Principle as a particular case. Indeed, such a principle says that we can extend any map $h : S_1 \times ... \times S_n \to S$ to a map $h^* : \mathcal{F}(S_1) \times ... \times \mathcal{F}(S_n) \to \mathcal{F}(S)$ by the formula

$$h^*(s_1,...,s_n)(x) = Sup\{s_1(x_1) \wedge ... \wedge s_n(x_n) : h(x_1,...,x_n) = x\}.$$

Now, define the operator H by setting, for $X_1 \in \mathcal{P}(S_1), ... , X_n \in \mathcal{P}(S_n)$,

$$H(X_1,...,X_n) = \{h(x_1,...,x_n) : x_1 \in X_1, ... , x_n \in X_n\}.$$

Then $H^* = h^*$. Indeed,

$$\begin{aligned}&H^*(s_1,...,s_n)(x)\\ &= Sup\{\lambda \in U : x \in H(C(s_1,\lambda),...,C(s_n,\lambda))\}\\ &= Sup\{\lambda \in U : x_1, ... ,x_n \text{ exist } s.t.\ x = h(x_1,...,x_n) \text{ and } s_1(x_1) \geq \lambda, ... ,s_n(x_n) \geq \lambda\}\\ &= Sup\{s_1(x_1) \wedge ... \wedge s_n(x_n) : h(x_1,...,x_n) = x\} = h^*(s_1,...,s_n)(x).\end{aligned}$$

2. AN EXTENSION PRINCIPLE FOR CLOSURE SYSTEMS

Let C be a closure system in a set S. Then, by identifying any set in C with the related characteristic function, we can claim that C is also a fuzzy closure system in S. Nevertheless, we will propose a way to extend C into a fuzzy closure system which is in accordance with the concept of canonical extension of a closure operator defined in the previous section. Also in this case, the way suggested is to identify the fuzzy subsets with the continuous chains. In fact, we can associate C with the class of chains whose elements belong to C or, if we will, the class of corresponding fuzzy subsets. Thus, given any class C of subsets of S, we set

$$C^* = \{s \in \mathcal{F}(S) : \mathrm{C}(s,\lambda) \in C \text{ for every } \lambda \neq 0\}, \tag{2.1}$$

and we say that C^* is the *canonical extension* of C.

Theorem 2.1. *Let C be any class of classical subsets of S. Then the class of crisp elements of C^* coincides with C and, therefore, C^* is an extension of C. Further,*

$$C^* \text{ is a fuzzy closure system} \Leftrightarrow C \text{ is a closure system.}$$

Proof. If s is a crisp element of C^*, then, trivially, s is the characteristic function of an element of C. Conversely, if $X \in C$ then, since all the cuts of c_X coincide with X, $c_X \in C^*$.

Assume that C^* is a fuzzy closure system and let $(X(i))_{i\in I}$ be any family of elements of C. Then, since $(c_{X(i)})_{i\in I}$ is a family of elements of C^*, $\bigcap_{i\in I} c_{X(i)} \in C^*$. As $\bigcap_{i\in I} c_{X(i)}$ is the characteristic function of $\bigcap_{i\in I} X(i)$, this demonstrates that $\bigcap_{i\in I} X(i) \in C$. Thus C is a closure system. Let C be a closure system and $(s_i)_{i\in I}$ any family of elements of C^*, then, since

$$\mathrm{C}(\textstyle\bigcap_{i\in I} s_i,\lambda) = \bigcap_{i\in I} \mathrm{C}(s_i,\lambda) \in C,$$

it is also $\bigcap_{i\in I} s_i \in C^*$. Thus C^* is a fuzzy closure system. □

The next theorem shows that the notions of canonical extension of a closure system and canonical extension of a closure operator are strictly related in accordance with the following diagrams :

$$\begin{array}{ccc} C & \longrightarrow & Co(C) \\ \downarrow & & \downarrow \\ C^* = Cs(Co(C)^*) & \longleftarrow & Co(C)^* \end{array} \qquad \begin{array}{ccc} J & \longrightarrow & Cs(J) \\ \downarrow & & \downarrow \\ J^* = Co(Cs(J^*)) & \longleftarrow & Cs(J^*) = Cs(J)^* \end{array}$$

Theorem 2.2. *Let C be a closure system and J a closure operator. Then*

$$C^* = Cs(Co(C)^*) \ \text{ and } \ Cs(J)^* = Cs(J^*). \tag{2.2}$$

Moreover,

$$J^* = Co(Cs(J^*)) \ \text{ and } \ Co(C)^* = Co(C^*). \tag{2.3}$$

Proof. In order to prove the first equation in (2.2), observe that if s is a fixed point for $Co(C)^*$, then, by Lemma 1.3,

$$\mathrm{C}(s,\mu) = \mathrm{C}(Co(C)^*(s),\mu) = \bigcap\{Co(C)(\mathrm{C}(s,\lambda)) : \lambda < \mu\} \in C.$$

As a consequence, $s \in C^*$. Conversely, if $s \in C^*$, then, every cut $\mathrm{C}(s,\mu)$ belongs to C and hence it is a fixed point for $Co(C)$. Thus

$$(Co(C)^*)(s)(x) = Sup\{\lambda \in U : x \in Co(C)(\mathrm{C}(s,\lambda))\}$$

$$= Sup\{\lambda \in U : x \in C(s,\lambda)\} = s(x).$$

This demonstrates that s is a fixed point for $Co(C)^*$. The remaining part of the proof is trivial. □

Note that from (2.3) it follows that, for every $s \in \mathcal{F}(S)$ and $x \in S$,

$$Co(C)^*(s)(x) = Sup\{\lambda \in U : x \in Co(C)(C(s,\lambda))\}. \tag{2.4}$$

The following proposition gives a very simple way to get the canonical extension of a closure system. Recall that, given a class $\mathcal{A}$ of fuzzy subsets, we denote the fuzzy closure system generated by $\mathcal{A}$ by $c(\mathcal{A})$.

Proposition 2.3. *Let C be a class of sets and set*

$$Q(C) = \{\lambda \vee X : X \in C, \lambda \in U, \lambda \neq 1\}. \tag{2.5}$$

Then, we have

$$c(Q(C)) \supseteq C^* \supseteq Q(C). \tag{2.6}$$

Furthermore, if C is a closure system, then

$$C^* = c(Q(C)), \tag{2.7}$$

i.e., C^* is the fuzzy closure system generated by $Q(C)$.

Proof. Given X in C and $\lambda \in U$, for every $\mu \in U$,

$$C(\lambda \vee X, \mu) = \begin{cases} S & \text{if } \lambda \leq \mu, \\ X & \text{if } \mu > \lambda. \end{cases}$$

Then $\lambda \vee X \in C^*$ and, therefore, $C^* \supseteq Q(C)$. Let $s \in C^*$, then, since every $C(s,\lambda)$ belongs to C and $s = \bigcap_{\lambda\neq 1}\lambda \vee C(s,\lambda)$, we have that $s \in c(Q(C))$. This demonstrates that $c(Q(C)) \supseteq C^*$.

Moreover, (2.6) yields $c(C^*) = c(Q(C))$ and (2.7) follows. □

3. CANONICAL EXTENSIONS AND CONTINUOUS DEFORMATIONS

The canonical extensions satisfy interesting continuity properties.

Theorem 3.1. *Let $f: U \rightarrow U$ be a continuous, order-preserving map and J an a-c-operator. Then, for every fuzzy subset s,*

$$J^*(f \circ s) = f \circ J^*(s) \cup J(\varnothing).$$

In particular, if $f(1) = 1$ or $J(\varnothing) = \varnothing$,

$$J^*(f \circ s) = f \circ J^*(s).$$

Proof. We must prove that, for every $x \in S$,

$$J^*(f \circ s)(x) = \begin{cases} 1 & \text{if } x \in J(\varnothing), \\ f(J^*(s)(x)) & \text{otherwise.} \end{cases} \tag{3.1}$$

Consider the case $x \in J(\varnothing)$. Then $x \in J(C(f \circ s, \lambda))$ for every $\lambda \in U$ and therefore, $J^*(f \circ s)(x) = 1$. Consider the case $x \notin J(\varnothing)$. Then, due to the continuity of f, $f(U)$ is an interval $[a,b]$. Moreover, since f is order preserving, for every $\lambda \in [a,b]$, the set $f^{-1}(\{\lambda\}))$ is a closed interval whose least element we denote by β_λ. Then

$$C(f \circ s,\lambda) = \begin{cases} C(s,\beta_\lambda) & \text{if } \lambda \in [a,b], \\ \varnothing & \text{if } \lambda > b, \\ S & \text{if } \lambda < a. \end{cases}$$

Indeed, it is evident that $C(f \circ s,\lambda) = \varnothing$ for $\lambda > b$ and $C(f \circ s,\lambda) = S$ for $\lambda < a$. Let $\lambda \in [a,b]$. Then,

$$x \in C(f \circ s,\lambda) \Leftrightarrow f(s(x)) \geq \lambda \Leftrightarrow s(x) \geq \beta_\lambda \Leftrightarrow x \in C(s,\beta_\lambda).$$

Thus,

$$\begin{aligned} J^*(f \circ s)(x) &= Sup\{\lambda \in U : x \in J(C(f \circ s,\lambda))\} \\ &= Sup(\{\lambda \in U : \lambda < a, x \in J(C(f \circ s,\lambda))\} \\ &\quad \cup \{\lambda \in U : \lambda \in [a,b], x \in J(C(f \circ s,\lambda))\}) \\ &= Sup\{\lambda \in [a,b] : x \in J(C(s, \beta_\lambda))\} \\ &= Sup\{f(\beta_\lambda) : x \in J(C(s,\beta_\lambda))\} \\ &= Sup\{f(\beta) : x \in J(C(s,\beta))\} \\ &= f(Sup\{\beta : x \in J(C(s,\beta))\}) = f(J^*(s)(x)). \end{aligned}$$

In order to prove the next theorem we need a characterization of the meet-preserving maps in [0,1]. Recall that a function $f : U \to U$ is called *upper semicontinuous* if $f^{-1}([\lambda,1])$ is a closed subset for every $\lambda \in U$.

Lemma 3.2. *Let $f : U \to U$ be a map. Then the following statements are equivalent:*

(i) *f is upper semicontinuous and order-preserving.*

(ii) *f is meet-preserving.*

(iii) $\lambda \leq f(1) \Rightarrow \exists \mu_\lambda \leq 1$ *s.t.* $f^{-1}([\lambda,1]) = [\mu_\lambda,1]$ *and* $\lambda > f(1) \Rightarrow f^{-1}([\lambda,1]) = \varnothing$.

Proof. (i) $\Rightarrow$ (ii). Let A be any subset of U. Then $f(Inf(A)) \leq f(a)$ for every $a \in A$ and therefore $f(Inf(A)) \leq Inf(f(A))$. Moreover, since f is upper semicontinuous, $f^{-1}([Inf(f(A)), 1])$ is a closed subset containing A and, hence, $Inf(A)$. Thus $f(Inf(A)) \geq Inf(f(A))$ and this proves (ii).

(ii) $\Rightarrow$ (iii). It is obvious that f is order-preserving and this entails that, for every $\lambda \leq f(1)$, the set $f^{-1}([\lambda,1])$ is an interval containing 1. Set $\mu_\lambda = Inf(f^{-1}([\lambda,1]))$. By hypothesis

$$f(\mu_\lambda) = f(Inf(f^{-1}([\lambda,1]))) = Inf(f(f^{-1}([\lambda,1]))) \geq \lambda,$$

and so, μ_λ is an element of the set $f^{-1}([\lambda,1])$, namely, the least element of such an interval. This proves that $f^{-1}([\lambda,1]) = [\mu_\lambda,1]$. It is obvious that $f^{-1}([\lambda,1]) = \varnothing$ for every $\lambda > f(1)$.

(iii) $\Rightarrow$ (i). It is evident that f is upper-semicontinuous. To prove that f is order-preserving, observe at first that $f(1)$ is the greatest value of f in U. Suppose

$\lambda \le \mu \le 1$. Then, since $f(\lambda) \le f(1)$, we have that $f^{-1}([f(\lambda),1])$ is an interval we denote by $[\mu_{f(\lambda)}, 1]$. Since $\lambda \in f^{-1}([f(\lambda), 1]) = [\mu_{f(\lambda)},1]$ and $\mu \ge \lambda$, we have that μ is an element in $f^{-1}([f(\lambda),1])$. Thus $f(\mu) \ge f(\lambda)$ and f is order-preserving. □

Theorem 3.3. *Let J be an a-c-operator and $f : U \to U$ a meet-preserving map such that $f(1) = 1$, then*

$$J^*(f \circ s) \subseteq f \circ J^*(s) \tag{3.2}$$

and, consequently,

$$s \in Cs(J^*) \Rightarrow f \circ s \in Cs(J^*). \tag{3.3}$$

Let C be any class of subsets, then

$$s \in C^* \Rightarrow f \circ s \in C^*. \tag{3.4}$$

Proof. By hypothesis and in accordance with (iii) of Lemma 3.2, for every $\lambda \in U$ a real number μ_λ exists such that $C(s,\mu_\lambda) = C(f \circ s,\lambda)$ where μ_λ is the least element μ such that $f(\mu) \ge \lambda$. Then,

$$J^*(f \circ s)(x) = Sup(\{\lambda \in U : x \in J(C(f \circ s,\lambda))\}) = Sup(\{\lambda \in U : x \in J(C(s,\mu_\lambda))\}).$$

Now, from the inequality $f(\mu_\lambda) \ge \lambda$ it follows that

$$\begin{aligned} Sup(\{\lambda \in U : x \in J(C(s,\mu_\lambda))\}) &\le Sup(\{f(\mu_\lambda) \in U : x \in J(C(s,\mu_\lambda))\}) \\ &\le Sup(\{f(\mu) \in U : x \in J(C(s,\mu))\}) \\ &\le f(Sup\{\mu \in U : x \in J(C(s,\mu))\}) = f(J^*(s)(x)). \end{aligned}$$

This proves (3.2). To prove (3.3) observe that if $s \in Cs(J^*)$ then

$$f \circ s \subseteq J^*(f \circ s) \subseteq f \circ J^*(s) = f \circ s.$$

Finally, in order to prove (3.4), observe that by Lemma 3.2 for every $\lambda \in U$, $\mu_\lambda \in U$ exists such that $f^{-1}([\lambda,1]) = [\mu_\lambda,1]$. Then, for every $s \in C^*$,

$$C(f \circ s,\lambda) = \{x \in S : f(s(x)) \ge \lambda\} = \{x \in S : s(x) \ge \mu_\lambda\} \in C$$

and this demonstrates that $f \circ s$ belongs to C^*. □

4. DUALIZING THE EXTENSION PRINCIPLE

Equation (1.8) shows that it is possible to refer to open cuts instead of closed cuts in defining the canonical extension of a classical operator. As a mathematical curiosity, the same question will now be considered for the canonical extension of a class C. To do this, set

$$C^o = \{s \in \mathcal{F}(S) : O(s,\lambda) \in C \text{ for every } \lambda \in U, \lambda \ne 1\}. \tag{4.1}$$

Proposition 4.1. *Let C be a closure system. Then $C^o \subseteq C^*$, but, in general, $C^o \ne C^*$. Let C be an interior system. Then $C^o \supseteq C^*$, but, in general, $C^o \ne C^*$.*

Proof. Let C be a closure system and assume that $s \in C^o$. Then, given $\lambda \ne 0$, since $C(s,\lambda) = \bigcap_{\mu<\lambda} O(s,\mu)$ and $O(s,\mu) \in C$ for every $\mu \in U$, we have that $C(s,\lambda) \in C$. This proves that $s \in C^*$. Let C be the class of closed subsets of the real line $\boldsymbol{R}$ and s a continuous fuzzy subset. Then it is evident that $s \notin C^o$ while $s \in C^*$. Let C be an interior system and suppose $s \in C^*$. Then given $\lambda \ne 1$, since $C(s,\mu) \in C^*$ for every $\mu \in U$, we have that $O(s,\lambda) = \bigcup_{\mu>\lambda} C(s,\mu) \in C$ and this proves that $s \in C^o$.

Hence $C^o \supseteq C^*$. Finally, the class C of all open subsets of $\boldsymbol{R}$ is an example of interior system such that $C^o \neq C^*$. In fact, if s is any continuous fuzzy subset, then $s \in C^o$ and $s \notin C^*$. □

The proof of the next proposition is trivial:

Proposition 4.2. *Let C be any class of classical subsets of S. Then C^o is an extension of C. More specifically, C coincides with the class of crisp elements of C^o. Further,*

$$C^o \text{ is a fuzzy interior system} \Leftrightarrow C \text{ is an interior system.}$$

We now prove that the following diagrams commute:

$$\begin{array}{ccc} C & \longrightarrow & C_d \\ \downarrow & & \downarrow \\ C^o & \longleftarrow & (C_d)^* \end{array} \qquad \begin{array}{ccc} I & \longrightarrow & Is(I) \\ \downarrow & & \downarrow \\ I^* & \longrightarrow & Is(I)^o \end{array} \qquad \begin{array}{ccc} O & \longrightarrow & Io(O) \\ \downarrow & & \downarrow \\ O^o & \longrightarrow & Io(O)^* \end{array}$$

Proposition 4.3. *Let C be a class of subsets, then*

$$C^o = (C_d)^*{}_d. \tag{4.2}$$

Moreover, for every interior operator I,

$$Is(I^*) = Is(I)^o, \tag{4.3}$$

for every interior system O,

$$Io(O^o) = Io(O)^*. \tag{4.4}$$

Proof. In order to prove (4.2) observe that

$$\begin{aligned} (C_d)^*{}_d &= \{s \in \mathcal{F}(S) : C(\text{-}s,\lambda) \in C_d \text{ for every } \lambda \in U\} \\ &= \{s \in \mathcal{F}(S) : \text{-}C(\text{-}s,\lambda) \in C \text{ for every } \lambda \in U\} \\ &= \{s \in \mathcal{F}(S) : O(s,1-\lambda) \in C \text{ for every } \lambda \in U\} \\ &= \{s \in \mathcal{F}(S) : O(s,\lambda) \in C \text{ for every } \lambda \in U\} = C^o. \end{aligned}$$

In order to prove (4.3) observe that, by Proposition 1.6, $(I^*)_d = (I_d)^*$ and by Proposition 2.5 of Chapter 2,

$$Is(I^*)_d = Cs(I^*{}_d),\ Is(I_d) = Cs(I)_d.$$

So

$$Is(I^*) = Cs(I^*{}_d)_d = Cs(I_d{}^*)_d = (Cs(I_d)^*)_d = (Is(I)_d)^*{}_d = Is(I)^o.$$

In order to prove (4.4) observe that, since O_d is a closure system,

$$Io(O^o) = Io((O_d)^*)_d = Co(O_d{}^*)_d = (Co(O_d)^*)_d = (Co(O_d))_d{}^* = Io(O)^*. \quad □$$

5. EXTENSION OF A COMPACT CLOSURE OPERATOR

In the case of compact operators, we can give a very simple formula for the canonical extension. At first, given a nonempty subset X of S and a fuzzy subset s of S, we set

$$Incl(X,s) = Inf_{x \in X}\, s(x).$$

Moreover, we set $Incl(\varnothing,s) = 1$. We say that $Incl(X,s)$ *is the inclusion degree of* X *in* s. A straightforward verification enables us to prove that, for any X, Y in $\mathcal{P}(S)$ and s, s' in $\mathcal{F}(S)$:

i) $X \subseteq Y \Rightarrow Incl(X,s) \geq Incl(Y,s)$; ii) $s \subseteq s' \Rightarrow Incl(X,s) \leq Incl(X,s')$

iii) $Incl(X \cup Y,s) = Incl(X,s) \wedge Incl(Y,s)$; iv) $X \subseteq C(s,\lambda) \Leftrightarrow Incl(X,s) \geq \lambda$.

Theorem 5.1. *Let* $J : \mathcal{P}(S) \to \mathcal{P}(S)$ *be an order-preserving operator. Then*

$$J^*(s)(x) = Sup\{Incl(X,s) : X \in \mathcal{P}(S), x \in J(X)\}. \tag{5.1}$$

Consequently, if J *is compact,*

$$J^*(s)(x) = Sup\{Incl(X,s) : X \in \mathcal{P}_f(S), x \in J(X)\}. \tag{5.2}$$

Moreover,

$$J^* \text{ is continuous} \Leftrightarrow J \text{ is compact.}$$

Proof. As J is monotone,

$$J(C(s,\lambda)) = \cup\{J(X) : X \subseteq C(s,\lambda)\} = \cup\{J(X) : Incl(X,s) \geq \lambda\}.$$

As a consequence,

$$\begin{aligned} J^*(s)(x) &= Sup\{\lambda \in U : x \in J(C(s,\lambda))\} \\ &= Sup\{\lambda \in U : X \in \mathcal{P}(S) \text{ exists such that } Incl(X,s) \geq \lambda \text{ and } x \in J(X)\} \\ &= Sup\{Incl(X,s) : X \in \mathcal{P}(S), x \in J(X)\}. \end{aligned}$$

Assume that J is compact. Then (5.2) is evident. To prove that J^* is continuous, let $(s_i)_{i\in I}$ be a directed family of fuzzy subsets and set $s = \cup_{i\in I} s_i$. We claim that, for any finite subset X,

$$Incl(X, \cup_{i\in I} s_i) = Sup_{i\in I}\, Incl(X,s_i).$$

Indeed, if $X = \{x_1,\ldots,x_n\}$,

$$\begin{aligned} Incl(X, \cup_{i\in I} s_i) &= (\cup_{i\in I} s_i)(x_1) \wedge \ldots \wedge (\cup_{i\in I} s_i)(x_n) \\ &= (Sup_{i\in I}\, s_i(x_1)) \wedge \ldots \wedge (Sup_{i\in I}\, s_i(x_n)) \\ &= Sup_{i(1),\ldots,i(n)}\, (s_{i(1)}(x_1) \wedge \ldots \wedge s_{i(n)}(x_n)). \end{aligned}$$

Since $(s_i)_{i\in I}$ is directed, $i \in I$ exists such that $s_{i(1)} \subseteq s_j,\ldots,s_{i(n)} \subseteq s_j$. Consequently,

$$Incl(X, \cup_{i\in I} s_i) = Sup_{i\in I}\, (s_i(x_1) \wedge \ldots \wedge s_i(x_n)) = Sup_{i\in I}\, Incl(X,s_j).$$

Thus,

$$\begin{aligned} J^*(s)(x) &= Sup\{Incl(X,s) : X \in \mathcal{P}(S), x \in J(X)\} \\ &= Sup\{Sup_{j\in I}\, Incl(X,s_j) : X \in \mathcal{P}(S), x \in J(X)\} \\ &= Sup_{j\in I}\{Sup\, Incl(X,s_j) : X \in \mathcal{P}(S), x \in J(X)\} = \cup_{j\in I} J^*(s_j)(x). \end{aligned}$$

Trivially, if J^* is continuous, then J is compact. □

Note that, in general,

$$O(J^*(s),\mu) \neq J(O(s,\mu)) \quad \text{and} \quad J(C(s,\mu)) \neq C(J^*(s),\mu).$$

For example, let J be the usual topological closure operator in the interval $[0,1]$ and s a continuous map. Then $J^*(s) = s$ and, while $O(J^*(s),\mu) = O(s,\mu))$ is an open set, $J(O(s,\mu))$ is closed. In the case of compact operators we have the following proposition:

Proposition 5.2. *Let* $J : \mathcal{P}(S) \to \mathcal{P}(S)$ *be a compact a-c-operator. Then, for any* $\mu \neq 1$,

$$O(J^*(s),\mu) = J(O(s,\mu)), \tag{5.3}$$

while, in general, $J(C(s,\mu)) \neq C(J^*(s),\mu)$.

Proof. $(C(s,\lambda))_{\lambda>\mu}$ is a directed family. Then, by Lemma 1.3,

$$O(J^*(s),\mu) = \cup_{\lambda>\mu} J(C(s,\lambda)) = J(\cup_{\lambda>\mu} C(s,\lambda)) = J(O(s,\mu)).$$

An example of compact closure operator J for which $J(C(s,\mu)) \neq C(J^*(s),\mu)$ is furnished by the natural extension of the classical propositional calculus proposed in Section 3 of Chapter 6. □

The next proposition shows that, in the case of algebraic closure systems, the extensions by closed cuts coincide with the extensions by open cuts.

Proposition 5.3. *Assume that* C *is a classical algebraic closure system. Then*

$$C^* = C^o. \tag{5.4}$$

Moreover, C^* *is an algebraic closure system.*

Proof. The inclusion $C^o \subseteq C^*$ was proven in Proposition 4.1. Assume that $s \in C^*$, that is $C(s,\lambda) \in C$ for every $\lambda \neq 0$. Then, since C is algebraic and, for every $\mu \neq 1$, $(C(s,\lambda))_{\lambda>\mu}$ is a directed family, we have that

$$O(s,\mu) = \cup_{\lambda>\mu} C(s,\lambda) \in C.$$

Consequently, $s \in C^o$ and (5.4) is proved.

Let $\mathcal{T}$ be a directed family of elements of C^* and $s = lim\ \mathcal{T}$. Then, for every $\lambda \in U$ such that $\lambda \neq 1$, $O(s,\lambda) = \cup\{O(s',\lambda) : s' \in \mathcal{T}\}$. Since the family $\{O(s',\lambda) : s' \in \mathcal{T}\}$ is directed and C is algebraic, this proves that $O(s,\lambda) \in C$. Consequently, $s \in C^o$ and therefore, $s \in C^*$. Thus, C^* is algebraic. □

Recall that if J is an operator, then $c(J)$ denotes the closure operator generated by J.

Theorem 5.4. *For every classical compact a-c-operator* J,

$$c(J^*) = c(J)^*,$$

i.e.,

$$\cup_{n\in N} (J^*)^n = (\cup_{n\in N} J^n)^*. \tag{5.5}$$

Proof. Observe at first that, for every $n \in N$,

$$O((J^*)^n(s),\lambda) = J^n(O(s,\lambda)). \tag{5.6}$$

Indeed, in the case $n = 1$ (5.6) coincides with (5.3). Assume (5.6) for the integer n, then

$$O((J^*)^{n+1}(s),\lambda) = O(J^*((J^*)^n(s)),\lambda) = J(O((J^*)^n(s),\lambda)) = J(J^n(O(s,\lambda))) = J^{n+1}(O(s,\lambda)).$$

Also,

$$O(c(J^*)(s),\lambda) = c(J)(O(s,\lambda)), \tag{5.7}$$

Indeed,

$$O(c(J^*)(s),\lambda) = O(\cup_{n\in N} (J^*)^n(s),\lambda) = \cup_{n\in N} (O((J^*)^n(s),\lambda))$$
$$= \cup_{n\in N} J^n(O(s,\lambda)) = c(J)(O(s,\lambda)).$$

Finally, taking (5.7) and (1.8) into account,

$$c(J^*)(s)(x) = Sup\{\lambda \in U : x \in O(c(J^*)(s),\lambda)\}$$

$$= Sup\{\lambda \in U : x \in c(J)(O(s,\lambda))\} = c(J)^*(s)(x).$$ □

6. EXTENSION OF A CRISP LOGIC

We can interpret the results in the previous sections in terms of fuzzy logic. Given a crisp abstract deduction system $(\mathcal{F}, \mathcal{D})$, we say that $(\mathcal{F}, \mathcal{D}^*)$ is *the canonical extension* of $(\mathcal{F}, \mathcal{D})$. To get a more expressive formula for $\mathcal{D}^*$, we can write $X \vdash x$ to denote that $x \in \mathcal{D}(X)$ and $x_1,\dots,x_n \vdash x$ to denote that $x \in \mathcal{D}(\{x_1,\dots,x_n\})$.

Definition 6.1. Let $(\mathcal{F}, \mathcal{D})$ be a crisp deduction system and v a fuzzy set of formulas. Then we say that a formula α *is a consequence of* v *to degree* λ, in brief $v \vdash_\lambda \alpha$, provided that $C(v,\lambda) \vdash \alpha$.

In others words, we set $v \vdash_\lambda \alpha$ provided that α can be proved by formulas that are true at least to degree λ in the valuation v.

Theorem 6.2. *The canonical extension $(\mathcal{F}, \mathcal{D}^*)$ of $(\mathcal{F}, \mathcal{D})$ can be defined by setting*

$$\mathcal{D}^*(v)(\alpha) = Sup\{\lambda \in U : v \vdash_\lambda \alpha\}. \tag{6.1}$$

Moreover, given a fuzzy subset τ of formulas,

$$\tau \text{ is a theory of } (\mathcal{F}, \mathcal{D}^*) \Leftrightarrow \text{ every closed cut of } \tau \text{ is a theory of } (\mathcal{F}, \mathcal{D}).$$

Proof. Equation (6.1) is evident. Let $\mathcal{T} = Cs(\mathcal{D})$ be the class of theories of $(\mathcal{F}, \mathcal{D})$. Then, since $Cs(\mathcal{D}^*) = Cs(\mathcal{D})^*$, the class of theories of $\mathcal{D}^*$ is $\mathcal{T}^*$. □

From this theorem and Proposition 2.3, it follows that, given a consistent theory T of $(\mathcal{F}, \mathcal{D})$ and $\lambda \neq 1$, the fuzzy subset $\lambda \vee T$ is a theory of $(\mathcal{F}, \mathcal{D}^*)$ we call a *λ-theory*. In a sense, the class of λ-theories is a basis for the class of theories.

Theorem 6.3. *Let $(\mathcal{F}, \mathcal{D}^*)$ be the canonical extension of a crisp deduction system $(\mathcal{F}, \mathcal{D}^*)$. Then $(\mathcal{F}, \mathcal{D}^*)$ is not logically compact and*

$$Inc(\mathcal{D}^*) = 1.$$

Moreover, every theory in $(\mathcal{F}, \mathcal{D}^*)$ is an intersection of λ-theories.

Proof. Let T be a consistent theory of $(\mathcal{F}, \mathcal{D})$. Then the class $\{\lambda \vee T : \lambda \neq 1\}$ is an inductive class of consistent theories whose union is not consistent. This proves that $(\mathcal{F}, \mathcal{D}^*)$ is not logically compact. The fact that $Inc(\lambda \vee T) = \lambda$ entails that $Inc(\mathcal{D}^*) = 1$.

The remaining part of the theorem follows from Proposition 2.3. □

Since $\mathcal{D}(\emptyset)$ is the set Tau of tautologies, by Theorem 3.1 we have:

Theorem 6.4. *Let $f: U \to U$ be a continuous, order-preserving map. Then*

$$\mathcal{D}^*(f \circ v) = f \circ \mathcal{D}^*(v) \cup Tau. \tag{6.2}$$

In particular, if $f(1) = 1$,

$$\mathcal{D}^*(f \circ v) = f \circ \mathcal{D}^*(v). \tag{6.3}$$

As an example, let $f(x) = \lambda \cdot x + \mu$ where $\lambda > 0$, $\mu \geq 0$ and $\lambda + \mu \leq 1$. Then

$$\mathcal{D}^*(\lambda \cdot v + \mu) = (\lambda \cdot \mathcal{D}^*(v)(x) + \mu) \cup Tau.$$

Hence, if v is the characteristic function of a set of formulas X and $\mu = 0$,

$$\mathcal{D}^*(\lambda \cdot X) = \lambda \cdot \mathcal{D}(X) \cup Tau,$$

or, equivalently,

$$\mathcal{D}^*(\lambda \wedge X) = \lambda \wedge \mathcal{D}(X) \cup Tau. \tag{6.4}$$

In the case $X = S$,

$$\mathcal{D}^*(s^{\lambda}) = \lambda \vee Tau. \tag{6.5}$$

From Theorem 3.3 we have:

Theorem 6.5. *Let $(\mathcal{F}, \mathcal{D})$ be a crisp deduction system and $f : U \to U$ a meet-preserving map such that $f(1) = 1$, then*

$$\tau \textit{ theory of } (\mathcal{F}, \mathcal{D}^*) \Rightarrow f \circ \tau \textit{ theory of } (\mathcal{F}, \mathcal{D}^*).$$

It is easy to see that, in the case of compact deduction systems, the following theorem holds:

Theorem 6.6. *Let $(\mathcal{F}, \mathcal{D})$ be a compact deduction system. Then*

$$\mathcal{D}^*(v)(\alpha) = Sup\{Incl(X, v) : X \in \mathcal{P}_f(\mathcal{F}) \textit{ such that } X \vdash \alpha\}. \tag{6.6}$$

Moreover, τ is a theory of $(\mathcal{F}, \mathcal{D}^*)$ iff every open cut (equivalently, closed cut) of τ is a theory of $(\mathcal{F}, \mathcal{D})$.

From Theorem 5.4 we have:

Theorem 6.7. *Let $(\mathcal{F}, J, \mathcal{A})$ be a step-by-step crisp deduction system whose associated deduction system is $(\mathcal{P}(\mathcal{F}), \mathcal{D})$. Then $(\mathcal{F}, J^*, \mathcal{A})$ is a fuzzy step-by-step deduction system whose associated deduction system is $(\mathcal{F}, \mathcal{D}^*)$.*

Corollary 6.8. *Let τ be a theory of a canonical extension $(\mathcal{F}, \mathcal{D}^*)$ and assume that $\alpha_1,...,\alpha_m \vdash \alpha$. Then*

$$\tau(\alpha) \geq \tau(\alpha_1) \wedge ... \wedge \tau(\alpha_m). \tag{6.7}$$

In particular,

$$\alpha \vdash \beta \Rightarrow \tau(\alpha) \leq \tau(\beta) \quad ; \quad \alpha \vdash \beta \textit{ and } \beta \vdash \alpha \Rightarrow \tau(\alpha) = \tau(\beta).$$

Proof. Let $\lambda = \tau(\alpha_1) \wedge ... \wedge \tau(\alpha_m)$. Then $\alpha_1,...,\alpha_m \in C(\tau,\lambda)$ and, because $C(\tau,\lambda)$ is a theory of $\mathcal{D}$, $\alpha \in C(\tau,\lambda)$. Hence, $\tau(\alpha) \geq \lambda = \tau(\alpha_1) \wedge ... \wedge \tau(\alpha_m)$. □

Let $(\mathcal{F}, \mathcal{D}, \mathcal{M})$ be a crisp logic, then we call *U-models* the elements of the set

$$Q(\mathcal{M}) = \{ \lambda \vee M : M \in \mathcal{M}, \lambda \in U, \lambda \neq 1\}.$$

Theorem 6.9. *Let $(\mathcal{F}, \mathcal{D}, \mathcal{M})$ be a crisp logic. Then, $(\mathcal{F}, \mathcal{D}^*, \mathcal{M}^*)$ is a fuzzy logic we call the canonical extension of $(\mathcal{F}, \mathcal{D}, \mathcal{M})$. Such logic is equivalent to*

$(\mathcal{F},\mathcal{D}^*,Q(\mathcal{M}))$, *i.e., the class of U-models defines an equivalent semantics.*

Proof. From the equality $\mathcal{D} = Co(\mathcal{M})$ it follows that $Co(\mathcal{M}^*) = Co(\mathcal{M})^* = \mathcal{D}^*$. This proves that $(\mathcal{F}, \mathcal{D}^*, \mathcal{M}^*)$ is a fuzzy logic. Moreover, in accordance with Proposition 2.3, $Co(\mathcal{M}^*) = Co(Q(\mathcal{M}))$. □

Proposition 6.10. *Let* $(\mathcal{F}, \mathcal{D}, \mathcal{M})$ *be a crisp logic such that every* $M \in \mathcal{M}$ *is maximal in* $\mathcal{M}$. *Then every U-model* $\lambda \vee M$ *is maximal in the class*

$$\{\tau \in \mathcal{F}(\mathcal{F}) : \tau \text{ is a theory of } \mathcal{D}^* \text{ such that } Inc(\tau) \leq \lambda\}.$$

Proof. It is evident that $Inc(\lambda \vee M) = \lambda$. Let τ be a theory containing $\lambda \vee M$ such that $Inc(\tau) \leq \lambda$. To prove that $\tau = \lambda \vee M$ observe that for every $\mu \leq \lambda$,

$$C(\tau,\mu) \supseteq C(\lambda \vee M,\mu) = \mathcal{F}.$$

On the other hand, if $\mu > \lambda$, since $Inc(\tau) \leq \lambda$, $C(\tau,\mu) \neq \mathcal{F}$. So, $C(\tau,\mu)$ is a consistent theory containing M. Since M is maximal, we can conclude that $C(\tau,\mu) = M$. □

7. CHARACTERIZATIONS OF THE CANONICAL EXTENSIONS

Observe that fuzzy closure operators exist which do not extend a classical closure operator. For instance, we can consider the operator

$$J(s)(x) = s(x) \vee \mu,$$

where μ is an element of U different from zero. Furthermore, extensions of a classical closure operator exist which are not canonical extensions. As an example, let C be a closure system in [0,1] containing 0 and 1 and different from [0,1]. Then the fuzzy operator J, defined by setting

$$J(s)(x) = Inf\{\lambda \in C : \lambda \geq s(x)\},$$

is different from the identity map. Since its restriction to $\mathcal{P}(S)$ is the identity map, we can conclude that J cannot be obtained by (1.1).

In the following theorem we show that each equality in Lemma 1.3 characterizes the canonical extensions.

Theorem 7.1. *Let* $(\mathcal{F}, \mathcal{D})$ *be a fuzzy deduction system extending a classical deduction system, i.e.,* $\mathcal{D}(v)$ *is crisp everywhere* v *is crisp. Then the following are equivalent:*

(i) $(\mathcal{F}, \mathcal{D})$ *is a canonical extension of a crisp deduction system.*

(ii) $O(\mathcal{D}(v),\mu) = \bigcup_{\lambda>\mu} \mathcal{D}(C(v,\lambda))$ *for any* $\mu \in U$.

(iii) $C(\mathcal{D}(v),\mu) = \bigcap_{\lambda<\mu} \mathcal{D}(C(v,\lambda))$ *for any* $\mu \in U$.

Proof. Implications (i) ⇒ (ii) and (i) ⇒ (iii) follow from Lemma 1.3. To prove that (ii) ⇒ (i), let J be the restriction of $\mathcal{D}$ to the crisp fuzzy subsets. Then

$$\begin{aligned}\mathcal{D}(v)(x) &= Sup\{\mu \in U : x \in O(\mathcal{D}(v),\mu)\}\\ &= Sup\{\mu \in U : x \in \mathcal{D}(C(v,\lambda)) \text{ for a suitable } \lambda > \mu\}\\ &= Sup\{\lambda \in U : x \in \mathcal{D}(C(v,\lambda))\}\\ &= Sup\{\lambda \in U : x \in J(C(v,\lambda))\} = J^*(v)(x).\end{aligned}$$

To prove (iii) $\Rightarrow$ (i) observe that

$$\begin{aligned}\mathcal{D}(v)(x) &= Sup\{\mu \in U : x \in C(\mathcal{D}(v),\mu)\}\\ &= Sup\{\mu \in U : x \in \mathcal{D}(C(v,\lambda)) \text{ for every } \lambda < \mu\}\\ &= Sup\{\lambda \in U : x \in \mathcal{D}(C(v,\lambda))\}\\ &= Sup\{\lambda \in U : x \in J(C(v,\lambda))\} = J^*(v)(x).\end{aligned}$$

Theorem 7.2. *A fuzzy deduction system* $(\mathcal{F},\mathcal{D})$ *is the canonical extension of a crisp deduction system iff*

(i) *every closed cut of a theory is a theory,*

(ii) τ *theory of* $(\mathcal{F},\mathcal{D}) \Rightarrow \lambda \vee \tau$ *theory of* $(\mathcal{F},\mathcal{D})$.

Proof. If $(\mathcal{F},\mathcal{D})$ is the canonical extension of a closure system, then it is evident that (i) and (ii) are satisfied. Let's assume (i) and (ii) are true, let $\mathcal{T}$ be the class of theories of $(\mathcal{F},\mathcal{D})$ and let C be the class of crisp elements of $\mathcal{T}$. Then, given $s \in C^*$, by definition $C(s,\lambda) \in C$ and by (ii) $\lambda \vee C(s,\lambda) \in \mathcal{T}$. Consequently, since $\mathcal{T}$ is a closure system and $s = \bigcap_{\lambda\in U}\lambda \vee C(s,\lambda)$, we have that $s \in \mathcal{T}$. This demonstrates that $C^* \subseteq \mathcal{T}$. Conversely, if $s \in \mathcal{T}$, since by (i) $C(s,\lambda) \in C$, we have that, for every $\lambda \in U$, $\lambda \vee C(s,\lambda) \in C^*$. Consequently, since C^* is a closure system and $s = \bigcap_{\lambda\in U}\lambda \vee C(s,\lambda)$, we have that $s \in C^*$. This demonstrates that $\mathcal{T} \subseteq C^*$. Thus $C^* = \mathcal{T}$ and we can conclude that $\mathcal{D}$ is the canonical extension of $Co(C)$.

Notice that in Condition (ii) we can confine ourselves to crisp relations.

Theorem 7.3. *Let* $(\mathcal{F},\mathcal{D})$ *be a fuzzy deduction system. Then the following are equivalent:*

(j) $(\mathcal{F},\mathcal{D})$ *is the canonical extension of a crisp deduction system.*

(jj) *For every meet-preserving map* $f: U \to U$ *such that* $f(1) = 1$,
τ *theory of* $(\mathcal{F},\mathcal{D}) \Rightarrow f\circ\tau$ *theory of* $(\mathcal{F},\mathcal{D})$.

(jjj) *For every order-preserving continuous map* $f: U \to U$ such that $f(1) = 1$,
τ *theory of* $(\mathcal{F},\mathcal{D}) \Rightarrow f\circ\tau$ *theory of* $(\mathcal{F},\mathcal{D})$.

Proof. Implication (j) $\Rightarrow$ (jj) was proven in Theorem 3.3. (jj) $\Rightarrow$ (jjj) is trivial. In fact, by Lemma 3.2 any continuous order-preserving map is meet-preserving. To prove (jjj) $\Rightarrow$ (j), we prove that (jjj) implies the conditions (i) and (ii) in Theorem 7.2. Indeed, given any element $\mu \in U$, consider the mapping $f_\mu : U \to U$ defined by setting $f_\mu(x) = 1$ if $x \geq \mu$ or, otherwise, $f_\mu(x) = 0$ otherwise. Then, it is easy to find a sequence $(f_n)_{n\in N}$ of order-preserving and continuous maps such that $f_n(1) = 1$ and $f_\mu = Inf\{f_n : n \in N\}$. Assume that τ is a theory. Then jjj) entails that the product $f_n\circ\tau$ is a theory. Since the class $\mathcal{T}$ of theories is a closure system, also $f_\mu\circ\tau = \bigcap_{n\in N}(f_n\circ\tau)$ is a theory. Now, $f_\mu\circ\tau$ is the characteristic function of $C(\tau,\mu)$, so $\mathcal{T}$ contains the closed cuts of its elements. In order to prove condition (ii), consider the continuous and order preserving map g_μ defined by $g_\mu(x) = \mu \vee x$ and observe that by (jjj) $g_\mu\circ\tau = \mu \vee \tau$ is a theory.

Theorem 7.4. *Let $(\mathcal{F}, \mathcal{D})$ be a fuzzy deduction system. Then the following are equivalent :*

(i) *$(\mathcal{F}, \mathcal{D})$ is a canonical extension of a crisp deduction system.*

(ii) *For every meet-preserving map f such that $f(1) = 1$,*

$$\mathcal{D}(f \circ v) \subseteq f \circ \mathcal{D}(v).$$

(iii) *For every order-preserving continuous map f such that $f(1) = 1$,*

$$\mathcal{D}(f \circ v) \subseteq f \circ \mathcal{D}(v).$$

Proof. Implication (i) $\Rightarrow$ (ii) follows from Theorem 3.3, (ii) $\Rightarrow$ (iii) is trivial. In order to prove (iii) $\Rightarrow$ (i) we prove condition (jjj) of Theorem 7.3. Indeed, let f be an order-preserving continuous map such that $f(1) = 1$ and let τ be a theory. Then from the inclusions

$$f \circ \tau \subseteq \mathcal{D}(f \circ \tau) \subseteq f \circ \mathcal{D}(\tau) = f \circ \tau,$$

it follows that $f \circ \tau = \mathcal{D}(f \circ \tau)$. Thus, $f \circ \tau$ is a theory. □

Very simple characterizations can be given for the canonical extensions of compact deduction systems. For instance, the following theorem shows that the canonical extensions of crisp logics can be characterized by the equation $O(\mathcal{D}(s),\mu) = \mathcal{D}(O(s,\mu))$.

Theorem 7.5. *Let $(\mathcal{F}, \mathcal{D})$ be a continuous fuzzy deduction system extending a crisp deduction system. Then $(\mathcal{F}, \mathcal{D})$ is a canonical extension if and only if,*

$$O(\mathcal{D}(s),\mu) = \mathcal{D}(O(s,\mu)) \tag{7.1}$$

for any $\mu \neq 1$.

Proof. Let $(\mathcal{F}, \mathcal{D})$ be a canonical extension. Then (7.1) follows from Proposition 5.2. Assume that (7.1) is satisfied by any $\mu \neq 1$. Then, by the continuity of $\mathcal{D}$,

$$O(\mathcal{D}(s),\lambda) = \mathcal{D}(O(s,\lambda)) = \mathcal{D}(\cup_{\mu>\lambda} C(s,\mu)) = \cup_{\mu>\lambda} \mathcal{D}(C(s,\mu)).$$

Consequently, by Theorem 7.1, $(\mathcal{F}, \mathcal{D})$ is a canonical extension. □

Theorem 7.6. *Let $(\mathcal{F}, \mathcal{D})$ be a continuous fuzzy deduction system extending a crisp deduction system. Then the following are equivalent:*

(i) *$(\mathcal{F}, \mathcal{D})$ is a canonical extension.*

(ii) *For every join-preserving function $f : U \to U$ such that $f(1) = 1$*

$$\mathcal{D}(f \circ v) = f \circ \mathcal{D}(v).$$

Proof. (i) $\Rightarrow$ (ii). Assume that $\mathcal{D}$ is the canonical extension of a crisp deduction operator H. Then, if α is a tautology,

$$\mathcal{D}(f \circ v)(\alpha) = 1 = f(1) = f(\mathcal{D}(v)(\alpha)).$$

Otherwise,

$$\begin{aligned}\mathcal{D}(f \circ v)(x) &= H^*(f \circ v)(x) = Sup\{f(v(\gamma_1)) \wedge \ldots \wedge f(v(\gamma_m)) : x \in H(\gamma_1,\ldots,\gamma_m)\} \\ &= Sup\{f(v(\gamma_1) \wedge \ldots \wedge v(\gamma_m)) : x \in H(\gamma_1,\ldots,\gamma_m)\} \\ &= f(Sup\{v(\gamma_1) \wedge \ldots \wedge v(\gamma_m) : x \in H(\gamma_1,\ldots,\gamma_m)\}) \\ &= f(H^*(v)(x)) = f(\mathcal{D}(v)(x)).\end{aligned}$$

(ii) $\Rightarrow$ (i). For every $\mu \in U$, let $f_\mu : U \to U$ be the join-preserving function defined by setting $f_\mu(x) = 1$ if $x > \mu$ or, otherwise, $f_\mu(x) = 0$. Then, since $f_\mu \circ v$ and $f_\mu \circ \mathcal{D}(v)$ are the characteristic functions of $O(v,\mu)$ and $O(\mathcal{D}(v),\mu)$ respectively, from (ii) we have $\mathcal{D}(O(v,\mu)) = O(\mathcal{D}(v),\mu)$. Thus, by Theorem 7.5, $\mathcal{D}$ is a canonical extension. □

8. DEGREE OF INCONSISTENCY OF A CANONICAL EXTENSION

We consider the question of the degree of inconsistency in the case of the canonical extensions.

Proposition 8.1. *Let v be a fuzzy subset of formulas. Then the following propositions are equivalent:*
(i) *v is inconsistent.*
(ii) $C(v,\lambda)$ *is inconsistent for every* $\lambda \neq 1$.
(iii) $O(v,\lambda)$ *is inconsistent for every* $\lambda \neq 1$.

Proof. (i) $\Leftrightarrow$ (ii). At first observe that for every formula α the set $\{\lambda \in U : \alpha \in \mathcal{D}(C(v,\lambda))\}$ is an interval containing 0. Then

v inconsistent $\Leftrightarrow$ $\mathcal{D}^*(v)(\alpha) = Sup\{\lambda \in U : \alpha \in \mathcal{D}(C(v,\lambda))\} = 1$ for every $\alpha \in \mathbb{F}$
$\Leftrightarrow$ $\{\lambda \in U : \alpha \in \mathcal{D}(C(v,\lambda))\} \supseteq [0,1)$ for every $\alpha \in \mathbb{F}$
$\Leftrightarrow$ $\mathcal{D}(C(v,\lambda)) = \mathbb{F}$ for every $\lambda \neq 1$
$\Leftrightarrow$ $C(v,\lambda)$ is inconsistent for every $\lambda \neq 1$.

In order to prove (ii) $\Rightarrow$ (iii) observe that if, $\lambda \neq 1$, then $\mu > \lambda$ exists such that $\mu \neq 1$ and $O(v,\lambda) \supseteq C(v,\mu)$. The inconsistency of $C(v,\mu)$ entails the inconsistency of $O(v,\lambda)$. Finally, to prove that (iii) $\Rightarrow$ (i) we apply (1.8). □

Proposition 8.2. *If* $C(v,1)$ *is inconsistent, then v is inconsistent but the converse implication does not hold.*

Proof. The first part of the proposition is trivial. Assume that an injective sequence $\alpha_1, \alpha_2, \ldots$ of contradictions exists, i.e., $\mathcal{D}(\alpha_i) = \mathbb{F}$ for every $i \in N$. Let $(\lambda_n)_{n\in N}$ be the family of all the rational numbers in U different from 0 and 1, and define an initial valuation v by setting $v(x) = \lambda_n$ if $x = \alpha_n$ or, otherwise, $v(x) = 0$. Then it is evident that, since every cut $C(v,\lambda)$, $\lambda \neq 1$, is inconsistent, v is inconsistent in spite of the consistency of $C(v,1) = \varnothing$. □

The following theorem gives some information about the degree of inconsistency.

Theorem 8.3. *Given an initial valuation v, the following hold:*
(i) $Inc(v) = Sup\{\lambda \in U : C(v,\lambda) \text{ inconsistent}\}$.
(ii) $Inc(v) = Inf\{\lambda \in U : C(v,\lambda) \text{ consistent}\}$.
(iii) $Inc(s^\lambda) = \lambda$.
Moreover, if $\mathcal{D}$ *is compact,*
(iv) $Inf\{v(\beta) : \beta \in \mathbb{F}\} \leq Inc(v) \leq Sup\{v(\beta) : \beta \in \mathbb{F}\}$.

Proof. In order to prove (i) and (ii), at first we observe that

$$Sup\{\lambda \in U : C(v,\lambda) \text{ inconsistent}\} = Inf\{\lambda \in U : C(v,\lambda) \text{ consistent}\}.$$

Indeed, the inconsistency of $C(v,\lambda_2)$ entails that for any $\lambda_1 \leq \lambda_2$, $C(v,\lambda_1)$ is inconsistent. Then, the set $\{\lambda \in U : C(v,\lambda) \text{ inconsistent}\}$ is an interval containing 0. As a consequence, the least upper bound of such an interval coincides with the greatest lower bound of its complement $\{\lambda \in U : C(v,\lambda) \text{ consistent}\}$. Now, for every formula α,

$$\mathcal{D}^*(v)(\alpha) = Sup\{\lambda \in U : C(v,\lambda) \vdash \alpha\} \geq Sup\{\lambda \in U : C(v,\lambda) \text{ inconsistent}\}$$

and this proves that

$$Inc(v) \geq Sup\{\lambda \in U : C(v,\lambda) \text{ inconsistent}\}.$$

Also, if $C(v,\lambda)$ is consistent, then a formula α exists which is not a consequence of $C(v,\lambda)$. This means that, if $C(v,\mu) \vdash \alpha$, then $\mu < \lambda$ and

$$\mathcal{D}^*(v)(\alpha) = Sup\{\mu \in U : C(v,\mu) \vdash \alpha\} \leq \lambda.$$

Thus

$$Inc(v) \leq \mathcal{D}^*(v)(\alpha) \leq Inf\{\lambda \in U : C(v,\lambda) \text{ consistent}\}.$$

(iii) is self-evident. To prove (iv) observe that

$$Inc(v) = Inf\{\mathcal{D}^*(v)(\alpha) : \alpha \in F\} \geq Inf\{v(\beta) : \beta \in F\}.$$

Moreover, if α is any formula which is not a tautology,

$$Inc(v) \leq \mathcal{D}^*(v)(\alpha) = Sup\{v(\alpha_1) \wedge \ldots \wedge v(\alpha_n) : \alpha_1,\ldots,\alpha_n \vdash \alpha\}$$
$$\leq Sup\{v(\beta) : \beta \in F\}.$$ □

Proposition 8.4. *If a contradiction χ exists, then, for every fuzzy subset v of formulas,*

$$Inc(v) = \mathcal{D}^*(v)(\chi). \tag{8.1}$$

Hence,

$$Inc(v) = Sup\{Incl(X,v) : X \text{ finite inconsistent subset of } Supp(v)\}.$$

Proof. Since $\chi \vdash \alpha$ for every formula α, we have $\mathcal{D}^*(v)(\chi) \leq \mathcal{D}^*(v)(\alpha)$ and this proves (8.1). The rest is evident. □

For every $\mu \in U$ we define the *μ-inconsistent theory* as the fuzzy subset $\mu \vee Tau$. In particular, the 0-inconsistent theory is the set of tautologies, the 1-inconsistent theory is the totally inconsistent theory. The following proposition shows the meaning of such a definition.

Proposition 8.5. *For every $\mu \in U$, $\mu \vee Tau$ is the smallest theory whose degree of inconsistency is greater than or equal to μ. Let X be any inconsistent set of formulas. Then,*

$$\mathcal{D}^*(\mu \wedge X) = \mu \vee Tau. \tag{8.2}$$

Proof. Obviously, the degree of inconsistency of $\mu \vee Tau$ is μ. Let τ be a theory whose degree of inconsistency is greater than or equal to μ. Then, since τ contains Tau and all its values are greater than or equal to μ, we have that τ contains $\mu \vee Tau$. Equation (8.2) follows from (6.4). □

Proposition 8.6. *Assume that f is continuous and order-preserving. Then, for every initial valuation v,*

$$Inc(f \circ v) = f(Inc(v)).$$

Moreover, let τ be a theory then, for every $\lambda \in U$, $\lambda \vee \tau$ is a theory such that

$$Inc(\lambda \vee \tau) = \lambda \vee Inc(\tau).$$

Consequently, no maximal theory exists.

Proof. Observe that

$$\begin{aligned} Inc(f \circ v) &= Inf\{\mathcal{D}^*(f \circ v)(\alpha) : \alpha \notin Tau\} \\ &= Inf\{f(\mathcal{D}^*(v)(\alpha)) : \alpha \notin Tau\} \\ &= f(Inf\{\mathcal{D}^*(v)(\alpha) : \alpha \notin Tau\}) \\ &= f(Inc(v)). \end{aligned}$$

Let τ be a theory and $\lambda \in U$. Then, by Theorem 7.1, $\lambda \vee \tau$ is a theory and $Inc(\lambda \vee \tau) = \lambda \vee Inc(\tau)$. Finally, let τ be any consistent theory, then, since $Inc(\tau) \neq 1$, there exists $\lambda \in U$ such that $Inc(\tau) < \lambda < 1$. Since

$$Inc(\lambda \vee \tau) = \lambda \vee Inc(\tau) = \lambda \neq 1,$$

$\lambda \vee \tau$ is a consistent theory containing τ and different from τ. □

We conclude this section by emphasizing an important property of the non-monotonic logic defined by the normalized operator $\mathcal{D}_n$. To this purpose, for every consistent initial valuation v, we define the *normalized valuation* v_n by setting

$$v_n(\alpha) = \begin{cases} 0 & \text{if } v(\alpha) \leq Inc(v), \\ \dfrac{v(\alpha) - Inc(v)}{1 - Inc(v)} & \text{otherwise.} \end{cases}$$

Theorem 8.7. *Given an initial valuation v,*

$$\mathcal{D}^*_n(v) = \mathcal{D}^*(v_n). \tag{8.3}$$

Hence, in the case $Inc(v) \neq 1$, $\mathcal{D}^*_n(v)$ is a totally consistent theory.

Proof. Define the function $f : U \to U$ by setting

$$f(x) = \begin{cases} 0 & \text{if } x \leq Inc(v), \\ \dfrac{x - Inc(v)}{1 - Inc(v)} & \text{otherwise.} \end{cases}$$

Then f is continuous, increasing and $f(1) = 1$. Consequently, since $f \circ v = v_n$ and $f \circ \mathcal{D}^* = \mathcal{D}^*_n$, (8.3) is a consequence of Proposition 6.4.

The remaining part of the proof is trivial. □

Corollary 8.8. *The class of theories of the non-monotonic logic defined by the normalized deduction operator* $\mathcal{D}^*_n$ *coincides with the class of all totally-consistent theories of the monotonic logic* $\mathcal{D}^*$.

Proof. The proof is obvious. □

9. CANONICAL SIMILARITY LOGIC.

In Ying [1994], a similarity-based logic is proposed which is related to the classical propositional calculus. The logical connectives are

$\to$ (implication),
f (a constant to denote *false*).

By denoting with *VAR* the set of propositional variables, the set $\mathbb{F}$ of formulas is defined by assuming that:

i) any element in *VAR* is a formula,
ii) f is a formula,
iii) if α and β are formulas then $\alpha \to \beta$ is a formula.

A 0-1-*assignment* is any map $m : VAR \to \{0,1\}$. A 0-1-assignment m is extended to the whole set of formulas by setting $m(f) = 0$ and

$$m(\alpha \to \beta) = \begin{cases} 1 & \text{if } m(\alpha) \leq m(\beta), \\ 0 & \text{otherwise.} \end{cases}$$

An assignment m *satisfies* the formula α if $m(\alpha) = 1$, m *is a model of the set* X of formulas if it satisfies all the formulas in X. We write $X \models \alpha$ if every model of X satisfies α. In any elementary textbook one proves that the resulting logic is axiomatizable and therefore, that the semantical relation $\models$ can be substituted with a syntactical relation $\vdash$ defined by a suitable set of axioms and inference rules. Moreover, M. Ying considers only *natural similarities* $R : \mathbb{F} \times \mathbb{F} \to U$, i.e., similarities obtained from a similarity $R_o : VAR \times VAR \to U$ on the set *VAR* by setting

(1) $R(\alpha \to \beta, \alpha' \to \beta') = R(\alpha,\alpha') \wedge R(\beta,\beta')$,
(2) $R(f,\alpha) = R(\alpha,f) = 1$ if $\alpha = f$,
(3) $R(f,\alpha) = R(\alpha,f) = 0$ if $\alpha \neq f$,
(4) $R(x,y) = R_o(x,y)$ if $x, y \in VAR$,
(5) $R(\alpha,\beta) = 0$ otherwise.

Finally, the *fuzzy consequence relation* is defined by setting

$$Con(X, \alpha) = Sup\{Incl(Y, SIM(X \cup Tau)) : Y \vdash \alpha\}, \tag{9.1}$$

where *Tau* is the set of tautologies. The meaning of $Con(X,\alpha)$ is evident, it represents the degree at which we can prove α by using formulas that are similar either to formulas in X or formulas in $\mathbb{A}$.

In order to compare Ying's definitions with abstract similarity logics, first we must extend these definitions in the following way (see Biacino and Gerla [1998]):

Definition 9.1. We call *canonical consequence relation associated with the classical propositional calculus and* R the fuzzy relation $Con : \mathcal{F}(\mathbb{F}) \times \mathbb{F} \to U$ defined by setting, for any initial valuation v and any formula α,

$$Con(v,\alpha) = Sup\{Incl(X, SIM(v \cup \mathbb{A})) : X \in \mathcal{P}_f(\mathbb{F}), X \vdash \alpha\} \qquad (9.2)$$

The proof of the following proposition is trivial:

Proposition 9.2. *Define* $H : \mathcal{F}(\mathbb{F}) \to \mathcal{F}(\mathbb{F})$ *by setting, for every initial valuation* v,

$$H(v) = SIM(v \cup \mathbb{A}). \qquad (9.3)$$

Then H *is a continuous closure operator such that*

$$Con(v, .\,) = \mathcal{D}^*(H(v)). \qquad (9.4)$$

We interpret $H(v)$ as the fuzzy subset of formulas which are similar either to a formula of v or to a logical axiom.

To come back to Ying's similarity logic we call *schema* a formula in which a propositional variable cannot occur in two different places. More precisely, by denoting with $Var(\alpha)$ the set of propositional variables of a formula α, the schemas are defined by assuming that:

i) any element in VAR is a schema,

ii) f is a schema,

iii) if α and β are schemas and $Var(\alpha) \cap Var(\beta) = \emptyset$, then $\alpha \to \beta$ is a schema.

We denote a schema whose propositional variables are $x_1,...,x_n$ by $\gamma(x_1,...,x_n)$. Let $\{y_1/x_1,...,y_n/x_n\}$ be a substitution associating each propositional variable x_i with the propositional variables y_i and let $\gamma(x_1,...,x_n)$ be a schema. Then $\gamma(y_1/x_1,...,y_n/x_n)$ is the formula obtained by replacing each x_i with y_i in $\gamma(x_1,...,x_n)$. We say that two formulas *have the same structure* if they are obtained by suitable substitutions from the same schema. Let R be a natural similarity and α and β two formulas with the same structure. Then, if γ is a schema and $\{y_1/x_1,...,y_n/x_n\}$ and $\{z_1/x_1,...,z_n/x_n\}$ are two substitution such that $\alpha = \gamma(y_1/x_1,...,y_n/x_n)$ and $\beta = \gamma(z_1/x_1,...,z_n/x_n)$, we have that

$$R(\alpha,\beta) = Inf\{R(y_i,z_i) : i = 1,...,n\}$$

Trivially, if α and β have different structures, then $R(\alpha,\beta) = 0$. These definitions enable us to prove that Ying's logic is an abstract similarity logic (see Biacino, Gerla, Ying [2000]).

Theorem 9.3. *Under Ying's hypothesis* $\mathcal{D}^* \circ H$ *is a (continuous) closure operator. Consequently, Ying's logic is an abstract similarity logic whose fuzzy consequence relation can be obtained by* (9.4).

Proof. Let s be a fuzzy set of formulas and set $s' = \mathcal{D}^*(H(s))$. By Theorem 9.4 in Chapter 1, to prove that $\mathcal{D}^* \circ H$ is a closure operator we have to prove that s' is a fixed point for H and therefore, for any $\alpha \in \mathbb{F}$,

$$SIM(s' \cup Tau)(\alpha) = Sup\{R(\alpha,\beta) \wedge s'(\beta) : \beta \in \mathbb{F}\} \vee SIM(Tau)(\alpha) \leq s'(\alpha).$$

In turn, as $s' \supseteq SIM(Tau)$, this is equivalent to proving that

$$s'(\alpha) \geq R(\alpha,\beta) \wedge s'(\beta),$$

i.e.,

$$Sup\{Incl(X,H(s)) : X \vdash \alpha\} \geq R(\alpha,\beta) \wedge Incl(Y,H(s)) \qquad (9.5)$$

for any finite set Y of formulas such that $Y \vdash \beta$. Now, in the case $R(\alpha,\beta) = 0$ (9.5) is obvious. Otherwise, there exists a schema $\gamma(x_1,\ldots,x_n)$ and two substitutions $\{y_1/x_1,\ldots, y_n/x_n\}$, $\{z_1/x_1,\ldots, z_n/x_n\}$ such that

$$\alpha = \gamma(y_1/x_1,\ldots, y_n/x_n) \ ; \ \beta = \gamma(z_1/x_1,\ldots, z_n/x_n).$$

Moreover, observe that if $Y \vdash \beta$ and $X = \{y_1 \to z_1, z_1 \to y_1\ldots, y_n \to z_n, z_n \to y_n\} \cup Y$, then $X \vdash \alpha$. In fact, let m be a 0-1-assignment satisfying X. Then m is a model of Y and therefore it satisfies β. Since m satisfies the set $\{y_1 \to z_1, z_1 \to y_1\ldots, y_n \to z_n, z_n \to y_n\}$ of implications, we also have that, $m(y_1) = m(z_1),\ldots,m(y_n) = m(z_n)$. Then m satisfies α.

By observing that,

$$Incl(\{y_i \to z_i\},H(s)\} = H(s)(y_i \to z_i) \geq SIM(Tau)(y_i \to z_i)$$
$$\geq R(y_i \to y_i, y_i \to z_i) = R(y_i,z_i),$$

we can conclude that

$$Incl(X,H(s)) = Incl(\{y_1 \to z_1, z_1 \to y_1\ldots, y_n \to z_n, z_n \to y_n\},H(s)\} \wedge Incl(Y,H(s))$$
$$\geq R(y_1,z_1) \wedge \ldots \wedge R(y_1,z_1) \wedge Incl(Y,H(s))$$
$$\geq R(\alpha,\beta) \wedge Incl(Y,H(s)).$$

Thus,

$$Sup\{Incl(X,H(s)) : X \vdash \alpha\} \geq R(\alpha,\beta) \wedge Incl(Y,H(s)).$$ □

Due to the presence of the canonical extension $\mathcal{D}^*$, we call Ying's logic *canonical similarity logic.*

10. FUZZY METALOGIC, FACTS AND PREFERENCES

We conclude with some final considerations on the fuzzy logics discussed in this chapter. First of all, differently from the truth-functional logics examined in Chapter 8,

in all these logics, fuzziness does not arise from a "fuzzy world" but from a fuzzyfication of some metalogic notions.

Indeed, consider the formula for the canonical extension of a crisp logic:

$$\mathcal{D}^*(v)(\alpha) = Sup\{Incl(X, v) : X \vdash \alpha\}.$$

Such a formula is a multivalued valuation of the (metalogical) claim: "*α is a consequence of v if there is a finite subset X of v able to prove α*". Indeed, usually the supremum operator *Sup* represents the interpretation of the existential quantifier. In such a case, the vagueness arises from the fuzzy subset v, i.e., from the notion of *"hypothesis"*. On the other hand, $\vdash$ is the classical deduction relation. In fact, the semantics and the deductive apparatus are classical.

Analogous considerations hold for similarity logic. Indeed, the formula

$$Con(v,\alpha) = Sup\{Incl(X, SIM(v \cup \mathcal{A})) : X \in \mathcal{P}_f(\mathcal{F}), X \vdash \alpha\}$$

is a multivalued valuation of the (metalogical) claim that

"α is a consequence of v if we can prove α from formulas which are either similar to formulas in v or to logical axioms".

In this case vagueness arises both from the notion of *"hypothesis"* and from the notion *"identity between formulas"*.

Secondly,

all logics considered in this chapter stem from such subjective notions as "preference" or "acceptability" and not concerned with the objective notions "truth" and "falseness".

Indeed, consider the following example. Let T be a system of axioms for set theory such that the choice axiom, denoted by CA, does not depend on T. Then we can consider the fuzzy subset of axioms v defined by setting

$$v(\alpha) = \begin{cases} 0.8 & \text{if } \alpha \text{ is } CA, \\ 1 & \text{if } \alpha \in T, \\ 0 & \text{otherwise.} \end{cases}$$

A simple calculation shows that:

$$\mathcal{D}^*(v)(\alpha) = \begin{cases} 0.8 & \text{if we can prove } \alpha \text{ only by using } CA, \\ 1 & \text{if } T \vdash \alpha, \\ 0 & \text{otherwise.} \end{cases}$$

Fuzziness in this case is not semantical in nature. Indeed, $v(\alpha)$ is a *degree of acceptability* for α and not a truth degree. In this sense, by recalling the Euclidean distinction between axiom and postulate, perhaps it's better to say v is the *fuzzy subset of the postulates we accept*. Thus, despite the fact that no vague predicate is considered in a classical set theory, in the metalanguage it is reasonable to consider a vague predicate as *"is acceptable"* and to represent it by a suitable fuzzy subset v. Equivalently, we can interpret $v(\alpha)$ as the *degree of preference* for α since the only reason we assign to CA the degree 0.8 instead of 1 is that we do not like to use CA.

For similarity-based logic it is completely evident that the information we elaborate is not a matter of fact. Indeed the valuation of a claim as

"detective story" is similar to *"thriller"*

depends more on the taste than on the actual nature of the books under consideration. Analogous considerations hold for the fuzzy logics related to fuzzy control (see Chapter 10). This leads to the following question:

is fuzzy logic a basis for a logic of the judgement values?

CHAPTER 4

APPROXIMATE REASONING

1. THE HEAP PARADOX

In accordance with Hilbert's approach, in this chapter we will show that it is possible to define the deduction operator of a fuzzy logic by a suitable extension of the notions of *inference rule* and *proof*. To justify such an extension we start from the famous paradox, the *"Heap Paradox"*, involving the vague predicate *"is small"*. A similar paradox is the *"Bald Man Paradox"*. An interesting treatment of such paradoxes by means of fuzzy logic can be found in Goguen [1968/69]. The paradox runs as follows:

(a) If you add one stone to a small heap, it remains small.

(b) A heap containing just one stone is small,

therefore:

(c) every heap is small.

To emphasize the formal structure of this argument, denote by *Small*(n) the sentence *"a heap with n stones is small"* and denote by *MP* the Modus Ponens rule. Then we can rewrite the above argument as follows. Assume, for any integer n, the implication

(a) $Small(n) \rightarrow Small(n+1)$

and the atomic formula

(b) $Small(1)$.

Then,

- from $Small(1)$ and $Small(1) \rightarrow Small(2)$ by *MP* we may state $Small(2)$;
- from $Small(2)$ and $Small(2) \rightarrow Small(3)$ by *MP* we may state $Small(3)$,

....

- from $Small(x)$ and $Small(x) \rightarrow Small(x+1)$ by *MP* we may state $Small(x+1)$.

In such a way, whatever the integer n is fixed, we may prove *Small*(n) by using $n-1$ times the *MP*-rule. On the other hand, a conclusion like *Small*(100.000.000.000) is contrary to our intuition about the notion *"small heap"* in spite of the fact that the reasoning is correct and the premises appear very reasonable.

Clearly, the core of such a paradox lies in the fuzziness of the predicate *"is small"* employed in our example. Indeed, it is easy to prove that:

Any vague "sufficiently graded" property originates a paradox similar to Heap Paradox.

Indeed, let $\mathcal{P}$ be any vague property in a set S and assume that the extension of such a property is represented by a fuzzy subset $s : S \rightarrow U$ whose values change from 0 to 1 in a graded way. More precisely, assume that a very small number ε and a sequence $x_1, x_2, \ldots$ of elements of S exist such that

$$s(x_1) = 1 \;,\; 0 < s(x_n) - s(x_{n+1}) < \varepsilon \;,\; lim_{n\to\infty} s(x_n) = 0.$$

Under these conditions

- $\mathcal{P}(x_1)$ is satisfied,

and all the implications

- $\mathcal{P}(x_n) \to \mathcal{P}(x_{n+1})$

seem very plausible. If we accept these formulas as a system of hypotheses in classical propositional logic, then we can obtain a paradox similar to Heap Paradox. Indeed, given any integer n, after a suitable number of applications of *MP* we can prove $\mathcal{P}(x_n)$. On the other hand, from the hypothesis $lim_{n\to\infty} s(x_n) = 0$, it follows that $\mathcal{P}$ looks to be false for a sufficiently large integer n. As an example, by considering the *"sufficiently graded"* property *"bald"*, we obtain the *Bald Man Paradox*. A more precise formalization of this phenomenon is possible in the framework of truth-functional logic. This supposes that $\to$ is interpreted by a continuous binary function $\to'$ such that $x \to' x$ is equal to 1. Obviously, such an argument is based on the hypothesis that sentences very near to the truth can be considered completely true. On the other hand, as only in the mathematical context the exact truth exists, this is a price that classical logic has to pay to be a tool in everyday human activity.

Now, a draconian way to solve the Heap Paradox is to claim that a predicate like *"is small"* is senseless and, consequently, a correct reasoning cannot use such a predicate. This is without doubt a solution, but it conflicts with the fact that vague notions as *"small"*, *"large"*, *"approximate solution"* are continuously used in everyday life and in scientific reasoning.

A second possibility is to deny that all the implications *Small*$(x) \to$ *Small*$(x+1)$ are true (it is unquestionable that *Small*(1) is true). But in this case, we must admit the existence of a *"critical"* number m such that *Small*$(m) \to$ *Small*$(m+1)$ is false and therefore *Small*(m) is true and *Small*$(m+1)$ is false. In other words,

while a heap with m stones is small, by adding just one stone this heap becomes not small !

Finally, the notion of approximate reasoning gives the most interesting (in our opinion) solution to this paradox. Indeed it is a fact that everyone is convinced that the implications *Small*$(n) \to$ *Small*$(n+1)$ are near-true but not completely true, in general. We can try to *"respect"* this conviction by assigning a truth value near to 1 to these formulas; the number 0.9, for instance. In accordance, we have to establish how the truth degree of a conclusion depends on the truth degrees of the premises in each inferential step. For example, we assume that,

IF you know that the formulas $\alpha \to \beta$ and α are true to the degrees λ_1 and λ_2,
THEN you can conclude that β is true to the degree $\lambda_1 \cdot \lambda_2$,

where λ_1 and λ_2 are elements in U and $\lambda_1 \cdot \lambda_2$ is the usual product. Under these hypotheses, the Heap Paradox argument can be restated as follows :

- since *Small*(1) [to degree 1] and *Small*$(1) \to$ *Small*(2) [to degree 0.9]
  we state
  Small(2) [to degree $1 \times 0.9 = 0.9$]

- since *Small*(2) [to degree 0.9] and $Small(2) \to Small(3)$ [to degree 0.9]
  we state
  Small(3) [to degree $0.9 \times 0.9 = 0.9^2$]

  . . .
- since *Small*(*n*) [to degree 0.9^{n-1}] and $Small(2) \to Small(3)$ [to degree 0.9]
  we state
  Small(*n*+1) [to degree $0.9^{n-1} \times 0.9 = 0.9^n$].

In particular, we can prove *Small*(100.000.000) at degree $0.9^{(100.000.000-1)}$, and this is not paradoxical. In fact, $0.9^{(100.000.000\ -1)}$ is a number very close to 0. In this way we get a formal representation of heap argument preserving its intuitive content but avoiding its paradoxical character.

Actually, things are slightly more complicated and also this new formalization gives rise to an (apparent) paradox. Indeed, it is reasonable to admit that the formulas *Small*(2),..., *Small*(10), say, are valued 1. In fact a heap with less than 11 stones is small. In this case several different proofs of the same formula are possible. For instance, consider the formula *Small*(3). Then, proceeding as above, we are able to demonstrate *Small*(3) to degree 0.81. Moreover, since *Small*(2) holds with degree 1 and $Small(2) \to Small(3)$ with degree 0.9, by the proposed fuzzy *MP* we can state *Small*(3) with degree $0.9 \times 1 = 0.9$. On the other hand, since *Small*(3) is an axiom with degree 1 we can also conclude that *Small*(3) holds with degree 1. So, the same deductive apparatus gives three different answers about the actual truth value of the formula *Small*(3); the numbers 0.81, 0.9 and 1. This forces a change in the interpretation of the evaluation part of *MP*. In fact, the meaning we assign to a rule is that:

- if you know that $\alpha \to \beta$ and α are true, at least to the degrees λ_1 and λ_2, then it is possible to conclude that β is true, at least to degree $\lambda_1 \cdot \lambda_2$.

In this way the above-quoted proofs of *Small*(3) state that

- *Small*(3) holds at least to degree 0.81,
- *Small*(3) holds at least to degree 0.9,
- *Small*(3) holds at least to degree 1,

respectively. These claims are not contradictory. In fact, they are equivalent to saying that *Small*(3) holds with degree 1. Thus, a fuzzy inference rule is not a tool to calculate the exact truth value of the conclusion. Instead, it is a way to calculate a constraint for its truth value by considering the available constraints on the truth values of the premises. In accordance, a fuzzy logic could be described as a tool to elaborate initial constraints for the truth values of the formulas (available information) to obtain new, improved constraints (derived information). In the next sections we give formal definitions in this direction (see also Chapter 5).

2. FUZZY INFERENCE RULES

We introduce the basic definitions concerning the deduction apparatus and the semantics of a fuzzy logic (see Pavelka [1979]) by starting from the crisp case. Let $\mathbb{F}$ be a set whose elements we call *formulas*. Then a *crisp inference rule* is any partial operation in $\mathbb{F}$, i.e., a map $r : D \to \mathbb{F}$ where $D \subseteq \mathbb{F}^n$, $n \in N$. We denote the domain of r by $Dom(r)$. A *Hilbert crisp deduction system*, in brief an *H-system*, is a pair $\mathcal{S} = (\mathbb{A}, \mathbb{R})$ such that $\mathbb{A}$ is a subset of $\mathbb{F}$, *the set of logical axioms*, and $\mathbb{R}$ a set of crisp inference rule. A *proof* π *of a formula* α *under the hypotheses* $\gamma_1,\dots,\gamma_h$ is any sequence $\alpha_1,\dots,\alpha_m$ of formulas such that $\alpha_m = \alpha$ and, for any $i = 1,\dots, m$,

- either $\alpha_i \in \{\gamma_1,\dots,\gamma_h\}$, ($\alpha$ is a hypothesis)
- or $\alpha_i \in \mathbb{A}$ (α is an axiom)
- or $\alpha_i = r(\alpha_{s(1)},\dots,\alpha_{s(n)})$, where $r \in \mathbb{R}$ and $s(1) < i, \dots , s(n) < i$ (α is obtained by an inference rule).

Given a set X of formulas, we write $X \vdash \alpha$ to denote that a proof of α exists whose hypotheses are contained in X. The *deduction operator* $\mathcal{D} : \mathcal{P}(\mathbb{F}) \to \mathcal{P}(\mathbb{F})$ *associated with* $(\mathbb{A}, \mathbb{R})$ is defined by setting, for any set X of formulas,

$$\mathcal{D}(X) = \{\alpha \in \mathbb{F} : X \vdash \alpha\}. \tag{2.1}$$

Theorem 2.1. *Let $\mathcal{D}$ be the deduction operator associated with an H-system. Then $\mathcal{D}$ is an algebraic closure operator. Conversely, let $\mathcal{D} : \mathcal{P}(\mathbb{F}) \to \mathcal{P}(\mathbb{F})$ be an algebraic closure operator. Then a crisp H-system exists whose deduction operator coincides with $\mathcal{D}$.*

Proof. The first part of the theorem is trivial. Let $\mathcal{D} : \mathcal{P}(\mathbb{F}) \to \mathcal{P}(\mathbb{F})$ be an algebraic closure operator and define, for any $n \in \mathbb{N}$, the n-ary rule r_i^n such that, for any $\alpha_1,\dots,\alpha_n$ in $\mathbb{F}$

- $r_i^n(\alpha_1,\dots,\alpha_n)$ is the i-th formula in $\mathcal{D}(\{\alpha_1,\dots,\alpha_n\})$ if $|\mathcal{D}(\{\alpha_1,\dots,\alpha_n\})| \geq i$,
- $r_i^n(\alpha_1,\dots,\alpha_n) = \alpha_1$ if $|\mathcal{D}(\{\alpha_1,\dots,\alpha_n\})| < i$.

Let $\mathcal{D}'$ be the deduction operator of the crisp H-system $\mathcal{S}$ whose set of logical axioms is $\mathcal{D}(\emptyset)$ and with the just defined inference rules. Then, since the theories of $\mathcal{S}$ coincide with the fixed points of $\mathcal{D}$, we have that $\mathcal{D}' = \mathcal{D}$. □

So, every crisp H-system $(\mathbb{A}, \mathbb{R})$ defines an abstract crisp deduction system $(\mathbb{F}, \mathcal{D})$. To characterize the theories of an H-system, we say that a set T of formulas is *closed under the n-ary rule r* if, for every $(\gamma_1,\dots,\gamma_n) \in Dom(r)$,

$$\gamma_1,\dots,\gamma_n \in T \Rightarrow r(\gamma_1,\dots,\gamma_n) \in T.$$

Proposition 2.2. *A set T of formulas is a theory of $(\mathbb{A}, \mathbb{R})$, i.e., a fixed point of the related deduction operator $\mathcal{D}$, iff*

(i) *T contains the set $\mathbb{A}$ of logical axioms,*
(ii) *T is closed under all the inference rules in $\mathbb{R}$.*

We call a *crisp Hilbert logic* any pair $(\mathcal{S}, \mathcal{M})$ where $\mathcal{S}$ is a crisp H-system and $\mathcal{M}$ a crisp semantics such that $(\mathbb{F}, \mathcal{D}, \mathcal{M})$ is a crisp abstract logic, i.e., $Lc = \mathcal{D}$.

By extending the above definitions, we define a *fuzzy Hilbert system*, in brief *fuzzy H-system*, a pair $\mathcal{S} = (a, \mathbb{R})$ where a is a fuzzy subset of $\mathbb{F}$, the *fuzzy subset of logical axioms*, and $\mathbb{R}$ is a set of fuzzy inference rules. In turn, a *fuzzy inference rule* is a pair $r = (r', r'')$, where

- r' is a partial n-ary operation on $\mathbb{F}$ whose domain we denote by $Dom(r)$;

- r'' is an n-ary operation on U preserving the least upper bound in each variable, that is

$$r''(x_1,..., Sup_{i\in I}\, y_i, ..., x_n) = Sup_{i\in I}\, r''(x_1, ..., y_i, ..., x_n). \tag{2.2}$$

In other words, an inference rule r consists

- of a syntactical component r' that operates on formulas (in fact, it is an inference rule in the usual sense);
- of a valuation component r'' which operates on truth values to calculate how the truth value of the conclusion depends on the truth values of the premises.

Condition (2.2) is also called *continuity condition* because it enables us to demonstrate that the deduction operator associated with a fuzzy H-system is continuous. We indicate an application of an inference rule r by

$$\frac{\alpha_1,...,\alpha_n}{r'(\alpha_1,...,\alpha_n)} \quad ; \quad \frac{\lambda_1,...,\lambda_n}{r''(\lambda_1,...,\lambda_n)} .$$

The meaning of such a rule is that:

IF you know that the formulas $\alpha_1,\ldots,\alpha_n$ are true (at least) to the degree $\lambda_1,\ldots,\lambda_n$,
THEN you can conclude that $r'(\alpha_1,\ldots,\alpha_n)$ is true (at least) to degree $r''(\lambda_1,\ldots,\lambda_n)$.

For instance, in Section 1 we considered a fuzzy rule in which r' is the classical MP, that is the function associating to any pair of formulas of type $\alpha \to \beta$ and α the formula β, and r'' is the usual product in U. More precisely, $Dom(r)$ is the set $\{(\alpha, \alpha \to \beta) : \alpha \in F, \beta \in F\}$, $r'(\alpha, \alpha \to \beta) = \beta$ and $r''(\lambda_1,\lambda_2) = \lambda_1 \cdot \lambda_2$.

A *proof* π of a formula α is a sequence $\alpha_1,...,\alpha_m$ of formulas such that $\alpha_m = \alpha$, together with a sequence of related *"justifications"*. This means that, given any formula α_i, we must specify whether

(i) α_i is assumed as a logical axiom; or
(ii) α_i is assumed as a proper axiom; or
(iii) α_i is obtained by a rule (in this case we must indicate also the rule and the formulas $\alpha_{s(1)},...,\alpha_{s(n)}$ used to obtain α_i).

Observe that we have only two proofs of α whose lengths is equal to 1. The formula α with the justification that α is assumed as a logical axiom and the formula α with the justification that α is assumed as a hypothesis. As in the classical case, for any $i \le m$, the initial segment $\alpha_1,...,\alpha_i$ is a proof of α_i we denote by π_i. Differently from the crisp case, it is necessary to indicate the justifications in a proof because different justifications of the same formula give rise to different valuations. Indeed, let $v : F \to U$ be any initial valuation and π a proof. Then the *valuation $Val(\pi,v)$ of π with respect to v* is defined by induction on the length m of π as follows: If the length of π is 1, then we set

$Val(\pi,v) = a(\alpha_m)$ if α_m is assumed as a logical axiom;

$Val(\pi,v) = v(\alpha_m)$ if α_m is assumed as a hypothesis.

Otherwise, we set

$$Val(\pi,v) = \begin{cases} a(\alpha_m) & \text{if } \alpha_m \text{ is assumed as a logical axiom,} \\ v(\alpha_m) & \text{if } \alpha_m \text{ is assumed as a proper axiom,} \\ r''(Val(\pi_{s(1)},v),\ldots,Val(\pi_{s(n)},v)) & \text{if } \alpha_m = r'(\alpha_{s(1)},\ldots, \alpha_{s(n)}) \end{cases}$$

where, $1 \le s(1) < m, \ldots, 1 \le s(n) < m$. If α is the formula proven by π, the meaning we assign to $Val(\pi,v)$ is that:

given the information v, π assures that α holds at least to degree $Val(\pi,v)$.

For instance, in the previous section v is a fuzzy set such that $v(Small(1)) = 1$ and $v(Small(n) \to Small(n+1)) = 0.9$, for every $n \in N$. A proof π_n of $Small(n)$ is given by the sequence

$<Small(1), Small(1) \to Small(2), Small(2), Small(2) \to Small(3), Small(3), \ldots, Small(n)>$,

where $Small(1)$ and the formulas $Small(n) \to Small(n+1)$ are assumed as proper axioms and the formulas like $Small(n)$, $n \neq 1$, are obtained by *MP*. The valuation of such a proof is obtained by recursion using the equations

$$Val(\pi_1,v) = 1 \text{ and } Val(\pi_i,v) = 0.9 \times Val(\pi_{i-2},v).$$

As observed in Section 1, given a formula α, it is possible that another proof π' of α exists such that $Val(\pi',v) > Val(\pi,v)$. This happens, for instance, if the assumptions used in π' are more true than the assumptions used in π. In other words, unlike the usual Hilbert systems, in a fuzzy *H*-system different proofs of the same formula α can give different contributions to the degree of validity of α. This suggests that, given a fuzzy set of axioms v (the available fuzzy information), in order to evaluate α we must refer to the whole set of proofs of α and calculate the least upper bound of the corresponding valuations.

Definition 2.3. Given a fuzzy *H*-system $\mathcal{S}$, we call *deduction operator* associated with $\mathcal{S}$ the operator $\mathcal{D} : \mathcal{F}(\mathbb{F}) \to \mathcal{F}(\mathbb{F})$ defined by setting,

$$\mathcal{D}(v)(\alpha) = Sup\{Val(\pi,v) : \pi \text{ is a proof of } \alpha\}, \tag{2.3}$$

for every initial valuation v and every formula α.

The meaning of $\mathcal{D}(v)(\alpha)$ is still

given the information v, we may prove that α holds at least to degree $\mathcal{D}(v)(\alpha)$,

but we also have that

$\mathcal{D}(v)(\alpha)$ is the best possible valuation we can draw from the information v.

A justification of the operator *Sup* in (2.3) will be given in Chapter 5 (see formula (4.3)). Observe that in computing $\mathcal{D}(v)(\alpha)$ we can confine ourselves only to *"normalized"* proofs $\pi = \alpha_1,\ldots,\alpha_n$ i.e. proofs such that $\alpha_1,\ldots,\alpha_n$ are pairwise different and two integers h and k exist such that $1 \le h \le k \le n$ and:

- $\alpha_1,\ldots,\alpha_h$ are the formulas justified as hypothesis;
- $\alpha_{h+1},\ldots,\alpha_k$ are the formulas justified as logical axiom;
- $\alpha_{k+1},\ldots,\alpha_n$ are the formulas obtained by an inference rule.

Definition 2.4. We say that a fuzzy subset of formulas $\tau : \mathbb{F} \to U$ is *closed under the fuzzy inference rule r* if, for every $(\alpha_1,\ldots,\alpha_n) \in Dom(r)$,

$$\tau(r'(\alpha_1,\ldots,\alpha_n)) \ge r''(\tau(\alpha_1),\ldots,\tau(\alpha_n)). \tag{2.4}$$

In other words, if $\alpha_1,\ldots,\alpha_n$ belong to τ, to degree $\tau(\alpha_1),\ldots,\tau(\alpha_n)$, then $r'(\alpha_1,\ldots,\alpha_n)$ belongs to τ at least to degree $r''(\tau(\alpha_1),\ldots,\tau(\alpha_n))$.

Proposition 2.5. *Given any initial valuation v, $\mathcal{D}(v)$ contains the fuzzy set $a : \mathbb{F} \to U$ of logical axioms and it is closed under all the inference rules.*

Proof. We prove that $\mathcal{D}(v)(\alpha) \geq a(\alpha)$ by observing that by assuming α as a logical axiom we obtain a proof π of α (whose length is one) such that $Val(\pi,v) = v(\alpha)$. To prove that $\mathcal{D}(v)$ is closed under any rule r, let $(\alpha_1,\dots,\alpha_n) \in Dom(r)$ and assume that π_i is a proof of α_i for $i = 1,\dots,n$. Then by adding $r'(\alpha_1,\dots,\alpha_n)$ to these proofs, we obtain a proof $\pi = <\pi_1,\dots,\pi_n, r'(\alpha_1,\dots,\alpha_n)>$ of $r'(\alpha_1,\dots,\alpha_n)$ such that

$$Val(\pi,v) = r''(Val(\pi_1,v),\dots,Val(\pi_n,v)).$$

Thus, taking (2.2) into account,

$$\begin{aligned} &r''(\mathcal{D}(v)(\alpha_1),\dots,\mathcal{D}(v)(\alpha_n)) \\ &\quad = r''(Sup\{Val(\pi_1,v) : \pi_1 \text{ is a proof of } \alpha_1\},\dots,Sup\{Val(\pi_n,v) : \pi_n \text{ is a proof of } \alpha_n\}) \\ &\quad = Sup\{r''(Val(\pi_1,v),\dots,Val(\pi_n,v)) : \pi_i \text{ is a proof of } \alpha_i\} \\ &\quad \leq Sup\{Val(\pi,v) : \pi \text{ is a proof of } r'(\alpha_1,\dots,\alpha_n)\} = \mathcal{D}(v)(r'(\alpha_1,\dots,\alpha_n)). \end{aligned}$$

The following theorem (see also Section 9 of Chapter 11) shows that the theory of continuous abstract deduction systems coincides with the theory of fuzzy Hilbert systems, i.e.,

approximate reasoning = theory of continuous fuzzy closure operators.

Theorem 2.6. *The deduction operator $\mathcal{D}$ associated with a fuzzy Hilbert system is a continuous closure operator. Conversely, let $\mathcal{D} : \mathcal{F}(\mathbb{F}) \to \mathcal{F}(\mathbb{F})$ be any continuous fuzzy closure operator. Then, a fuzzy Hilbert system exists whose associated deduction operator coincides with $\mathcal{D}$.*

Proof. Let $\mathcal{D}$ be the deduction operator associated with a fuzzy Hilbert system. In order to prove that $\mathcal{D}$ is a closure operator, observe that, given an initial valuation v and a formula α, we can consider the proof of α obtained by assuming α as a proper axiom. Then, $\mathcal{D}(v)(\alpha) \geq v(\alpha)$. Since the valuation part of any inference rule is monotone, $\mathcal{D}$ is monotone. Finally, to prove that $\mathcal{D}(\mathcal{D}(v)) \subseteq \mathcal{D}(v)$ we prove that $Val(\pi,\mathcal{D}(v)) \leq \mathcal{D}(v)(\alpha)$ for every proof π of α. The proof is by induction on the length of π. In fact, if $Val(\pi, \mathcal{D}(v)) = \mathcal{D}(v)(\alpha)$ or $Val(\pi, \mathcal{D}(v)) = a(\alpha)$ such an inequality is evident. Assume that

$$Val(\pi, \mathcal{D}(v)) = r''(Val(\pi_{s(1)}, \mathcal{D}(v)),\dots,Val(\pi_{s(n)}, \mathcal{D}(v)))$$

and let $\alpha_{s(1)},\dots, \alpha_{s(n)}$ be the formulas proved by $\pi_{s(1)},\dots, \pi_{s(n)}$, respectively. Then, since $\mathcal{D}(v)$ is closed under the inference rules and by induction hypothesis

$$Val(\pi_{s(1)}, \mathcal{D}(v)) \leq \mathcal{D}(v)(\alpha_{s(1)}),\dots,Val(\pi_{s(n)},\mathcal{D}(v)) \leq \mathcal{D}(v)(\alpha_{s(n)}),$$

we have that

$$\begin{aligned} Val(\pi,\mathcal{D}(v)) &= r''(Val(\pi_{s(1)},\mathcal{D}(v)),\dots,Val(\pi_{s(n)},\mathcal{D}(v))) \\ &\leq r''(\mathcal{D}(v)(\alpha_{s(1)}),\dots,\mathcal{D}(v)(\alpha_{s(n)})) \\ &\leq \mathcal{D}(v)(r'(\alpha_{s(1)},\dots,\alpha_{s(n)})) = \mathcal{D}(v)(\alpha). \end{aligned}$$

The proof that $\mathcal{D}$ is continuous is evident (for a complete proof see Section 5).

Assume that $\mathcal{D} : \mathcal{F}(\mathbb{F}) \to \mathcal{F}(\mathbb{F})$ is a continuous fuzzy closure operator and, for any $s_f \in \mathcal{F}_f(\mathbb{F})$ such that $s_f \neq \emptyset$ and $\alpha \in \mathbb{F}$, define $r = (r',r'')$ by setting

$$r'(\alpha_1,\ldots,\alpha_n) = \begin{cases} \alpha & \text{if } \{\alpha_1,\ldots,\alpha_n\} = Supp(s_f), \\ \text{undefined} & \text{otherwise.} \end{cases}$$

and

$$r''(\lambda_1,\ldots,\lambda_n) = \begin{cases} \mathcal{D}(s_f)(\alpha) & \text{if } \lambda_1 > s_f(\alpha_1), \ldots, \lambda_n > s_f(\alpha_n) \\ 0 & \text{otherwise.} \end{cases}$$

It is evident that r'' satisfies (2.2) and therefore that $r = (r', r'')$ is a fuzzy inference rule. Let $\mathcal{S}$ be the fuzzy Hilbert system whose fuzzy inference rules are the just-defined rules and whose fuzzy subset of logical axioms is $\mathcal{D}(\varnothing)$. To demonstrate that $\mathcal{D}$ coincides with the deduction operator of $\mathcal{S}$, we prove that the fixed points of $\mathcal{D}$ coincide with the theories of $\mathcal{S}$. Let τ be a theory of $\mathcal{S}$, then, by Theorem 6.3 in Chapter 2, to prove that τ is a fixed point of $\mathcal{D}$ it is sufficient to prove that $\tau \supseteq \mathcal{D}(s_f)$ for any $s_f \in \mathcal{F}_f(\mathbb{F})$ such that $s_f \ll \tau$. Indeed, if $s_f = \varnothing$, then $\tau \supseteq \mathcal{D}(s_f)$ by hypothesis. In the case $s_f \neq \varnothing$, let $\alpha_1,\ldots,\alpha_n$ be the elements in $Supp(s_f)$ and $\alpha \in \mathbb{F}$. Moreover, consider the inference rule associated with s_f and α. Then, since $\tau(\alpha_1) > s_f(\alpha_1),\ldots, \tau(\alpha_n) > s_f(\alpha_n)$, we have that

$$\tau(\alpha) = \tau(r'(\alpha_1,\ldots,\alpha_n)) \geq r''(\tau(\alpha_1),\ldots,\tau(\alpha_n)) = \mathcal{D}(s_f)(\alpha).$$

This proves that $\tau \supseteq \mathcal{D}(s_f)$ and therefore that τ is a fixed point of $\mathcal{D}$.

Let τ be a fixed point of $\mathcal{D}$, then $\tau = \mathcal{D}(\tau) \supseteq \mathcal{D}(\varnothing)$. Moreover, let (r',r'') be any rule and assume that (r',r'') is defined by s_f such that $Supp(s_f) = \{\alpha_1,\ldots,\alpha_n\} \neq \varnothing$ and by $\alpha \in \mathbb{F}$. We claim that, for every $x_1,\ldots,x_n \in \mathbb{F}$,

$$\tau(r'(\alpha_1,\ldots,\alpha_n)) \geq r''(\tau(\alpha_1),\ldots,\tau(\alpha_n)).$$

Indeed, if $r''(\tau(\alpha_1),\ldots,\tau(\alpha_n)) \neq 0$ then $\tau(\alpha_1) > s_f(\alpha_1),\ldots, \tau(\alpha_n) > s_f(\alpha_n)$. Consequently,

$$\tau(r'(\alpha_1,\ldots,\alpha_n)) = \tau(\alpha) = \mathcal{D}(\tau)(\alpha) \geq \mathcal{D}(s_f)(\alpha) = r''(\tau(\alpha_1),\ldots,\tau(\alpha_n)).$$ □

In accordance with such a theorem, any fuzzy *H*-system $\mathcal{S}$ defines an abstract fuzzy deduction system $(\mathbb{F}, \mathcal{D})$. A simple characterization of the theories of $(\mathbb{F}, \mathcal{D})$ is given in the next theorem:

Theorem 2.7. *A fuzzy subset τ of formulas is a theory of a fuzzy H-system iff τ contains the fuzzy subset a of logical axioms and it is closed under any rule in $\mathbb{R}$.*

3. FUZZY HILBERT LOGIC AND HOMOMORPHISMS

The basic notion of fuzzy Hilbert logic is given in the following:

Definition 3.1. We call a *fuzzy Hilbert logic* any pair $(\mathcal{S}, \mathcal{M})$ where $\mathcal{M}$ is a fuzzy semantics and $\mathcal{S}$ a fuzzy *H*-system whose deduction operator $\mathcal{D}$ coincides with the logical consequence operator *Lc* associated with $\mathcal{M}$.

We say that a fuzzy semantics $\mathcal{M}$ is *axiomatizable* if a fuzzy *H*-system $\mathcal{S}$ exists such that $(\mathcal{S}, \mathcal{M})$ is a fuzzy Hilbert logic. A fuzzy *H*-system $\mathcal{S} = (a, \mathbb{R})$ is *sound* with respect to $\mathcal{M}$ if every element $m \in \mathcal{M}$ is a theory in $\mathcal{S}$, i.e., m is closed under the inference rules and it contains the fuzzy set a of axioms. Equivalently, $\mathcal{S}$ is sound with respect to $\mathcal{M}$ if $\mathcal{D} \leq Lc$.

Proposition 3.2. *A fuzzy semantics $\mathcal{M}$ is axiomatizable by the fuzzy H-system $\mathcal{S}$ iff*
(a) *$\mathcal{S}$ is sound with respect to $\mathcal{M}$,*
(b) *every theory τ is an intersection of elements of $\mathcal{M}$.*

Proof. Assume that the logical consequence operator Lc and the deduction operator $\mathcal{D}$ coincide. Then, since every $m \in \mathcal{M}$ is a theory of $\mathcal{D}$, $\mathcal{S}$ is sound. Let τ be a theory, then $\tau = \mathcal{D}(\tau) = Lc(\tau)$ and therefore τ is the intersection of the models containing τ. Conversely, assume (a) and (b) and let v be a set of formulas. Then by (a) $\mathcal{D}(v) \subseteq Lc(v)$. On the other hand, since $\mathcal{D}(v)$ is the intersection of a family $(m_i)_{i\in I}$ of elements of $\mathcal{M}$,

$$\mathcal{D}(v) = \bigcap_{i\in I} m_i \supseteq \bigcap\{m \in \mathcal{M} : m \supseteq v\} = Lc(v).$$ □

The deduction operator of a given fuzzy H-system is invariant with respect to a large class of functions. To show this, we define a *homomorphism* of a fuzzy H-system $\mathcal{S}$ as any map $h : U \to U$ such that
(i) $a(\alpha) = h(a(\alpha))$ for every $\alpha \in Supp(a)$;
(ii) $r''(h(\lambda_1),...,h(\lambda_n)) = h(r''(\lambda_1,...,\lambda_n))$ for every $r = (r',r'') \in \mathbb{R}$ and $\lambda_1,...,\lambda_n \in U$;
(iii) h is join-preserving.

Theorem 3.3. *Let h be a homomorphism of $\mathcal{S}$ and v an initial valuation. Then*

$$\mathcal{D}(h \circ v) = h \circ \mathcal{D}(v), \tag{3.1}$$

and, consequently,

$$\tau \text{ theory in } \mathcal{S} \Rightarrow h \circ \tau \text{ theory in } \mathcal{S}. \tag{3.2}$$

Proof. In order to prove (3.1) let $\pi = \alpha_1,...,\alpha_m$ be a proof. We claim that

$$Val(\pi, h\circ v) - h(Val(\pi,v)). \tag{3.3}$$

Indeed, if α_m is assumed as a hypothesis, then

$$Val(\pi, h\circ v) = h(v(\alpha_m)) = h(Val(\pi,v)).$$

If α_m is assumed as a logical axiom, then,

$$Val(\pi, h\circ v) = a(\alpha_m) = h(a(\alpha_m)) = h(Val(\pi,v)).$$

This shows that the proposition holds for $m = 1$, so we can proceed by induction on m. Suppose $m \neq 1$ and $\alpha_m = r'(\alpha_{s(1)},...,\alpha_{s(n)})$. Then

$$\begin{aligned} Val(\pi, h\circ v) &= r''(Val(\pi_{s(1)}, h\circ v),...,Val(\pi_{s(n)}, h\circ v)) \\ &= r''(h(Val(\pi_{s(1)},v)),...,h(Val(\pi_{s(n)},v))) \\ &= h(r''(Val(\pi_{s(1)},v),...,Val(\pi_{s(n)},v)) = h(Val(\pi,v)). \end{aligned}$$

The remaining part of the proposition is trivial. □

Let $\mathcal{M}$ be a fuzzy semantics. Then we define *homomorphism* of $\mathcal{M}$ any map $h : U \to U$ such that

$$m \in \mathcal{M} \Rightarrow h \circ \tau \in \mathcal{M}.$$

As a consequence of Theorem 3.3, we have the following proposition:

Proposition 3.4. *Let* ($\mathcal{S}$, $\mathcal{M}$) *be a fuzzy Hilbert logic, then every homomorphism of* $\mathcal{S}$ *is a homomorphism of* $\mathcal{M}$.

Assume that a is crisp and that all the valuation parts of the rules coincide with $\wedge$. Then h is a homomorphism if h is join-preserving and $h(1) = 1$. This enables us to demonstrate the following theorem:

Theorem 3.5. *Let* $\mathcal{S}$ *be an H-system whose system of axioms is crisp and such that all the valuation parts of the rules coincide with the minimum. Then its deduction operator* $\mathcal{D}$ *is a canonical extension of a classical deduction operator.*

Proof. For any $\lambda \in U$, define the map h_λ by setting $h_\lambda(x) = 0$ if $x \leq \lambda$ and $h_\lambda(x) = 1$ if $x>\lambda$. Then h_λ is a homomorphism of $\mathcal{S}$. Consequently, since $h_\lambda \circ s$ is the characteristic function of the open cut $O(s,\lambda)$,

$$\mathcal{D}(O(s,\lambda)) = O(\mathcal{D}(s),\lambda).$$

Thus, in accordance with Theorem 7.5 in Chapter 3, $\mathcal{D}$ is the canonical extension of a classical deduction operator. □

4. DEGREE OF CONSISTENCY AND NON-MONOTONICITY

Let $\mathcal{S}$ be a fuzzy H-system. Then the *degree of inconsistency of an initial valuation* v and the *degree of inconsistency of* $\mathcal{S}$ are defined with reference to the associated deduction system. So,

$$Inc(v) = Inf\{\mathcal{D}(v)(\alpha) : \alpha \in \mathcal{F}\}, \tag{4.1}$$

and

$$Inc(\mathcal{S}) = Sup\{Inc(v) : v \text{ is consistent}\}. \tag{4.2}$$

The notion of degree of inconsistency emphasizes that there are proofs π which are not useful. Indeed, given a fuzzy subset v of formulas, if π is a proof of a formula α and $Val(\pi,v) \leq Inc(v)$, then such a proof is useless.

The next two propositions show the existence of a large class of fuzzy H-systems admitting theories with any prefixed degree of inconsistency. In the following we say that an initial valuation v *strengthens* α, if $\mathcal{D}(v)(\alpha) > Tau(\alpha)$.

Proposition 4.1. *Let* $\mathcal{S}$ *be a fuzzy H-system such that each valuation component* r'' *of its inference rules satisfies*

$$r''(\lambda_1,\ldots,\lambda_n) \leq \lambda_1 \wedge \ldots \wedge \lambda_n. \tag{4.3}$$

Then, if the initial valuation v *strengthens* α,

$$\mathcal{D}(v)(\alpha) \leq Sup\{v(\beta) : \beta \in \mathcal{F}\}.$$

Moreover,

$$\mathcal{D}(s^\lambda) = \lambda \vee Tau.$$

Proof. Firstly, we prove that, for every proof π of α,

$$Val(v,\pi) \leq Tau(\alpha) \vee Sup\{v(\beta) : \beta \in \mathcal{F}\},$$

and therefore that

$$\mathcal{D}(v)(\alpha) \leq Tau(\alpha) \vee Sup\{v(\beta) : \beta \in \mathcal{F}\}, \tag{4.4}$$

We shall proceed by induction on the length of π. Let α be the formula proved by π. Then, if α is assumed as a hypothesis or as a logical axiom, such an inequality is trivial. Suppose that α is obtained by the inference rule $r = (r',r'')$ and therefore that $\alpha = r'(\alpha_{s(1)},\dots,\alpha_{s(n)})$. Set $c = Sup\{v(\beta) : \beta \in F\}$ and let $\pi_{s(1)},\dots,\pi_{s(n)}$ be the proofs in π of $\alpha_{s(1)},\dots,\alpha_{s(n)}$, respectively. Then, since r'' is order-preserving, by induction hypothesis,

$$Val(v,\pi) = r''(Val(v,\pi_{s(1)}),\dots,Val(v,\pi_{s(n)})) \leq r''(Tau(\alpha_{s(1)}) \vee c,\dots, Tau(\alpha_{s(n)}) \vee c).$$

As a consequence, if $Tau(\alpha_{s(1)}) \geq c,\dots, Tau(\alpha_{s(n)}) \geq c$, we have

$$Val(v,\pi) \leq r''(Tau(\alpha_{s(1)}),\dots,Tau(\alpha_{s(n)})) \leq Tau(r'(\alpha_{s(1)},\dots,\alpha_{s(n)})) = Tau(\alpha).$$

If an index $j \in \{1,..,n\}$ exists such that $Tau(\alpha_{s(j)}) < c$, then by (4.3)

$$Val(v,\pi) \leq (Tau(\alpha_{s(1)}) \vee c) \wedge \dots \wedge (Tau(\alpha_{s(n)}) \vee c)$$
$$\leq [Tau(\alpha_{s(1)}) \wedge \dots \wedge Tau(\alpha_{s(n)})] \vee c \leq Tau(\alpha_{s(j)}) \vee c = c.$$ □

Observe that Condition (4.3) is satisfied everywhere $r''(\lambda_1,\dots,\lambda_n) = \lambda_1 \odot \dots \odot \lambda_n$, where $\odot$ is a continuous triangular norm.

Theorem 4.2. *Let S be a fuzzy H-system whose inference rules satisfy condition* (4.3). *Then*

$$Inf\{v(\beta) : \beta \in F\} \leq Inc(v) \leq Sup\{v(\beta) : \beta \in F\}. \tag{4.5}$$

In particular, for any $\lambda \in U$, $Inc(s^\lambda) = \lambda$ and therefore $Inc(S) = 1$.

Proof. It is evident that,

$$Inc(v) = Inf\{\mathcal{D}(v)(\alpha) : \alpha \in F\} \geq Inf\{v(\beta) : \beta \in F\}.$$

Besides, by (4.4),

$$\mathcal{D}(v)(\alpha) \leq (Sup\{v(\beta) : \beta \in F\}) \vee Tau(\alpha)$$

and, hence,

$$Inc(v) \leq Sup\{v(\beta) : \beta \in F\} \vee Inf\{Tau(\beta) : \beta \in F\}.$$

Since $Inc(v) \geq Inc(\emptyset) = Inf\{Tau(\beta) : \beta \in F\}$, (4.5) is completely proven.

The rest is evident. □

5. STEP-BY-STEP DEDUCTION AND CONTINUITY

Any Hilbert deduction system S defines a step-by-step deduction system in a natural way. Let $r = (r',r'')$ be an inference rule in R. Then we define the operator K_r by setting

$$K_r(v)(\alpha) = Sup\{r''(v(\alpha_1),\dots,v(\alpha_n)) : r'(\alpha_1,\dots,\alpha_n) = \alpha\}. \tag{5.1}$$

In other words, $K_r(v)$ is the fuzzy subset of formulas we can derive from v by one application of r. Also, we define the operator K by setting

$$K(v)(\alpha) = Sup\{r''(v(\alpha_1),\dots,v(\alpha_n)) : (r',r'') \in R \text{ and } r'(\alpha_1,\dots,\alpha_n) = \alpha\} \tag{5.2}$$

or, equivalently,

$$K(v) = \cup_{r \in R} K_r(v). \tag{5.3}$$

Then, $K(v)$ is the fuzzy subset of formulas we can derive from v by just one application of some inference rule in R.

Proposition 5.1. *Let $\mathcal{S} = (a, \mathbb{R})$ be a fuzzy H-system and define K by* (5.3). *Then $(\mathbb{F}, K, a)$ is a step-by-step fuzzy deduction system whose associated deduction system coincides with that of $\mathcal{S}$.*

Proof. In order to prove that K is continuous, observe that, as every valuation component r'' is an order-preserving map, K is order-preserving, too. Let $\mathcal{T}$ be a directed family of fuzzy subsets, $v = lim\,\mathcal{T}$, and α a formula. Then, by observing that for every $s_1,...,s_n \in \mathcal{T}$ an element $s \in \mathcal{T}$ exists such that $s_1 \subseteq s,...,s_n \subseteq s$, we have:

$$
\begin{aligned}
K(v)(\alpha) &= Sup\{r''(v(\alpha_1),...,v(\alpha_n)) : (r',r'') \in \mathbb{R} \text{ and } r'(\alpha_1,...,\alpha_n) = \alpha\} \\
&= Sup\{r''(s_1(\alpha_1),...,s_n(\alpha_n)) : (r',r'') \in \mathbb{R},\ r'(\alpha_1,...,\alpha_n) = \alpha \text{ and } s_1,...,s_n \in \mathcal{T}\} \\
&= Sup\{r''(s(\alpha_1),...,s(\alpha_n)) : (r',r'') \in \mathbb{R},\ r'(\alpha_1,...,\alpha_n) = \alpha \text{ and } s \in \mathcal{T}\} \\
&= \bigcup\{K(s)(\alpha) : s \in \mathcal{T}\}.
\end{aligned}
$$

In accordance with (7.2) in Chapter 1, define the immediate consequence operator H by setting,

$$H(v) = K(v) \cup v \cup a. \tag{5.4}$$

and observe that τ is a fixed point of H iff $\tau \supseteq a$ and $\tau \supseteq K(\tau)$. Then, by Theorem 2.7, the fixed points of H are the theories of $\mathcal{S}$ and this demonstrates that $\mathcal{D}$ is the closure operator generated by H. □

Proposition 5.1 entails the following theorem:

Theorem 5.2. *The deduction operator $\mathcal{D}$ of a fuzzy H-system is a continuous closure operator. Consequently, the class of its theories is an algebraic closure system.*

Note that, while all the deduction operators of the H-systems are continuous, there are H-systems whose deduction operator is not logically compact. Examples are given by those H-systems arising as the canonical extension of a crisp system (see Theorem 6.3 in Chapter 3). On the other hand, the following proposition furnishes a condition for logical compactness:

Proposition 5.3. *Let $\mathcal{M}$ be an axiomatizable fuzzy semantics. Then*

$$Contr(\mathcal{M}) \neq \emptyset \Rightarrow \mathcal{M} \text{ logically compact.}$$

Proof. This is a direct consequence of Proposition 7.4 in Chapter 2. □

We conclude this section with an interesting notion proposed in Pavelka [1978].

Definition 5.4. A fuzzy H-system is with *optimal proofs* if, for every fuzzy subset v of formulas and $\alpha \in \mathbb{F}$, a proof π exists such that $\mathcal{D}(v)(\alpha) = Val(\pi,v)$.

Clearly, the fuzzy H-systems are not necessarily with optimal proofs.

Proposition 5.5. *The deduction operator of a fuzzy H-system with optimal proofs is p-compact.*

6. BUILDING UP FUZZY HILBERT SYSTEMS BY INEQUALITIES

A very large class of fuzzy H-systems can be obtained starting from suitable systems of inequalities as follows.

Theorem 6.1. *Let $\mathbb{F}$ be a set, $a : \mathbb{F} \to U$ a fuzzy subset of $\mathbb{F}$ and let $\mathcal{M}$ be the class of all proper fuzzy subsets s of $\mathbb{F}$ containing a and satisfying a family of inequalities of type*

$$s(p(x_1,...,x_h)) \geq q(s(p_1(x_1,...,x_h)),...,s(p_k(x_1,...,x_h))) \tag{6.1}$$

where,

- $p, p_1, \ldots, p_k$ are partial operations defined on a subset D of $\mathbb{F}^h$,

- q is a function from U^k to U preserving the least upper bounds in each variable.

Then $\mathcal{M}$ is axiomatizable. More precisely, a fuzzy H-system exists whose class of consistent theories is $\mathcal{M}$ and whose fuzzy subset of logical axioms is a. If q is continuous, then $\mathcal{M}$ is closed under ultraproducts.

Proof. First, observe that an inequality like (6.1) cannot be considered directly as the inequality expressing the closure under an inference rule whose premises are the formulas $x_1,...,x_n$. In fact, while the formula $p(x_1,\ldots,x_h)$ depends on the formulas $x_1,...,x_h$, in the right side of (6.1) there are the formulas $p_1(x_1,\ldots, x_h)$, ... , $p_k(x_1,...,x_h)$. We can eliminate such an inconvenience by considering inference rules whose antecedents are not only the formulas $x_1,...,x_n$ but also the formulas $p_1(x_1,\ldots, x_h)$, ... , $p_k(x_1,...,x_h)$. More precisely, we define the *normalization function* $n : \boldsymbol{R} \to U$ by setting, for every $\lambda \in \boldsymbol{R}$,

$$n(\lambda) = \begin{cases} 1 & \text{if } \lambda \geq 1, \\ \lambda & \text{if } 0 \leq \lambda < 1, \\ 0 & \text{otherwise.} \end{cases}$$

It is evident that n is an increasing continuous function preserving the least upper bounds. Consider the fuzzy rule $r = (r', r'')$ whose domain is

$$\{(x_1,..., x_h, y_1,...,y_k) : (x_1,...,x_h) \in D, y_1 = p_1(x_1,...,x_h),\ldots, y_k = p_k(x_1,...,x_h)\},$$

and such that

$$r'(x_1,...,x_h,y_1,...,y_k) = p(x_1,...,x_h) \quad ; \quad r''(\lambda_1,...,\lambda_h,\mu_1,...,\mu_k) = n(q(\mu_1,...,\mu_k)).$$

Consider a fuzzy H-system whose fuzzy subset of logical axioms is a and whose fuzzy inference rules are the rules obtained in such a way for any considered inequality as in (6.1). In addition, for any inequality as in (6.1) we consider a collapsing rule enabling to claim that if $\alpha_1,...,\alpha_k$ where proven to degree $\lambda_1,...,\lambda_k$ such that $q(\lambda_1,...,\lambda_k) > 1$, then every formula can be proven to degree 1. Formally, this is achieved by adding to the deduction apparatus, for every formula α, a rule $r = (r', r'')$ defined by setting $r'(\alpha_1,...,\alpha_k) = \alpha$, and

$$r''(\lambda_1,...,\lambda_k) = \begin{cases} 1 & \text{if } q(\lambda_1,...,\lambda_k) > 1, \\ 0 & \text{otherwise.} \end{cases}$$

Trivially, a proper fuzzy set τ of formulas is closed under the collapsing rules iff $q(\tau(y_1),\dots,\tau(y_k)) \leq 1$. Then, for $y_1,\dots,y_k \in F$

$$n(q(\tau(y_1),\dots,\tau(y_k)) \geq q(\tau(y_1),\dots,\tau(y_k)).$$

Now, let τ be a consistent theory of the so-defined H-system. Then for every inference rule r and every $(x_1,\dots, x_h, y_1,\dots,y_k) \in Dom(r)$,

$$\begin{aligned}\tau(p(x_1,\dots,x_h)) &= \tau(r'(x_1,\dots,x_h,y_1,\dots,y_k))\\ &\geq r''(\tau(x_1),\dots,\tau(x_h),\tau(y_1),\dots,\tau(y_k)) = n(q(\tau(y_1),\dots,\tau(y_k))\\ &\geq q(\tau(y_1),\dots,\tau(y_k)).\end{aligned}$$

Therefore, since $y_1 = p_1(x_1,\dots,x_h), \dots , y_k = p_k(x_1,\dots,x_h)$, we have

$$\tau(p(x_1,\dots,x_h)) \geq n(q(\tau(p_1(x_1,\dots,x_h)),\dots,\tau(p_k(x_1,\dots,x_h)))).$$

Conversely, let τ be a proper fuzzy subset containing a and satisfying (6.1). Then, since $\tau(p(x_1,\dots,x_h)) \in U$, we also have that

$$\tau(p(x_1,\dots,x_h)) \geq n((q(\tau(p_1(x_1,\dots,x_h))),\dots,\tau(p_k((x_1,\dots,x_h))).$$

Hence, τ is closed under the rule r. It is evident that τ is closed with respect to the collapsing rules. Thus, we may conclude that τ is a theory.

The closure under ultraproducts follows from Proposition 8.2 in Chapter 2. □

Under the hypotheses of Proposition 6.1, the deduction operator $\mathcal{D}$ enables us to get the smallest function $\mathcal{D}(s)$ greater than or equal to s satisfying all the inequalities (6.1). Furthermore, we may obtain $\mathcal{D}(s)$ as the fixed point of a suitable operator J. Indeed, by setting

$$J'(s)(z) = Sup\{q^i(s(p^i_1(x_1,\dots,x_{n(i)})),\dots,s(p^i_{k(i)}(x_1,\dots,x_{n(i)}))) : i \in I \text{ and } p^i(x_1,\dots,x_{n(i)}) = z\}$$

and

$$J(s) = J'(s) \cup s \cup a,$$

we have

$$\mathcal{D}(s) = \cup_{n\in N} J^n(s).$$

As an example, let $\boldsymbol{B}$ be a Boolean algebra. Then a *generalized necessity* is any map $n : \boldsymbol{B} \to U$ such that, for every $x, y \in \boldsymbol{B}$,

$$n(1) = 1 \quad ; \quad n(x \wedge y) = n(x) \wedge n(y)$$

(see Section 6 in Chapter 6). Equivalently, we require that $n(1) = 1$ and

$$n(x \wedge y) \geq n(x) \wedge n(y) \quad ; \quad n(x) \geq n(x \wedge y). \tag{6.2}$$

Now, the first inequality in (6.2) can be obtained from (6.1) setting $h = k = 2$ and assuming that p, p_1, p_2 and q are the functions defined by setting, for $x_1,x_2 \in \boldsymbol{B}$ and $\lambda_1, \lambda_2 \in U$.

$$p(x_1,x_2) = x_1 \wedge x_2 \quad ; \quad p_1(x_1,x_2) = x_1 \quad ; \quad p_2(x_1,x_2) = x_2 \quad ; \quad q(\lambda_1,\lambda_2) = \lambda_1 \wedge \lambda_2.$$

The second inequality in (6.2) is obtained by setting $h = 2$, $k = 1$ and assuming that the functions p, p_1 and q are defined by

$$p(x_1,x_2) = x_1 \quad ; \quad p_1(x_1,x_2) = x_1 \wedge x_2 \quad ; \quad q(\lambda) = \lambda.$$

So, the generalized necessities define an H-system whose set of logical axioms is $\{1\}$ (to obtain the equality $n(1) = 1$) and with the following inference rules:

$$\frac{x_1, x_2}{x_1 \wedge x_2} \qquad ; \qquad \frac{\lambda_1, \lambda_2}{\lambda_1 \wedge \lambda_2}$$

$$\frac{x_1, x_2, y}{x_1} \quad ; \quad \frac{\lambda_1, \lambda_2, \lambda_3}{\lambda_3}$$

where the second rule is defined in the set $\{(x_1, x_2, y) : y = x_1 \wedge x_2\}$. No collapsing rule is assumed. Consequently, the class of generalized necessities is a closure system closed under ultraproducts. It is possible to find more simple rules (see Chapter 6).

7. ANY FUZZY HILBERT SYSTEM IS EQUIVALENT TO A CRISP SYSTEM

Let $\mathcal{S} = (a, \mathbb{R})$ be a fuzzy H-system and, as usual, denote by $\mathbb{F}$, $\mathcal{D}$ and $\mathcal{T}$ the set of formulas, the deduction operator and the class of theories of $\mathcal{S}$, respectively. We claim that $\mathcal{S}$ can be represented as a crisp H-system provided that an infinitary inference rule is admitted. In order to prove this, recall that in Section 11 of Chapter 2 we denote by $\mathbb{F}_s$ the set $\{(\alpha,\lambda) : \alpha \in \mathbb{F}, \lambda \in U\}$ of signed formulas and we consider the map $E : \mathcal{F}(\mathbb{F}) \to \mathcal{P}(\mathbb{F}_s)$ associating any fuzzy set v of formulas with its hypograph $E(v) = \{(\alpha,\lambda) \in \mathbb{F}_s : v(\alpha) \geq \lambda\}$. It is easy to verify that such a function is an isomorphism between $\mathcal{F}(\mathbb{F})$ and the lattice $G(\mathbb{F}) = E(\mathcal{F}(\mathbb{F}))$. Furthermore, we associate the deduction operator $\mathcal{D}$ of a fuzzy logic with a crisp closure operator $\mathcal{D}_s : G(\mathbb{F}) \to G(\mathbb{F})$ defined by setting, for every X in $G(\mathbb{F})$,

$$\mathcal{D}_s(X) = E(\mathcal{D}(E^{-1}(X))). \tag{7.1}$$

In such a way an abstract crisp logic is achieved which is equivalent to the considered fuzzy logic.

Given a fuzzy inference rule $r = (r',r'')$, we define *the crisp inference rule associated with r* as the rule r_c defined on the domain

$$Dom(r_c) = \{((\alpha_1,\lambda_1),\ldots,(\alpha_n,\lambda_n)) \in \mathbb{F}_s^n : (\alpha_1,\ldots,\alpha_n) \in Dom(r)\}$$

by setting

$$r_c((\alpha_1,\lambda_1),\ldots,(\alpha_n,\lambda_n)) = (r'(\alpha_1,\ldots,\alpha_n), r''(\lambda_1,\ldots,\lambda_n)).$$

This means that in place of the fuzzy rule:

$$\frac{\alpha_1,\ldots,\alpha_n}{r'(\alpha_1,\ldots,\alpha_n)} \quad ; \quad \frac{\lambda_1,\ldots,\lambda_n}{r''(\lambda_1,\ldots,\lambda_n)}$$

we have the crisp rule:

$$\frac{(\alpha_1,\lambda_1),\ldots,(\alpha_n,\lambda_n)}{(r'(\alpha_1,\ldots,\alpha_n), r''(\lambda_1,\ldots,\lambda_n))}$$

The *crisp H-system associated with* $\mathcal{S} = (a, \mathbb{R})$ is defined as the crisp H-system $\mathcal{S}_c$ such that

- its set of formulas is $\mathbb{F}_s$
- its set of logical axioms is $\mathbb{A} = E(a)$
- its set $\mathbb{R}_c$ of inference rules is $\{r_c : r \in \mathbb{R}\}$.

We denote the related deduction operator by $\mathcal{D}_c$ (as we will show in the following, $\mathcal{D}_c$ does not coincide with $\mathcal{D}_s$).

Theorem 7.1. *Let τ be a fuzzy subset of formulas. Then*

τ is a theory of $\mathcal{S}$ $\Leftrightarrow$ $E(\tau)$ is a theory of $\mathcal{S}_c$.

Proof. Let τ be a theory of $\mathcal{S}$. Then, since E is order-preserving and $a \subseteq \tau$, $E(\tau)$ contains the set $\mathcal{A} = E(a)$ of logical axioms of $\mathcal{S}_c$. Let $(\alpha_1,\lambda_1),...,(\alpha_n,\lambda_n)$ be signed formulas in $E(\tau)$ such that $((\alpha_1,\lambda_1),...,(\alpha_n,\lambda_n))$ belongs to $Dom(r_s)$. Then, $\tau(\alpha_1) \geq \lambda_1,..., \tau(\alpha_n) \geq \lambda_n$ and $(\alpha_1,...,\alpha_n) \in Dom(r)$. By recalling that τ is closed under r and that r'' is increasing with respect to any component, we have

$$\tau(r'(\alpha_1,...,\alpha_n)) \geq r''(\tau(\alpha_1),...,\tau(\alpha_n)) \geq r''(\lambda_1,...,\lambda_n).$$

Hence, the signed formula $(r'(\alpha_1,...,\alpha_n), r''(\lambda_1,...,\lambda_n))$ belongs to $E(\tau)$ and this proves that $E(\tau)$ is a theory of $\mathcal{S}_c$.

Conversely, assume that $E(\tau)$ is a theory of $\mathcal{S}_c$. Then $E(\tau) \supseteq E(a)$ and, therefore, $\tau \supseteq a$. Let $r = (r',r'')$ be any inference rule in $\mathcal{S}$, $(\alpha_1,...,\alpha_n)$ an element of $Dom(r)$ and set $\lambda_1 = \tau(\alpha_1),...,\lambda_n = \tau(\alpha_n)$. Then, $((\alpha_1,\lambda_1),...,(\alpha_n,\lambda_n)) \in Dom(r_c)$ and $(\alpha_1,\lambda_1), ...,(\alpha_n,\lambda_n)$ are elements in $E(\tau)$. Since $E(\tau)$ is closed under r_c, we have that

$$r_c((\alpha_1,\lambda_1),...,(\alpha_n,\lambda_n)) = (r'(\alpha_1,...,\alpha_n), r''(\lambda_1,...,\lambda_n)) \in E(\tau)$$

and, hence, that $\tau(r'(\alpha_1,...,\alpha_n)) \geq r''(\tau(\alpha_1),...,\tau(\alpha_n))$. Thus, τ is a theory of $\mathcal{S}$. □

Observe that there are theories in $\mathcal{S}_c$ which are not hypographs, i.e., images of a fuzzy subset of formulas. Therefore Theorem 7.1 does not establish a one-one correspondence between the fuzzy theories of $\mathcal{S}$ and the crisp theories of $\mathcal{S}_c$. Now, E maps the lattice $\mathcal{F}(\mathbb{F})$ of all fuzzy subsets of $\mathbb{F}$ on the class $G(\mathbb{F})$ of the subsets X of $\mathbb{F}_s$ such that:

(i) $(x,\mu) \in X$ and $\lambda \leq \mu \Rightarrow (x,\lambda) \in X$,

(ii) if $\{\lambda \in U : (x,\lambda) \in X)\}$ is nonempty, then it admits a greatest element.

So, the class of theories of $\mathcal{S}$ corresponds to the class of theories of $\mathcal{S}_c$ satisfying both (i) and (ii). This suggests the consideration of further inference rules corresponding to conditions (i) and (ii). Unluckily, (ii) imposes to admit infinitary inference rules.

Definition 7.2. Given a fuzzy H-system $\mathcal{S}$, the *crisp (infinitary) H-system associated with* $\mathcal{S}$ is the system $\mathcal{S}_\infty$ achieved by adding the following rules to the rules in $\mathcal{S}_c$:

- for every $\lambda \in U$ the rule r_λ defined by setting

$$Dom(r_\lambda) = \{(\alpha,\mu) \in \mathbb{F}_s : \lambda \leq \mu\} \quad \text{and} \quad r_\lambda(\alpha,\mu) = (\alpha,\lambda).$$

- the infinitary inference rule r_∞ defined by

$$r_\infty((\alpha,\lambda_n)_{n\in N}) = (\alpha, Sup_{n\in N}\, \lambda_n)$$

for every $\alpha \in \mathbb{F}$ and every sequence $(\lambda_n)_{n\in N}$ of elements of U.

The rule r_λ says that if a formula α holds at least to degree μ and $\lambda \leq \mu$, then α holds at least to degree λ. The rule r_∞ says that if for every $\lambda_n \in U$, $\lambda_n \neq 0$, the formula α holds (at least) to degree λ_n, then α holds (at least) to degree $Sup_{n\in N}\, \lambda_n$. Due to the presence of an infinitary rule, we avoid the notion of proof for $\mathcal{S}_\infty$. Indeed, we call *theory* of $\mathcal{S}_\infty$ any set of signed formulas closed under the inference rules of $\mathcal{S}_\infty$. We call *deduction operator* of $\mathcal{S}_\infty$ the closure operator associated with the class of theories of $\mathcal{S}_\infty$.

Theorem 7.3. *The deduction operator of $\mathcal{S}_\infty$ coincides with the operator $\mathcal{D}_s$ defined in* (7.1). *Then, since*

$$\mathcal{D}(s) = E^{-1}(\mathcal{D}_s(E(s))), \tag{7.2}$$

we can reduce the fuzzy H-system $\mathcal{S}$ to the crisp (infinitary) H-system $\mathcal{S}_\infty$.

Proof. It is sufficient to prove that $E(\mathcal{T})$ coincides with the class of all the theories of $\mathcal{S}_\infty$. If τ is a theory in $\mathcal{S}$, then we have just proven that $E(\tau)$ is a theory of $\mathcal{S}_c$. As $E(\tau)$ satisfies (i) and (ii), $E(\tau)$ is closed under the rules r_λ and the rule r_∞. Thus $E(\tau)$ is a theory of $\mathcal{S}_\infty$. Conversely, let T be a theory of $\mathcal{S}_\infty$, then, since T is closed under the rules r_λ and r_∞, a fuzzy subset τ of formulas exists such that $E(\tau) = T$. Because T is a theory of $\mathcal{S}_c$, by Theorem 7.1 τ is a theory of $\mathcal{S}$. □

Note that while $\mathcal{D}_s \geq \mathcal{D}_c$, in general $\mathcal{D}_s \neq \mathcal{D}_c$. To establish the exact connection between $\mathcal{D}_s$ and $\mathcal{D}_c$, recall that $G(\mathcal{F})$ is a non-algebraic closure system and $Co(G(\mathcal{F}))$ a non-algebraic closure operator. For every $X \subseteq \mathcal{F}$, denote by $<X>$ the element of $G(\mathcal{F})$ generated by X.

Theorem 7.4. $\mathcal{D}_s = Co(G(\mathcal{F})) \circ \mathcal{D}_c$, *i.e., for every* $X \subseteq \mathcal{F}_s$,

$$\mathcal{D}_s(X) = <\mathcal{D}_c(X)>. \tag{7.3}$$

As a consequence, for every fuzzy subset of formulas s,

$$\mathcal{D}(s) = E^{-1}(<\mathcal{D}_c(E(s))>). \tag{7.4}$$

Proof. We will prove that $<\mathcal{D}_c(X)>$ is a theory of $\mathcal{S}_\infty$ containing X. Indeed, let T_1 be the closure of $\mathcal{D}_c(X)$ with respect to the rules r_λ, that is,

$$T_1 = \{(\alpha,\lambda) \in \mathcal{F}_s : \exists(\alpha,\mu) \in \mathcal{D}_c(X), \lambda \leq \mu\}.$$

Let T_2 be the closure of T_1 with respect to the rule r_∞, that is,

$$T_2 = \{(\alpha,\lambda) : \lambda = Sup_{n\in N}\, \mu_n \text{ where } ((\alpha,\mu_n))_{n\in N} \text{ is a sequence of elements of } T_1\}.$$

It is evident that $T_2 = <\mathcal{D}_c(X)>$. We claim that T_1 is a theory of $\mathcal{S}_c$. Indeed, since $T_1 \supseteq \mathcal{D}_c(X)$, T_1 contains the set $\mathcal{A}$ of logical axioms. Let r be an inference rule of $\mathcal{S}$ and $(\alpha_1,\lambda_1),\ldots,(\alpha_n,\lambda_n)$ elements of T_1 belonging to the domain of r_c. Then $\mu_1,\ldots,\mu_n$ exist such that $\mu_1 \geq \lambda_1,\ldots,\mu_n \geq \lambda_n$ and $(\alpha_1, \mu_1),\ldots,(\alpha_n, \mu_n)$ are in $\mathcal{D}_c(X)$. Since $\mathcal{D}_c(X)$ is closed under r_c, $(r'(\alpha_1,\ldots,\alpha_n), r''(\mu_1,\ldots,\mu_n)) \in \mathcal{D}_c(X)$. Since $r''(\mu_1,\ldots,\mu_n) \geq r''(\lambda_1,\ldots,\lambda_n)$ this entails that $(r'(\alpha_1,\ldots,\alpha_n), r''(\lambda_1,\ldots,\lambda_n)) \in T_1$.

We claim that T_2 is a theory of $\mathcal{S}_\infty$. Indeed, it is evident that T_2 is closed under the rules r_λ and r_∞. Let r be an inference rule and $(\alpha_1,\lambda_1),\ldots,(\alpha_n,\lambda_n)$ elements of T_2 belonging to the domain of r_c, that is $(\alpha_1,\ldots,\alpha_n) \in Dom(r)$. Then, there exist sequences $(\alpha_1, \mu^1{}_m)_{m\in N},\ldots,(\alpha_n, \mu^n{}_m)_{m\in N}$ of elements of T_1 such that

$$\lambda_1 = Sup_{m\in N}\, \mu^1{}_m, \ \ldots\ , \lambda_n = Sup_{m\in N}\, \mu^n{}_m.$$

The closure of T_1 under r_c, entails that, for every $s(1),\ldots,s(n)$ in $\boldsymbol{N}$, $(r'(\alpha_1,\ldots,\alpha_n), r''(\mu^1{}_{s(1)},\ldots,\mu^n{}_{s(n)})) \in T_1$. Thus

$$(r'(\alpha_1,\ldots,\alpha_n), r''(\lambda_1,\ldots,\lambda_n)) = (r'(\alpha_1,\ldots,\alpha_n), Sup\{r''(\mu^1{}_{s(1)},\ldots,\mu^n{}_{s(n)}) : s(1),\ldots,s(n) \in \boldsymbol{N}\})$$

is an element of T_2 and this proves that T_2 is closed under r_c. Since every theory of $\mathcal{S}_\infty$ containing X contains $< \mathcal{D}_c(X) >$, (7.3) is proved.

The rest is evident. □

Remark. We can ask:

is there anything in fuzzy logic (defined via fuzzy subsets) which cannot be translated into a corresponding crisp logic (defined via signed formulas) ?

The first observation is that, due to the presence of an infinitary rule in the system $\mathcal{S}_\infty$, the deduction operator $\mathcal{D}_s$ is not compact. This is in accordance with the fact that $\mathcal{D}_s$ is the product of the compact operator $\mathcal{D}_c$ and the non-compact operator $Co(G(\mathbb{F}))$. Then, non-compactness is a price we must pay to represent a fuzzy Hilbert system by a crisp Hilbert system based on signed formulas. Furthermore, as we will see in Chapter 11, the set $E(v)$ of signed formulas we associate with a recursively enumerable fuzzy set of axioms v need not be recursively enumerable (see the remark in Section 2 of Chapter 11). Consequently, the proposed translation does not preserve effective enumerability. On the other hand, this notion is basic for logic since it expresses the effectiveness of the deduction process.

8. BALD MEN, ŁUKASIEWICZ CONJUNCTION AND INDUCTION PRINCIPLE

As we have observed in Section 1, Bald Man Paradox is similar to Heap Paradox. Indeed, assume that $Bald(n)$ represents the claim *"a man with n hairs is bald"*. Then $Bald(1)$ is true and, for any $n \in N$, $Bald(n) \rightarrow Bald(n+1)$ is plausible and this enables us to prove $Bald(n)$ for any $n \in N$. But, if we try to apply fuzzy logic, we become aware that

in order to solve the Bald Man Paradox, the product does not yield an adequate valuation part for fuzzy Modus Ponens.

Indeed, in Section 1 we showed that the paradoxical argument can be formalized in a fuzzy proof of $Small(100.000.000)$ whose valuation was $\delta = 0.9^{(100.000.000-1)}$, where $\delta \neq 0$. This entails that

"the truth degree of Small(100.000.000) is greater or equal to δ"

and this is compatible with the idea that *Small(100.000.000)* is very near to false but not completely false. When we try to apply the same apparatus to Bald Man Paradox, we can show that $Bald(100.000.000)$ is also proved to degree δ. But, a man with 100.000.000 hairs is surely not bald and therefore $Bald(100.000.000)$ is completely false. This means that the claim

"the truth degree of Bald(100.000.000) is greater or equal to δ"

cannot be accepted.

This argument shows that Bald Man Paradox appears different from Heap Paradox when we try to solve it by fuzzy logic. However, if we consider the *"reverse form"* of Heap Paradox, the product gives the same difficulties we meet with the Bald Man paradox. Note that Heap Paradox proceeds by "addition" but that one might proceed also in reverse, by "subtraction". Namely, let *"Big"* be the negation of *"Small"*, i.e., $Big(n)$ means that a heap with n stones is big, and admit that:

- for a sufficiently large integer b, $Big(b)$ holds
- all the implications $Big(n) \rightarrow Big(n-1)$ are plausible.

Then by using $b-1$ times $M.P$, we can conclude that $Big(1)$ holds.

By using fuzzy logic formalisms, assume that $Big(b)$ is true to degree $\lambda \neq 0$ and that the formulas $Big(n) \rightarrow Big(n-1)$ holds at least to a degree $\mu \neq 0$. Then by proceeding as in Section 1, we obtain that $Big(1)$ can be proved to degree $\lambda \cdot (\mu)^{b-1} \neq 0$ and therefore, that its truth degree is different from zero. As in the case of Bald Man Paradox, this is completely unsatisfactory. Indeed, in spite of the fact that $\lambda \cdot (\mu)^{b-1}$ is very near to zero, $Big(1) = \neg\, Small(1)$ is true to degree 0, exactly.

Thus, we can conclude that:

The usual product is not even adequate for the "reverse form" of Heap Paradox.

An answer to these questions is given by *Łukasiewicz conjunction* $\odot$. In fact, define, as usual, the power x^n by setting

$$x^0 = 1 \text{ and } x^n = x^{n-1} \odot x.$$

We claim that:

Łukasiewicz conjunction is able to solve both Bald Man Paradox and Heap Paradox.

In fact, as it is easy to prove, for every $x \neq 1$, an integer n exists such that $x^n = 0$. This solves Bald Man Paradox (the number $0.9^{(100.000.000-1)}$ is exactly equal to zero). Moreover, it is easy to prove that, for every integer n an element $x \neq 0$ exists such that $x^n = 0$. Therefore, in the case of the *"subtraction"* form of Heap Paradox, given the integer b, by a suitable choice of λ and μ, we can obtain $\lambda \odot (\mu)^{b-1} = 0$.

We conclude this section by emphasizing that the same argument on the basis of Heap Paradox enables us to show a surprising feature of fuzzy logic which is:

the falsity of the induction principle whenever we consider vague predicates.

In fact, the *induction principle* is expressed by the two formulas $\alpha(1)$ and $((\forall n(\alpha(n) \rightarrow \alpha(n+1)) \rightarrow \forall n(\alpha(n))$, or, equivalently, by

$$\alpha(1) \rightarrow ((\forall n(\alpha(n) \rightarrow \alpha(n+1)) \rightarrow \forall n\alpha(n),$$

where α is any formula expressing a property in the set N of natural numbers. Moreover, assume that the induction principle holds at degree μ, and that α represents a vague property for which $\forall n(\alpha(n))$ is false and $\forall n(\alpha(n) \rightarrow \alpha(n+1))$ is true to degree $\nu \neq 0$ (as in the Heap Paradox). Then, by *MP* we can prove the formula $\forall n\alpha(n)$ to degree $\nu \odot \mu$. Since such a formula is false, $\nu \odot \mu. = 0$. Then, in the case that $\odot$ is either the minimum or the product operation, we can conclude that $\mu = 0$. In the case that $\odot$ is the Łukasiewicz conjunction, it is $\nu + \mu \leq 1$ and therefore $\mu \leq 1-\nu$. Assume that a suitable choice of the vague property α enables us to obtain ν as close to 1 as we want. Then, we can also conclude that $\mu = 0$.

CHAPTER 5

LOGIC AS MANAGMENT OF CONSTRAINTS ON THE TRUTH VALUES

1. HEAP PARADOX BY NEGATIVE INFORMATION

Let U be any complete lattice. Then, as observed in Chapter 4, in fuzzy logic it would be misleading to consider an initial valuation $v : \mathbb{F} \to U$ as a fuzzy subset of $\mathbb{F}$. As a matter of fact, for any formula α, the number $v(\alpha)$ is not the truth value of α but a constraint on its actual truth value, namely a constraint like

"the truth value of α is greater than or equal to $v(\alpha)$".

Now, it is rather natural to admit also that the available information about an unknown fuzzy world can be expressed by constraints of a more general kind. For example, constraints as

"the truth value of α is between 0.3 and 0.5",

"the truth value of β is less than 0.7".

The general form of such information is

"the truth value of α belongs to X",

where X is a subset of U (see Gerla [1999]).

Such a general approach can also be justified by giving a different form to Heap paradox. Again, we assume that a heap containing just one stone is small, i.e., that

Small(1) is true.

Moreover, we express directly our opposition to accept the existence of a *critical* number n such that a heap with n stones is small and a heap with $n+1$ stones is not small. Indeed, we claim that

$Small(n) \wedge \neg Small(n+1)$ *is "impossible".*

Then,

- since *Small*(1) holds and $Small(1) \wedge \neg Small(2)$ is impossible, the formula *Small*(2) holds,
- since *Small*(2) holds and $Small(2) \wedge \neg Small(3)$ is impossible, the formula *Small*(3) holds,

. . .

- since $Small(n-1)$ holds and $Small(n-1) \wedge \neg Small(n)$ is impossible, the formula $Small(n)$ holds.

In such an argument the presence of the expression *"impossible"* means that we admit negative information as a premise in an inferential step. This is not usual in formalized logic where inference rules in which the premises are true formulas are preferred. Instead, in our opinion, in the everyday arguments it is rather frequent to infer from negative information. Clearly, for classical logic the distinction is not essential because the claim that a formula is false is equivalent to the claim that its negation is true. Now, the better way to express negative information is to consider *"signed"* formulas. This enables us to formalize the just exposed argument by the classical *"inference rule"*

$$\frac{(A, 1)\ ,\ (A \wedge \neg B, 0)}{(B\ , 1)}$$

and by assuming the signed formulas:

(a) (*Small*(1), 1)

(b) (*Small*(*n*) $\wedge \neg$ *Small*(*n*+1), 0).

Indeed,

from (*Small*(1), 1) and (*Small*(1) $\wedge \neg$ *Small*(2), 0)
we prove (*Small*(2), 1),

from (*Small*(2), 1) and (*Small*(2) $\wedge \neg$ *Small*(3), 0)
we prove (*Small*(3), 1),

. . .

from (*Small*(*n*−1), 1) and (*Small*(*n*−1) $\wedge \neg$ *Small*(*n*), 0)
we prove (*Small*(*n*), 1).

Now, an answer to this paradox can be obtained by observing that everyone is convinced that the formulas *Small*(*n*) $\wedge \neg$ *Small*(*n*+1) are very far from the truth but, in general, not completely false. In other words, the truth values of these formulas are either equal to 0 or, in any case, *"very near" to* 0. This can be expressed in a precise way by enlarging the notion of initial valuation or, equivalently, the notion of signed formula. In accordance, we need change the notion of fuzzy inference rule and fuzzy reasoning. Indeed we call *signed formula* a pair (A, X) where A is a formula and X a constraint for its truth value. Moreover, we propose the rule

$$\frac{(A, [\lambda, 1])\ \ ,\ (A \wedge \neg B, [0, \mu])}{(B,\ [0 \vee (\lambda - \mu), 1])}$$

which means that:

IF	A is approximately true (to degree λ),
AND	$A \wedge \neg B$ is approximately false (to degree μ),
THEN	B is approximately true (to degree $\lambda - \mu$).

This rule is sound with respect to Łukasiewicz semantics (see Definition 6.1 in Chapter 8). Indeed, let x and y be the truth values of the formulas A and B, respectively. Then from $x \geq \lambda$ and $(x + (1-y) - 1) \vee 0 \leq \mu$ it follows that $x - y \leq \mu$ and therefore $y \geq \lambda - \mu$.

By assuming the signed formulas,

(*Small*(1), 1) , (*Small*(*n*) $\rightarrow$ *Small*(*n*+1), [0, 0.0001]),

we can consider the following approximate reasoning:

- from
  (*Small*(1), [1,1]) and (*Small*(1)$\wedge\neg$*Small*(2), [0, 0.0001]),
  we state (*Small*(2), [0.9999, 1]);
- since
  (*Small*(2), [0.9999, 1]) and (*Small*(2)$\wedge\neg$*Small*(3), [0, 000.1]),

we state (*Small*(3), [0.9998, 1]);

...

- since

(*Small*(10.000), [0.0001,1]) and (*Small*(10.000) $\wedge\neg$*Small*(10.001), [0, 0.0001]), we state (*Small*(10.0001), [0, 1]).

Thus, as in Chapter 4, for any integer n we can give a correct proof π of the formula *Small*(n) but, for n > 10.000, no information about the truth value of *Small*(n) is furnished by π.

2. CONSTRAINTS ON THE TRUTH VALUES

We define a *constraint* as any subset X of U and we say that a truth value $\lambda \in U$ *satisfies* X if $\lambda \in X$. A constraint X is an *improvement* of a constraint Y, in brief $X \geq Y$, provided that $X \subseteq Y$. As a matter of fact, in case U is infinite, the whole class $\mathcal{P}(U)$ of constraints is too big and no language should be able to represent its elements. So, it should be more convenient to refer only to particular constraints, for example interval-constraints. In any case it is useful to assume that the class of constraints we will consider is a closure system in U and hence a complete lattice.

Definition 2.1. Let U be a lattice and $\mathcal{C}$ a closure system in U. Then we say that $\mathcal{C}$ is a *constraint frame*. Given a subset X of U, let $<X>$ denote the constraint generated by X, i.e.

$$<X> = \cap\{ Y \in \mathcal{C}: Y \supseteq X \}. \tag{2.1}$$

As an example, we can assume that $\mathcal{C}$ is the set of closed intervals and therefore, that $<X>$ is the least interval containing X.

We consider $\mathcal{C}$ as a lattice with respect to the improvement relation. This means that the joins coincide with the intersections while the meet of a family $(X_i)_{i\in I}$ of constraints coincides with the element of $\mathcal{C}$ generated by the union $\cup_{i\in I}X_i$. This determines the correspondent notion of $\mathcal{C}$-subset of formulas.

Definition 2.2. Let U be a lattice, $\mathcal{C}$ a constraint frame in U and $\mathcal{F}$ the set of formulas of a given language. We define an *initial $\mathcal{C}$-valuation* or a *$\mathcal{C}$-system of axioms* as any $\mathcal{C}$-set of formulas, i.e., any map $\mathcal{V} : \mathcal{F} \to \mathcal{C}$. We denote the class $\mathcal{C}^{\mathcal{F}}$ of initial $\mathcal{C}$-valuations by $Val(\mathcal{C})$.

Then $Val(\mathcal{C})$ is a complete lattice with respect to the relation $\leq$ defined by setting, for every $\mathcal{V}_1$ and $\mathcal{V}_2$ in $Val(\mathcal{C})$,

$$\mathcal{V}_1 \leq \mathcal{V}_2 \Leftrightarrow \mathcal{V}_1(\alpha) \supseteq \mathcal{V}_2(\alpha) \text{ for every } \alpha \in \mathcal{F}.$$

If $\mathcal{V}_1 \leq \mathcal{V}_2$ we say that $\mathcal{V}_2$ is an *improvement* of $\mathcal{V}_1$. The least element in this lattice is the initial $\mathcal{C}$-valuation $\mathcal{V}_0$ defined by setting $\mathcal{V}_0(\alpha) = U$ for every formula α. We call it the *zero valuation* since such a valuation is satisfied by any model and therefore it corresponds to zero information. We call *plausible* a $\mathcal{C}$-valuation $\mathcal{V}$ such that $\mathcal{V}(\alpha) \neq \emptyset$ for every formula α.

The information carried on by $\mathcal{V}$ is that, given any formula α, the actual truth value of α satisfies $\mathcal{V}(\alpha)$.

Definition 2.3. Let $\mathcal{C}$ be a constraint frame in a complete lattice U, and $\mathcal{M}$ a U-semantics. Then, given an initial valuation $\mathcal{V}$, we say that $m \in \mathcal{M}$ is *a model of* $\mathcal{V}$, in brief $m \models \mathcal{V}$, if $m(\alpha)$ satisfies $\mathcal{V}(\alpha)$ for every formula α. We call *satisfiable* an initial $\mathcal{C}$-valuation admitting a model.

We say that two valuations are *logically equivalent* if they admit the same models.

Given a U-semantics $\mathcal{M}$ and an initial valuation $\mathcal{V}$ it is natural to search for the best improvement of $\mathcal{V}$ which is logically equivalent to $\mathcal{V}$. This leads to define the logical consequence operator $Lc : Val(\mathcal{C}) \to Val(\mathcal{C})$ as follows:

Definition 2.4. Given a fuzzy semantics $\mathcal{M}$, and a constraint frame $\mathcal{C}$, the *logical consequence operator* $Lc : Val(\mathcal{C}) \to Val(\mathcal{C})$ is defined by setting, for every initial valuation $\mathcal{V}$,

$$Lc(\mathcal{V})(\alpha) = <\{m(\alpha) : m \in \mathcal{M}, m \models \mathcal{V}\}>. \tag{2.2}$$

The following theorem, whose proof is trivial, shows that such a definition is the best one.

Theorem 2.5. *The logical consequence operator* $Lc: Val(\mathcal{C}) \to Val(\mathcal{C})$ *is a closure operator. Moreover, for any initial valuation* $\mathcal{V} : \mathcal{F} \to \mathcal{C}$, $Lc(\mathcal{V}) : \mathcal{F} \to \mathcal{C}$ *is the best improvement of* $\mathcal{V}$ *logically equivalent to* $\mathcal{V}$.

We have to look at an initial valuation $\mathcal{V}$ as a system of axioms (the available rough information), and at the derived valuation $Lc(\mathcal{V})$ as the set of its logical consequences (the information we can derive). In particular, $Lc(\mathcal{V}_0)$ is called the *$\mathcal{C}$-set of tautologies*, we denote by *Taut*. Then, $Taut : \mathcal{F} \to \mathcal{C}$ is defined by setting, for every formula α,

$$Taut(\alpha) = <\{m(\alpha) : m \in \mathcal{M}\}>. \tag{2.3}$$

Note. From a formal point of view, Lc is not the logical consequence operator of the U-semantics $\mathcal{M}$. Indeed, $\mathcal{M}$ is not a class of $\mathcal{C}$-valuations. Nevertheless, define $c : U \to \mathcal{C}$ by setting, for every $x \in U$,

$$c(x) = <\{x\}> \tag{2.4}$$

and, given $m \in \mathcal{M}$, set m' equal to $c \circ m$, i.e.,

$$m'(\alpha) = \bigcap\{X \in \mathcal{C} : m(\alpha) \in X\}.$$

Then, m' is the information on m which is possible to represent via the constraint frame $\mathcal{C}$. Moreover, we can set

$$\mathcal{M}' = \{m' \in Val(\mathcal{C}) : m \in \mathcal{M}\}.$$

Then $\mathcal{M}'$ is an abstract semantics in the lattice $Val(\mathcal{C})$ whose logical consequence operator is Lc.

We will give some basic examples of constraint frames.

Lower-bound constraints. Assume that $\mathcal{C}$ is the class of all lower-bound constraints, i.e.,

$$\mathcal{C} = \{[\lambda,1] \in \mathcal{P}(U) : \lambda \in U\}.$$

Such a class is a closure system. In fact, given any family $([\lambda_i, 1])_{i\in I}$ of elements in $\mathcal{C}$, the intersection $\bigcap_{i\in I} [\lambda_i, 1] = [Sup_{i\in I}\, \lambda_i, 1]$ is an element of $\mathcal{C}$. In such a way, in the case $U = [0,1]$, we restate the notions considered in Chapters 2, 3 and 4. Indeed, define $c : U \to \mathcal{C}$ by (2.4), i.e. set $c(x) = [x,1]$ for any $x\in U$. Then c is an isomorphism between U and $\mathcal{C}$ enabling us to identify any U-subset s with the $\mathcal{C}$-subset $\mathcal{V}_v$ defined by setting $\mathcal{V}_v(\alpha) = c(v(\alpha)) = [v(\alpha),1]$ for every formula α. Moreover, $m \in \mathcal{M}$ is a model of $\mathcal{V}_v$ iff m is a model of v in the sense of Chapter 2 and,

$$Lc(\mathcal{V}_v)(\alpha) = [Inf\{m(\alpha) : m \in \mathcal{M}, m \supseteq v\},1] = c(Lc(v)(\alpha)).$$

Interval-constraints. Another interesting class of constraints is obtained by setting $\mathcal{C}$ equal to the set of interval-constraints, i.e.,

$$\mathcal{C}= \{[a, b] : a, b \in U \text{ and } a \leq b\} \cup \{\emptyset\}.$$

Indeed, to prove that $\mathcal{C}$ is a closure system observe that if $([a_i, b_i])_{i\in I}$ is a family of intervals such that $a_i \leq b_i$, then, $\bigcap_{i\in I} [a_i, b_i]$ is the interval $[Sup_{i\in I}\, a_i, Inf_{i\in I}\, b_i]$. In this case $c(x) = \{x\}$ for every formula x and an initial valuation $\mathcal{V}$ is defined by a pair (v_1, v_2) of U-subsets of formulas such that $\mathcal{V}(\alpha) = [v_1(\alpha), v_2(\alpha)]$. In other words, the information carried on by $\mathcal{V}$ is that α is at least true with degree $v_1(\alpha)$ and at most true with degree $v_2(\alpha)$. As we will observe in the next sections, in the case of a balanced semantics it is indifferent to consider either lower-bound constraints or interval-constraints.

Interesting types of $\mathcal{C}$-valuations are the *atoms*. Namely, the atoms in $\mathcal{C}$ (with respect to the improvement relation) coincide with the one-element intervals and therefore, the atoms in the class of plausible valuations are the initial valuations $\mathcal{V}$ such that every $\mathcal{V}(\alpha)$ is a singleton. Given a fuzzy logic, these valuations are either satisfiable with just one model or are not satisfiable, i.e., are either inconsistent or categorical.

It is easy to see that the logical consequence operator is defined by

$$Lc(\mathcal{V})(\alpha) = [J(v_1,v_2)(\alpha)\, , J'(v_1,v_2)(\alpha)] \tag{2.5}$$

where

$$J(v_1,v_2) = \bigcap\{m \in \mathcal{M} : v_1 \subseteq m \subseteq v_2\},$$

and

$$J'(v_1,v_2) = \bigcup\{m \in \mathcal{M} : v_1 \subseteq m \subseteq v_2\}.$$

In a sense, $J(v_1,v_2)(\alpha)$ expresses a necessity and $J'(v_1,v_2)(\alpha)$ a possibility. In accordance,

$$Taut(\alpha) = [Inf\{m(\alpha) : m \in \mathcal{M}\}\ ;\ Sup\{m(\alpha) : m \in \mathcal{M}\}]. \tag{2.6}$$

Segment constraints. It is also possible to consider the frame

$$\mathcal{C}= \{\{\lambda\} : \lambda \in U\} \cup \{\emptyset\} \cup U.$$

In this case, given a plausible valuation $\mathcal{V}$, for each formula α, either we have complete information on its truth value, i.e.,$\mathcal{V}(\alpha) = \{\lambda\}$), or no information, i.e.,

$\mathcal{V}(\alpha) = \{\lambda\}$. This suggests the identification of a plausible valuation $\mathcal{V}$ with the partial function $v : D \to U$, where $D = \{\alpha \in \mathcal{F} : \mathcal{V}(\alpha) \neq U\}$ and $v(\alpha) = \lambda$ provided that $\mathcal{V}(\alpha) = \{\lambda\}$. In accordance, an initial valuation $v_1 : D_1 \to U$ is an improvement of an initial valuation $v_2 : D_2 \to U$ provided that v_1 is an extension of v_2, i.e., $D_1 \supseteq D_2$, and $v_1(\alpha) = v_2(\alpha)$ for every $\alpha \in D_2$. Also, a model m satisfies v if m is an extension of v, or, equivalently, v is a *segment* of m. In this case the $\mathcal{C}$-set of tautologies is defined by setting

$$Taut(\alpha) = \begin{cases} \{\lambda\} & \text{if } m(\alpha) = \lambda \text{ for every } m \in \mathcal{M}, \\ U & \text{otherwise.} \end{cases} \tag{2.7}$$

3. EXAMPLES: ZADEH LOGIC, BOOLEAN LOGIC, PROBABILITY LOGIC

To illustrate the different meaning of the proposed systems of constraints, we consider some simple examples of fuzzy semantics. These are truth-functional semantics (see Chapter 8). We assume that $\mathcal{F}$ is the set of formulas of a propositional calculus.

Zadeh logic. Set $U = [0,1]$ and $\mathcal{M}_Z$ equal to the class of truth-functional valuations m in U such that

$$m(\alpha \wedge \beta) = m(\alpha) \wedge m(\beta),$$
$$m(\alpha \vee \beta) = m(\alpha) \vee m(\beta),$$
$$m(\neg\alpha) = \sim m(\alpha).$$

We call *Zadeh logic* the resulting logic. As we will show in Proposition 6.1 of Chapter 8, in the case of lower bound constraints,

$$Taut(\alpha) = \begin{cases} [1/2, 1] & \text{if } \alpha \text{ is a classical tautology,} \\ [0, 1] & \text{otherwise.} \end{cases}$$

In the case of segment constrains, *Taut* is the map constantly equal to U. This means that the notion of tautology is not defined and therefore that the expressive power of the segment constraint frames is not adequate for such a logic. More useful are the interval-constraints. Indeed, in this case we can prove that

$$Taut(\alpha) = \begin{cases} [1/2, 1] & \text{if } \alpha \text{ is a classical tautology,} \\ [0, 1/2] & \text{if } \alpha \text{ is a contradiction,} \\ [0, 1] & \text{otherwise.} \end{cases}$$

Boolean logic. Crisp classical logics are obtained by assuming that $U = \{0,1\}$ and that $\mathcal{C}$ is the class of lower-bound constraints $\{\{1\},\{0,1\}\}$. In this case an initial valuation $\mathcal{V} : \mathcal{F} \to \mathcal{C}$ is identified with the set

$$T = \{\alpha \in \mathcal{F} : \mathcal{V}(\alpha) = \{1\}\}$$

of formulas. Indeed, in a crisp logic to give a set T of axioms is equivalent to assuming that each formula in T is true in the actual world and that we have no information about the remaining formulas. Such a choice means that we only consider the constraints $\{0,1\}$ (no information) and $\{1\}$ (positive information).

Nevertheless, it should be possible to consider also the constraint $\{0\}$, i.e., to assume that $\mathcal{C}= \{\varnothing, \{0\}, \{1\}, \{0,1\}\}$ obtaining a symmetric logic in which we are able to handle both positive and negative information. In this case, every plausible initial valuation $\mathcal{V}$ should be identified with the pair (T_1, T_2) such that,

$$T_1 = \{\alpha \in \mathcal{F} : \mathcal{V}(\alpha) = \{1\}\} \quad ; \quad T_2 = \{\alpha \in \mathcal{F} : \mathcal{V}(\alpha) = \{0,1\}\}.$$

So, $\mathcal{V}$ claims that the formulas in T_1 are true, the formulas in T_2 are possible and the remaining formulas are false.

A more interesting example can be achieved by assuming that $U = (\boldsymbol{B}, \vee, \wedge, -. \mathbf{0}, \mathbf{1})$ is any complete Boolean algebra (see. e.g. Rasiowa [1974], Section 2 in Chapter 13 and Tarski [1956]). Let $\mathcal{F}$ be the class of formulas of classical propositional logic and $\mathcal{M}$ the class of all *Boolean models* i.e., the truth-functional valuations $m : \mathcal{F} \to U$ such that

$$m(\alpha \wedge \beta) = m(\alpha) \wedge m(\beta) \quad ; \quad m(\alpha \vee \beta) = m(\alpha) \vee m(\beta) \quad ; \quad m(\neg\alpha) = -m(\alpha).$$

If $m \in \mathcal{M}$, then $m(\alpha) = 0$ for any contradiction α and $m(\alpha) = 1$ for every tautology α. Moreover, if α is logically equivalent to β, then $m(\alpha) = m(\beta)$. For instance, we can imagine that U is the Boolean algebra $\mathcal{P}(S)$ where S is the set of past cases we have stored and that, given $m \in \mathcal{M}$ and $\alpha \in \mathcal{F}$, $m(\alpha)$ is the set of past cases in which α was satisfied. By considering lower bound constraints, the meaning of an initial valuation $v : \mathcal{F} \to U$ is that, given a formula α, α is true at least in all the cases that are in the set $v(\alpha)$. In accordance, m is a model of v iff $m(\alpha) \supseteq v(\alpha)$ for every formula α. Furthermore,

$$Lc(v)(\alpha) = \bigcap\{m(\alpha) : m \geq v\},$$

and therefore,

$$Taut(\alpha) = \bigcap\{m(\alpha) : m \text{ is a Boolean model}\}.$$

It is obvious that

$$Taut(\alpha) = \begin{cases} S & \text{if } \alpha \text{ is a classical tautology,} \\ \varnothing & \text{otherwise.} \end{cases}$$

Observe that the theories of a Boolean logic are not necessarily Boolean valuations.

Assume that $\mathcal{C}$ is the class of interval-constraints in U. Then an initial $\mathcal{C}$-valuation $\mathcal{V}$ is defined by a pair (v_1,v_2) of Boolean functions such that $v_1 \leq v_2$ and its meaning is that, given any formula α, α is at least true in all the cases in $v_1(\alpha)$ and at most true in the cases that are in $v_2(\alpha)$. In other words, α is surely true in all the cases in $v_1(\alpha)$ and surely false in all the cases in the complement of $v_2(\alpha)$. Moreover,

$$Lc(\mathcal{V})(\alpha) = [J(v_1,v_2)(\alpha)\, , J'(v_1,v_2)(\alpha)]$$

where

$$J(v_1,v_2)(\alpha) = \bigcap\{m(\alpha) : v_2 \geq m \geq v_1\}, \quad J'(v_1,v_2)(\alpha) = \bigcup\{m(\alpha) : v_2 \geq m \geq v_1\}.$$

Finally, a direct inspection shows that

$$Taut(\alpha) = \begin{cases} [S,S] & \text{if } \alpha \text{ is a tautology,} \\ [\varnothing,\varnothing] & \text{if } \alpha \text{ is a contradiction,} \\ \mathcal{P}(S) & \text{otherwise.} \end{cases}$$

This means that if α is a tautology of classical propositional calculus, then α is satisfied in any case. If α is a contradiction, then there is no case in which α is satisfied.

Probability logic. Probability logic will be considered in Chapter 9. The semantics of such a logic is the class $\mathcal{M}$ of all valuations $m : \mathbb{F} \to [0,1]$ such that

$m(\alpha) = 1$ for every classical tautology α,

α logically equivalent to $\beta \Rightarrow m(\alpha) = m(\beta)$,

$m(\alpha \vee \beta) = m(\alpha) + m(\beta)$ provided that α is inconsistent with β.

Then in both the cases of interval-constraints and segment constrains

$$Taut(\alpha) = \begin{cases} [1,1] & \text{if } \alpha \text{ is a classical tautology,} \\ [0,0] & \text{if } \alpha \text{ is a contradiction,} \\ [0,1] & \text{otherwise.} \end{cases}$$

In the case of lower bound constraints,

$$Taut(\alpha) = \begin{cases} 1 & \text{if } \alpha \text{ is a classical tautology,} \\ 0 & \text{otherwise.} \end{cases}$$

4. HILBERT SYSTEMS FOR CONSTRAINTS

A further step is to define a deduction apparatus able to manage the constraints. To this aim, we define a $\mathbb{C}$-inference rule as a pair $r = (r',r'')$ such that r' is a partial n-ary operation in $\mathbb{F}$, and r'' is an n-ary operation in $\mathbb{C}$. The meaning is that:

IF the truth values of the formulas $\alpha_1,\ldots,\alpha_n$ satisfy the constraints $X_1,\ldots,X_n$,

THEN you can conclude that the truth value of $r'(\alpha_1,\ldots,\alpha_n)$ satisfies $r''(X_1,\ldots,X_n)$.

We assume that r'' satisfies the following continuity condition:

$$r''(X_1,\ldots, lim\, \mathcal{T},\ldots, X_n) = lim\ \{r''(X_1,\ldots, X,\ldots, X_n) : X \in \mathcal{T}\} \tag{4.1}$$

for every directed class $\mathcal{T}$ of constraints. Since we refer to the lattice $\mathbb{C}$, a class $\mathcal{T}$ of constraints is directed provided that

$$X \in \mathcal{T},\ Y \in \mathcal{T} \Rightarrow \text{ there is } Z \in \mathcal{T} \text{ such that } Z \subseteq X \text{ and } Z \subseteq Y.$$

In other words, two constraints in $\mathcal{T}$ are always compatible. Further, since the inductive limits in $\mathbb{C}$ are the intersections, (4.1) is equivalent to

$$r''(X_1,\ldots,\bigcap\mathcal{T},\ldots,X_n) = \bigcap\{r''(X_1,\ldots,X,\ldots,X_n) : X \in \mathcal{T}\}. \tag{4.2}$$

A *$\mathbb{C}$-Hilbert system* is given by a set $\mathbb{R}$ of $\mathbb{C}$-inference rules and a fixed $\mathbb{C}$-set of formulas $\mathcal{A}$, we call *logical axioms system*. Moreover, we assume that in $\mathbb{R}$ there is the following *concatenation rule* $r_o = (r_o', r_o'')$ defined by setting

$$r_o'(\alpha,\alpha) = \alpha \quad ; \quad r_o''(X,Y) = X \cap Y.$$

The definitions of *proof* and of *valuation* $Val(\pi,\mathcal{V})$ *of a proof* π *with respect to* $\mathcal{V}$ run as in Chapter 4. In particular, a deduction operator $\mathcal{D}$ is defined by setting

$$\mathcal{D}(\mathcal{V})(\alpha) = \bigcap\{Val(\pi,\mathcal{V}) : \pi \text{ is a proof of } \alpha\}. \quad (4.3)$$

It is possible to prove that $\mathcal{D}$ is a closure operator in the lattice $Val(\mathcal{C})$ and therefore that any $\mathcal{C}$-Hilbert system is associated with an abstract deduction system in the lattice $Val(\mathcal{C})$. Observe that, in the case of lower bound constraints, (4.3) gives a justification of the operator *Sup* in Definition 2.3 of Chapter 4.

The definitions and the results proved in the previous sections for the fuzzy *H*-systems are easily extended to the $\mathcal{C}$-Hilbert systems. We only emphasize that the concatenation rule enables us to demonstrate the following proposition:

Proposition 4.1. *Given two proofs* π' *and* π'' *of the same formula* α, *we can concatenate them by obtaining a proof* π *of* α *such that*

$$Val(\pi,\mathcal{V}) = Val(\pi',\mathcal{V}) \cap Val(\pi'',\mathcal{V}).$$

Consequently, the set of valuations of the proofs of a given formula is directed.

To show the role of such a property, consider the claim that $\mathcal{D}(\mathcal{V})$ is closed under any inference rule r, i.e.,

$$\mathcal{D}(\mathcal{V})(r'(\alpha_1,\ldots,\alpha_n)) \subseteq r''(\mathcal{D}(\mathcal{V})(\alpha_1),\ldots,\mathcal{D}(\mathcal{V})(\alpha_n)) \quad (4.4)$$

for every $\alpha_1,\ldots,\alpha_n$ in $Dom(r)$. Its proof runs as the corresponding proof for fuzzy Hilbert systems. The only difference is that the necessary sequence of statements

$$r''(\mathcal{D}(\mathcal{V})(\alpha_1),\ldots,\mathcal{D}(\mathcal{V})(\alpha_n))$$
$$= r''(\bigcap\{Val(\pi_1,\mathcal{V}) : \pi_1 \text{ is a proof of } \alpha_1\},\ldots,\bigcap\{Val(\pi_n,\mathcal{V}) : \pi_n \text{ is a proof of } \alpha_n\})$$
$$= \bigcap\{r''(Val(\pi_1,\mathcal{V}),\ldots,Val(\pi_n,\mathcal{V})) : \pi_i \text{ is a proof of } \alpha_i\}$$
$$\supseteq \bigcap\{Val(\pi,\mathcal{V}) : \pi \text{ is a proof of } r'(\alpha_1,\ldots,\alpha_n)\} = \mathcal{D}(\mathcal{V})(r'(\alpha_1,\ldots,\alpha_n)),$$

can be justified only by taking (4.2) into account and by the fact that each set $\{Val(\pi_i,\mathcal{V}) : \pi_i \text{ is a proof of } \alpha_i\}$ is directed.

We conclude this section by defining a $\mathcal{C}$-*Hilbert logic* as a pair $(\mathcal{S}, \mathcal{M})$ where $\mathcal{S}$ is a $\mathcal{C}$-Hilbert system and $\mathcal{M}$ a *U*-semantics, such that the deduction operator associated with $\mathcal{S}$ and the logical consequence operator defined by (2.2) coincide.

5. FUZZY LOGICS WITH A NEGATION

In the literature on approximate reasoning only lower bound constraints are considered. This is not restrictive in case the language under consideration is equipped with a suitable *"negation"* as a logical connective. We say that a fuzzy semantics $\mathcal{M}$ is *with a negation* if a logical connective $\neg : \mathbb{F} \to \mathbb{F}$ exists such that, for every $m \in \mathcal{M}$ and $\alpha \in \mathbb{F}$, $m(\alpha) \leq 1 - m(\neg\alpha)$, i.e.,

$$m(\alpha) + m(\neg\alpha) \leq 1.$$

Such a condition is a multi-valued translation of the classical condition

$$m \models \alpha \Rightarrow m \not\models \neg\, \alpha.$$

We say that an abstract fuzzy logic is *with a negation* if its semantics is with a negation.

Proposition 5.1. *Let $\mathcal{M}$ be a fuzzy semantics with a negation. Then, for every formula α and $m \in \mathcal{M}$,*

$$m(\alpha) \wedge m(\neg\alpha) \leq 0.5$$

and therefore,

$$Inc(\mathcal{M}) \leq 0.5.$$

Proof. Suppose $m(\alpha) > 0.5$. Then $m(\neg\alpha) \leq 1 - m(\alpha) < 1 - 0.5 = 0.5$. □

Observe that from Proposition 5.1 it follows that, if v is satisfiable, then

$$Lc(v)(\alpha) \wedge Lc(v)(\neg\alpha) \leq 0.5.$$

The following proposition shows that in a fuzzy logic with a negation each (lower-bound) initial valuation determines an equivalent interval-constraint valuation. Here, $v^{\perp}$ denotes the fuzzy subset defined by the equation

$$v^{\perp}(x) = 1 - v(\neg x).$$

Proposition 5.2. *Let v be a fuzzy set of formulas. Then the following are equivalent:*

(i) *m is a model of v.*
(ii) *m satisfies the interval valuation $(v, v^{\perp})$.*
(iii) *m satisfies the interval valuation $(Lc(v), Lc(v)^{\perp})$.*

Proof. Let m be a model of v. Then $m(\alpha) \geq v(\alpha)$ and $m(\neg\alpha) \geq v(\neg\alpha)$ for every formula α. Consequently, $m(\alpha) \leq 1 - m(\neg\alpha) \leq 1 - v(\neg\alpha) = v^{\perp}(\alpha)$ and this proves that $m(\alpha) \in [v(\alpha), v^{\perp}(\alpha)]$. The rest is evident. □

We may give a natural interpretation of $Lc(v)$ and $Lc(v)^{\perp}$. In fact, while the number $Lc(v)(\alpha)$ is the truth degree of the claim "*α is a consequence of v*", $Lc(v)^{\perp}(\alpha)$ is the truth degree of "*$\neg\alpha$ is not a consequence of v*". Equivalently, $Lc(v)^{\perp}(\alpha)$ is the degree of consistency of α with v. So, while $Lc(v)$ is the fuzzy set of formulas which are consequences of the fuzzy information v, $Lc(v)^{\perp}$ is the fuzzy set of formulas consistent with v. This is in accordance with the fact that $Lc(v) \subseteq m \subseteq Lc(v)^{\perp}$.

Definition 5.3. We say that a formula α is *decidable* in v if

$$Lc(v)(\alpha) + Lc(v)(\neg\alpha) = 1.$$

An initial valuation v is *complete* if every formula is decidable in v.

If α is decidable in v, then for every model m of v, $m(\alpha) = Lc(v)(\alpha)$. As a consequence, if v is complete, then $Lc(v) = Lc(v)^{\perp}$ and $Lc(v)$ coincides with the unique model of v. As we will see in the next sections, while every complete theory is maximal, the converse does not hold.

Definition 5.4. A fuzzy semantics is *balanced* if all the models are complete, i.e.,

$$m(\alpha) + m(\neg\alpha) = 1$$

for every $m \in \mathcal{M}$ and α formula. A fuzzy logic is balanced if its semantics is balanced. An initial valuation v is *balanced* if $v(\neg(\neg(\alpha))) = v(\alpha)$.

Observe that for a balanced semantics we have the equality

$$Contr(\mathcal{M})(\alpha) = Taut(\mathcal{M})(\neg\alpha),$$

a multi-valued translation of the classical equivalence

$$\alpha \text{ is a contradiction} \Leftrightarrow \neg\alpha \text{ is a tautology.}$$

In fact,

$$Contr(\mathcal{M})(\alpha) = \bigcap\{-m(\alpha) : m \in \mathcal{M}\} = \bigcap\{m(\neg\alpha) : m \in \mathcal{M}\} = Taut(\mathcal{M})(\neg\alpha).$$

If a semantics is balanced it is rather natural to confine ourselves only to balanced valuations.

Theorem 5.5. *Assume that $\mathcal{M}$ is balanced. Then the expressive power of the initial valuations (i.e., of the lower bound constraints) is equal to the expressive power of the valuations by interval-constraints. Namely, given an initial valuation v*

$$m \text{ is a model of } v \Leftrightarrow m \text{ satisfies } (v, v^{\perp}).$$

Given an interval-constraint (v_1, v_2) such that both v_1 and v_2 are balanced

$$m \text{ satisfies } (v_1, v_2) \Leftrightarrow m \text{ is a model of } v_1 \cup v_2^{\perp}.$$

Proof. The first equivalence was given in Proposition 5.2. Assume that m satisfies (v_1, v_2). Then, in particular, m is a model of v_1. Moreover, since $m(\alpha) \leq v_2(\alpha)$, it is $m(\alpha) = 1 - m(\neg\alpha) \geq 1 - v_2(\neg\alpha) = v_2^{\perp}(\alpha)$. Thus m is a model of $v_1 \cup v_2^{\perp}$. Conversely, let m be a model of $v_1 \cup v_2^{\perp}$. Then, for every formula α, $m(\alpha) \geq v_1(\alpha)$. Since $m(\neg\alpha) \geq v_2^{\perp}(\neg\alpha) = 1 - v_2(\neg(\neg\alpha))$, we have that $m(\alpha) = 1 - m(\neg\alpha) \leq v_2(\neg(\neg\alpha)) = v_2(\alpha)$. This proves that m satisfies (v_1, v_2). □

Proposition 5.6. *Assume that $\mathcal{M}$ is balanced and closed under ultraproducts and let v be a satisfiable initial valuation. Then, for every formula α, two models m, $m' \in \mathcal{M}$ of v exist such that*

$$Lc(v)(\alpha) = m(\alpha) \text{ and } Lc(v)^{\perp}(\alpha) = m'(\alpha).$$

Proof. Equality $Lc(v)(\alpha) = m(\alpha)$ is a direct consequence of Theorem 8.1 in Chapter 2. Let m' be a model of v such that $m'(\neg\alpha) = Lc(v)(\neg\alpha)$. Then

$$m'(\alpha) = 1 - m'(\neg\alpha) = 1\text{-}Lc(v)(\neg\alpha) = Lc(v)^{\perp}(\alpha).$$ □

Recall that in a classical logic the extension T_α of of a set T of formulas via a formula α is defined by setting

$$T_\alpha = \begin{cases} T & \text{if } \alpha \text{ is inconsistent with } T, \\ T \cup \{\alpha\} & \text{otherwise.} \end{cases}$$

Trivially, if T is consistent then T_α is consistent. Furthermore, α is decidable in T_α. This definition gives a useful tool to complete any consistent theory. In order to extend this notion to fuzzy logics, given a fuzzy set v of formulas, we call *extension of v by α* the fuzzy set v_α coinciding with v in the case $1 - v(\neg\alpha) < v(\alpha)$. Otherwise v_α is defined by setting, for every formula x,

$$v_\alpha(x) = \begin{cases} v(x) & \text{if } x \neq \alpha, \\ 1 - v(\neg\alpha) & \text{if } x = \alpha. \end{cases}$$

Proposition 5.7. *Assume that $\mathcal{M}$ is a balanced fuzzy semantics closed under ultraproducts and let τ be a satisfiable theory of $\mathcal{M}$. Then for every formula α, the extension τ_α is satisfiable and α is decidable in τ_α, i.e.,*

$$\tau_\alpha(\alpha) + \tau_\alpha(\neg\alpha) = 1. \tag{5.1}$$

Proof. If τ is satisfiable, then a model $m \in \mathcal{M}$ exists such that

$$1 - \tau(\neg\alpha) \geq 1 - m(\neg\alpha) \geq m(\alpha) \geq \tau(\alpha).$$

Then $\tau_\alpha(\alpha) = 1 - \tau(\neg\alpha)$ while $\tau_\alpha(\neg\alpha) = \tau(\neg\alpha)$. This proves (5.1). In order to prove τ_α is satisfiable, recall that by Proposition 5.6 a model m' of τ exists such that $m'(\alpha) = \tau^\perp(\alpha) = \tau_\alpha(\alpha)$. Then, m' is also a model of τ_α. □

Theorem 5.8. *Assume that $\mathcal{M}$ is balanced and closed under ultraproducts, let τ be a satisfiable theory, and let $(\alpha(n))_{n\in N}$ be an enumeration of all the formulas in F. Then by setting*

$$\tau_0 = \tau \;;\; \tau_{n+1} = Lc((\tau_n)_{\alpha(n+1)}) \;;\; m = \bigcup_{n\in N} \tau_n,$$

we get a satisfiable complete theory extending τ (i.e., a model of τ).

Proof. By Proposition 5.7 every τ_n is satisfiable, so, since Theorem 8.1 in Chapter 2 proves that $\mathcal{M}$ is logically compact, m is satisfiable. It is obvious that every formula is decidable in m. □

6. REFUTATION PROCEDURES.

Recall that if T is a system of axioms in a classical logic and α a formula, then the following are equivalent:

(i) $T \models \alpha$.

(ii) $T \cup \{\neg\alpha\}$ is not satisfiable.

This equivalence gives rise to deduction techniques called *refutation procedures*. In this section we will try to extend such techniques to fuzzy logic. Namely, given a formula α and an initial valuation v, we give a notion of degree of inconsistency of α with v in such a way that the following are equivalent:

(j) α is a consequence of v to degree λ.

(jj) The degree of inconsistency of $\neg\alpha$ with v is λ.

In the following we denote by $\{\alpha\}^\lambda$ the singleton assuming in α the value λ.

Definition 6.1. Let v be an initial valuation and α a formula. Then the *degree of consistency of α with v* is the number

$$Cons(v,\alpha) = Sup\{\lambda \in U : v \cup \{\alpha\}^\lambda \text{ is satisfiable}\}. \tag{6.1}$$

The *degree of inconsistency of* α *with* v is

$$Inc(v,\alpha) = 1 - Cons(v,\alpha).$$

Trivially,

$$Inc(v,\alpha) = Inf\{\lambda \in U : v \cup \{\alpha\}^{1-\lambda} \text{ is satisfiable}\}. \tag{6.2}$$

and

$$v \text{ is not satisfiable} \Leftrightarrow Inc(v,\alpha) = 1 \text{ for any } \alpha \in F$$
$$\Leftrightarrow Cons(v, \alpha) = 0 \text{ for any } \alpha \in F.$$

Proposition 6.2. *The following equalities hold:*

$$Cons(v,\alpha) = Inf\{\lambda \in U : v \cup \{\alpha\}^{\lambda} \text{ is not satisfiable}\}. \tag{6.3}$$
$$Inc(v,\alpha) = Sup\{\lambda \in U : v \cup \{\alpha\}^{1-\lambda} \text{ is not satisfiable}\}. \tag{6.4}$$

Proof. The set

$$\{\lambda \in U : v \cup \{\alpha\}^{\lambda} \text{ is satisfiable}\}$$

is an interval. Indeed, if $\mu < \lambda$ and $v \cup \{\alpha\}^{\lambda}$ is satisfiable, then $v \cup \{\alpha\}^{\mu}$ is satisfiable. Consequently

$$Sup\{\lambda \in U : v \cup \{\alpha\}^{\lambda} \text{ is satisfiable}\} = Inf\{\lambda \in U : v \cup \{\alpha\}^{\lambda} \text{ is not satisfiable}\}.$$

We can demonstrate (6.4) in a similar way. □

The next proposition shows that, in a sense, any balanced semantics admits a refutation procedure.

Theorem 6.3. *Let* $\mathcal{M}$ *be a balanced semantics. Then*

$$Lc(v)(\alpha) = Inc(v, \neg\alpha) \tag{6.5}$$

and

$$Lc(v)^{\perp}(\alpha) = Cons(v, \alpha). \tag{6.6}$$

Proof. Suppose $\lambda < Lc(v)(\alpha)$. Then, for every model m of v we have that $m(\alpha) > \lambda$ and $m(\neg\alpha) = 1 - m(\alpha) < 1 - \lambda$. This means that $v \cup \{\neg\alpha\}^{1-\lambda}$ is not satisfiable and hence that

$$Lc(v)(\alpha) = Sup\{\lambda \in U : \lambda < Lc(v)(\alpha)\}$$
$$\leq Sup\{\lambda \in U : v \cup \{\neg\alpha\}^{1-\lambda} \text{ is not satisfiable}\} = Inc(v,\neg\alpha).$$

Assume that $v \cup \{\neg\alpha\}^{1-\lambda}$ is not satisfiable. Then, for every model m of v, $m(\neg\alpha) < 1-\lambda$ and $m(\alpha) = 1 - m(\neg\alpha) > \lambda$. Hence $Lc(v)(\alpha) \geq \lambda$ and this demonstrates that $Lc(v)(\alpha) \geq Inc(v,\neg\alpha)$. (6.6) is a direct consequence of (6.5). □

From (6.5) and (6.6) it follows that if v is a fuzzy set representing the available information about an unknown world m, then the better interval-constraint we may get for the actual truth degree $m(\alpha)$ is furnished by the degree of inconsistency of $\neg\alpha$ and the degree of consistency of α with v. Of course, the basic problem is to find effective methods to compute the fuzzy relations *Inc* and *Cons*.

Theorem 6.4. *Let* $\mathcal{M}$ *be a balanced fuzzy semantics and Lc the associated logical consequence operator. Then,*

$$\mathcal{M} \text{ logically compact} \Leftrightarrow Lc \text{ is continuous}.$$

Proof. Assume that $\mathcal{M}$ is logically compact and let v be the inductive limit of a directed family $(v_i)_{i\in I}$. Then, since $v \cup \{\neg\alpha\}^{1-\lambda}$ is the inductive limit of the directed family $(v_i \cup \{\neg\alpha\}^{1-\lambda})_{i\in I}$ and $Sat(\mathcal{M})$ is inductive, we get that

$$v \cup \{\neg\alpha\}^{1-\lambda} \text{ is satisfiable } \Leftrightarrow v_i \cup \{\neg\alpha\}^{1-\lambda} \text{ is satisfiable for every } i \in I.$$

As a consequence,

$$\begin{aligned} Lc(v)(\alpha) &= Inc(v, \neg\alpha) = Sup\{\lambda \in U : v \cup \{\neg\alpha\}^{1-\lambda} \text{ is not satisfiable}\} \\ &= Sup\{\lambda \in U : v_i \cup \{\neg\alpha\}^{1-\lambda} \text{ is not satisfiable for a suitable } i \in I\} \\ &= Sup_{i\in I}(Sup\{\lambda \in U : v_i \cup \{\neg\alpha\}^{1-\lambda} \text{ is not satisfiable}\}) \\ &= Sup_{i\in I}\,(Inc(v_i, \neg\alpha) : i \in I) = Sup_{i\in I}\, Lc(v_i)(\alpha). \end{aligned}$$

Conversely, assume that Lc is continuous and let $(v_i)_{i\in I}$ be a directed family of satisfiable initial valuations. We prove that the limit $\bigcup_{i\in I} v_i$ is satisfiable. To do this, it is sufficient to prove that $Lc(\bigcup_{i\in I} v_i)$ is different than the map constantly equal to 1. Now, let α be any formula and assume that

$$Lc(\textstyle\bigcup_{i\in I} v_i)(\alpha) = Lc(\bigcup_{i\in I} v_i)(\neg\alpha) = 1.$$

Now, the continuity of Lc entails that $Lc(\bigcup_{i\in I} v_i) = \bigcup_{i\in I} Lc(v_i)$. So, we also have:

$$Sup_{i\in I}\, Lc(v_i)(\alpha) = Sup_{i\in I}\, Lc(v_i)(\neg\alpha) = 1.$$

From this it follows that i and $j \in I$ exist such that

$$Lc(v_i)(\alpha) \geq 2/3 \text{ and } Lc(v_j)(\neg\alpha) \geq 2/3.$$

Let $h \in I$ such that $v_i \subseteq v_h$ and $v_j \subseteq v_h$. Then $Lc(v_i) \subseteq Lc(v_h)$ and $Lc(v_j) \subseteq Lc(v_h)$ and hence $Lc(v_h)(\alpha) \geq 2/3$ and $Lc(v_h)(\neg\alpha) \geq 2/3$. Since v_h is satisfiable, this entails that a balanced model m exists such that

$$m(\alpha) \geq Lc(v_h)(\alpha) \geq 2/3 \text{ and } m(\neg\alpha) \geq Lc(v_h)(\neg\alpha) \geq 2/3,$$

which is absurd. □

7. EQUIVALENCE TO A CRISP LOGIC

In this section we will show that it is possible to extend the representation theorems in Section 11 of Chapter 2 to the logic of constraints proposed in this chapter. The first step is to extend the notion of a signed formula.

Definition 7.1. Given a constraint frame $\mathcal{C}$, a *signed formula* is a pair (α, X) where α is a formula and $X \in \mathcal{C}$ a constraint. We denote the set $\mathbb{F} \times \mathcal{C}$ of signed formulas by $\mathbb{F}_s$.

The information carried on by a signed formula (α, X) is that the truth value of α satisfies X. We will show that it is possible to represent any constraint-based fuzzy logic by a crisp logic. To show this, we define the operator $K : \mathcal{P}(\mathbb{F}_s) \to Val(\mathcal{C})$ associating any set T of signed formulas with the $\mathcal{C}$-subset $K(T) : \mathbb{F} \to \mathcal{C}$ defined by setting

$$K(T)(x) = \bigcap\{X \in \mathcal{C} : (x, X) \in T\}, \tag{7.1}$$

for any $x \in \mathbb{F}$. Moreover, we define the operator $E : Val(\mathcal{C}) \to \mathcal{P}(\mathbb{F}_s)$ by setting

$$E(\mathcal{V}) = \{(\alpha, X) \in \mathbb{F}_s : \mathcal{V}(\alpha) \subseteq X\}, \tag{7.2}$$

for any $\mathcal{C}$-valuation $\mathcal{V}$. These operators enable us to transform any deduction operator in the lattice $Val(\mathcal{C})$ into a crisp deduction operator in $\mathbb{F}_s$. Indeed, given a deduction operator $\mathcal{D}$ in the lattice $Val(\mathcal{C})$, we can consider the operator $\mathcal{D}_s$ in $\mathbb{F}_s$ defined by setting, for every $T \subseteq \mathbb{F}_s$,

$$\mathcal{D}_s(T) = E(\mathcal{D}(K(T))). \tag{7.3}$$

Successively, the formula

$$\mathcal{D}(\mathcal{V}) = E^{-1}(\mathcal{D}_s(E(\mathcal{V}))) \tag{7.4}$$

enables us to get $\mathcal{D}$ from $\mathcal{D}_s$. By following this line, we can restate all the results in Section 11 of Chapter 2. As an example, we can demonstrate the following representation theorem:

Theorem 7.2. *Let $\mathcal{C}$ be a constraint frame and $(Val(\mathcal{C}), \mathcal{D})$ an abstract logic. Then $\mathcal{D}_s : \mathcal{P}(\mathbb{F}_s) \rightarrow \mathcal{P}(\mathbb{F}_s)$ is a closure operator and therefore, $(\mathbb{F}_s, \mathcal{D}_s)$ is a crisp deduction system "equivalent" to $(Val(\mathcal{C}), \mathcal{D})$.*

We can also associate any fuzzy semantics $\mathcal{M}$ with the crisp semantics $\mathcal{M}_s = E(\mathcal{M})$ (see Section 12 of Chapter 2). This means that we identify any model m with the class of signed formulas it satisfies. Then, we can associate any abstract fuzzy logic $(\mathbb{F}, \mathcal{D}, \mathcal{M})$ with the equivalent abstract crisp logic $(\mathbb{F}_s, \mathcal{D}_s, \mathcal{M}_s)$.

There are several different ways to obtain the operator Lc_s. As an example, we say that $m \in \mathcal{M}$ *satisfies* a signed formula (α,X), in brief $m \models (\alpha, X)$, if $m(\alpha) \in X$. We say that m is a *model of* a set T of signed formulas, in brief $m \models T$, provided that $m \models (\alpha, X)$ for every $(\alpha,X) \in T$. We write $T \models (\alpha, X)$ if every model of T satisfies (α, X). Then:

$$Lc_s(T) = \{(\alpha, X) \in \mathbb{F}_s : T \models (\alpha, X)\}.$$

Equivalently, define the maps

$$mod : \mathcal{P}(\mathbb{F}_s) \rightarrow \mathcal{P}(\mathcal{M}) \quad \text{and} \quad th : \mathcal{P}(\mathcal{M}) \rightarrow \mathcal{P}(\mathbb{F}_s)$$

where, given $T \in \mathcal{P}(\mathbb{F}_s)$

$$mod(T) = \{m \in \mathcal{M} : m \models T\},$$

and, given $C \subseteq \mathcal{M}$,

$$th(C) = \{(\alpha, X) \in \mathbb{F}_s : m \models (\alpha, X) \text{ for every } m \in C\}.$$

Then, *mod* and *th* define a Galois connection such that

$$Lc = mod \circ th.$$

We say that $Lc_s(\emptyset)$ is the set of tautologies. Then a signed formula (α,X) is a *tautology* if it is satisfied by any model $m \in \mathcal{M}$. Given any formula α, the signed formula (α,U) is an example of tautology. We say that (α, X) is a *contradiction* if no model $m \in \mathcal{M}$ satisfies (α, X). Given any formula α, the signed formula $(\alpha, \emptyset)$ is an example of contradiction.

Finally, we can transform any fuzzy inference rule into a crisp rule. It is sufficient to write

$$\frac{(\alpha_1,X_1), \ldots ,(\alpha_n,X_n)}{(r'(\alpha_1,\ldots,\alpha_n), r''(X_1,\ldots,X_n))}$$

instead of

$$\frac{\alpha_1,\ldots,\alpha_n}{r'(\alpha_1,\ldots,\alpha_n)} \quad ; \quad \frac{X_1,\ldots,X_n}{r''(X_1,\ldots,X_n)} .$$

In accordance, we can transform any fuzzy Hilbert system based on constraints into a suitable crisp Hilbert system on signed formulas.

8. TABLEAUX METHOD

The notion of a formula signed by a constraint suggests the search for semantic tableau systems (see for example Hähnle [1993]). We confine ourselves only to sketching some ideas and results.

We start from a propositional language, a related truth-functional semantics $\mathcal{M}$ (see Chapter 8) and a constraint frame $\mathbb{C}$. As usual, we denote the class of all signed formulas by $\mathbb{F}_s$. Moreover, to translate the signed formulas in $\mathbb{F}_s$ into formulas of classical logic, we consider a classical propositional calculus whose set of propositional variables is $\mathbb{F}_s$. We call a *Boolean combination* of signed formulas any formula of such a calculus. An example is the formula $(\alpha,X) \wedge (\neg(\beta,Y) \vee (\alpha,Z))$. We say that a Boolean combination is *positive* if the negation $\neg$ does not occur in it. Any model m defines a (classical) valuation of the signed formulas and therefore of the Boolean combinations. In fact, we assign to any signed formula (α,X) the truth value 1 if $m(\alpha) \in X$, we assign the truth value 0 if $m(\alpha) \notin X$. We say that two Boolean combinations are *equivalent* if they assume the same truth value for any valuation defined by an element m in $\mathcal{M}$. As an example,

$(\alpha, X) \wedge (\alpha, Y)$ is equivalent to $(\alpha, X \cap Y)$,

$(\alpha, X) \vee (\alpha, Y)$ is equivalent to $(\alpha, X \cup Y)$,

and so on. Further equivalences follow from the structure of $\mathcal{M}$.

Definition 8.1. We say that a truth-functional semantics $\mathcal{M}$ is *tableaux-manageable* provided that, given any n-ary logical connective h, a signed formula like $(h(\alpha_1,\ldots,\alpha_n), X)$ is equivalent to any positive Boolean combination $q((\alpha_1,X_1), \ldots ,(\alpha_n,X_n))$ of signed formulas like $(\alpha_1,X_1), \ldots ,(\alpha_n,X_n)$.

Notice that if $\mathbb{C}$ is closed with respect to the complement, then any Boolean combination can be reduced to a positive combination. Indeed, in this case $\neg(\alpha,X)$ is equivalent to $(\alpha, -X)$. We assume that the Boolean combinations are in a disjunctive normal form. More precisely, in the case of positive combination, it is not restrictive to assume that $q((\alpha_1,X_1), \ldots ,(\alpha_n,X_n))$ is a Boolean combination $A_1 \vee \ldots \vee A_t$, where each A_i has the form $(\alpha_1, X_1^i) \wedge \ldots \wedge (\alpha_n, X_n^i)$ in which all the formulas $\alpha_1, \ldots ,\alpha_n$ occur. Indeed, if α_j does not occur in A_i, we can add the tautology (α_j,U) by obtaining an equivalent combination.

A simple geometrical characterization of the tableaux-manageable semantics is the following: Given a constraint frame $\mathbb{C}$, we call *n-rectangle* any Cartesian product of n elements in $\mathbb{C}$. Moreover, if h is a logic connective, then we denote its interpretation in U by $\underline{h}$.

Theorem 8.2. *A truth-functional semantics $\mathcal{M}$ is tableaux-manageable iff given any n-ary logical connective h, and $X \in \mathcal{C}$, $\underline{h}^{-1}(X)$ is a finite union of n-rectangles.*

Proof. Assume that $\mathcal{M}$ is tableaux-manageable. Then, with the notations we have just exposed,

$$\underline{h}^{-1}(X) = (X_1^{\ 1} \times \ldots \times X_n^{\ 1}) \cup \ldots \cup (X_1^{\ \prime} \times \ldots \times X_n^{\ \prime})$$

Conversely, let $(h(\alpha_1,\ldots,\alpha_n), X)$ be any signed formula and assume that $\underline{h}^{-1}(X)$ is equal to $(X_1^{\ 1} \times \ldots \times X_n^{\ 1}) \cup \ldots \cup (X_1^{\ \prime} \times \ldots \times X_n^{\ \prime})$ where the $X_i^{\prime}$ are suitable elements in $\mathcal{C}$. Then, for every model m,

$(h(\alpha_1,\ldots,\alpha_n),X)$ satisfied by $m \Leftrightarrow m((h(\alpha_1,\ldots,\alpha_n)) \in X$

$\Leftrightarrow \underline{h}(m(\alpha_1),\ldots,m(\alpha_n)) \in X \Leftrightarrow (m(\alpha_1),\ldots,m(\alpha_n)) \in \underline{h}^{-1}(X)$.

This last condition is equivalent to saying that m satisfies the positive Boolean combination

$$((\alpha_1, X_1^{\ 1}) \wedge \ldots \wedge (\alpha_n, X_n^{\ 1})) \vee \ldots \vee ((\alpha_1, X_1^{\prime}) \wedge \ldots \wedge (\alpha_n, X_n^{\prime})). \quad \square$$

The following proposition shows some interesting examples:

Proposition 8.3. *If U is finite and $\mathcal{C} = \mathcal{P}(U)$, then any fuzzy semantics is tableaux-manageable. Zadeh semantics with closed intervals as constraints is tableaux-manageable. Łukasiewicz semantics is not tableaux-manageable.*

Proof. Assume that U is finite. Then, from the fact that a *"point"* $(\lambda_1,\ldots,\lambda_n)$ is, trivially, a rectangle, we deduce that any fuzzy semantics is tableaux-manageable.

Let $\mathcal{M}_Z$ be the Zadeh semantics. Then, since

- $(\alpha \wedge \beta$, [a,b]) is equivalent to $((\alpha$, [a,b]) $\wedge$ $(\beta$, [a,1])) $\vee$ $((\alpha$, [a,1]) $\wedge$ $(\beta$, [a,b])),
- $(\alpha \vee \beta$, [a,b]) is equivalent to $((\alpha$, [a,b]) $\wedge$ $(\beta$, [0,b])) $\vee$ $((\beta$, [a,b]) $\wedge$ $(\alpha$, [0,b])),
- $(\neg\alpha$, [a,b]) is equivalent to $(\alpha$, [1 − b, 1 − a]),

we can conclude that $\mathcal{M}$ is manageable.

Finally observe that in Łukasiewicz semantics, disjunction is interpreted in [0,1] by truncated addition, i.e., $x \oplus y = Min\{x + y, 1\}$. Moreover, given a and b in [0,1] with $a \leq b$,

$$\{(x, y) \in [0,1] \times [0,1] : a \leq x \oplus y \leq b\}$$

is the set of points in the square [0,1] × [0,1] between the lines $x + y = a$ and $x + y = b$. Since such a set is not a finite union of rectangles, Łukasiewicz semantics is not tableaux manageable. $\square$

If $\mathcal{M}$ is tableaux manageable, then for any logical connective h we can consider a *rewriting rule*

$$(h(\alpha_1,\ldots,\alpha_n), X) \mapsto q((\alpha_1,X_1), \ldots ,(\alpha_n,X_n)),$$

where $q((\alpha_1,X_1), \ldots ,(\alpha_n,X_n))$ is a Boolean combination equivalent to $(h(\alpha_1,\ldots,\alpha_n), X)$.

Such a system of rewriting rules enables us to transform any signed formula (α,X) into an equivalent positive Boolean combination of signed formulas (p,X) with p propositional variables.

We illustrate this procedure by starting from Zadeh logic. Indeed, in this case we have the following very simple system of rewriting rules:

$(\alpha \wedge \beta, [a,b]) \mid\!\rightarrow ((\alpha, [a,b]) \wedge (\beta, [a,1])) \vee ((\alpha, [a,1]) \wedge (\beta, [a,b])),$
$(\alpha \vee \beta, [a,b]) \mid\!\rightarrow ((\alpha, [a,b]) \wedge (\beta, [0,b])) \vee ((\beta, [a,b]) \wedge (\alpha, [0,b])),$
$(\neg\alpha, [a,b]) \mid\!\rightarrow (\alpha, [1-b, 1-a]),$

with the derived rules

$(\alpha \wedge \beta, [a,1]) \mid\!\rightarrow (\alpha, [a,1]) \wedge (\beta, [a,1]),$
$(\alpha \vee \beta, [a,1]) \mid\!\rightarrow (\alpha, [a,1]) \vee (\beta, [a,1]),$
$(\alpha \wedge \beta, [0,b]) \mid\!\rightarrow (\alpha, [0,b]) \vee (\beta, [0,b]),$
$(\alpha \vee \beta, [0,b]) \mid\!\rightarrow (\alpha, [0,b]) \wedge (\beta, [0,b]).$

For example, we can start from the formula $(p \wedge \neg p, [0.7,1])$ by obtaining the sequence

$(p \wedge \neg p, [0.7,1])$; $(p, [0.7,1]) \wedge (\neg p, [0.7,1])$; $(p, [0.7,1]) \wedge (p, [0, 0.3])$;
$(p, [0.7,1] \cap [0, 0.3])$; $(p,\varnothing)$

of equivalent Boolean combinations. Since no model satisfies $(p,\varnothing)$, we can conclude that no model satisfies $(p \wedge \neg p, [0.7,1])$.

If we start from $(\neg p \wedge (\neg q \vee p), [0.7,1])$, then we get the sequence

$(\neg p \wedge (\neg q \vee p), [0.7,1])$; $(\neg p, [0.7,1]) \wedge (\neg q \vee p, [0.7,1])$;
$(\neg p, [0.7,1]) \wedge (\neg q, [0.7,1]) \vee (p, [0.7,1])$;
$(\neg p, [0.7,1]) \wedge ((q, [0, 0.3]) \vee (p, [0.7,1]))$;
$(p, [0,0.3]) \wedge ((q, [0, 0.3]) \vee (p, [0.7,1]))$;
$(p, [0,0.3]) \wedge (q, [0, 0.3]) \vee ((p, [0,0.3]) \wedge (p, [0.7,1]))$;
$(p, [0,0.3]) \wedge (q, [0, 0.3]) \vee ((p, [0,0.3] \cap [0.7,1]))$;
$(p, [0,0.3]) \wedge (q, [0, 0.3]) \vee (p, \varnothing)$;
$(p, [0,0.3]) \wedge (q, [0, 0.3])$.

Consequently, we can conclude that $(\neg p \wedge (\neg q \vee p), [0.7,1])$ is satisfied in the rectangle $[0, 0.3] \times [0, 0.3]$.

Let $\mathcal{M}$ be a tableaux-manageable fuzzy semantics. Then a *tableau proof* is a finite labelled tree, whose node labels are signed formulas, constructed as follows: Let h be a logical connective and assume that $(h(\alpha_1,\ldots,\alpha_n), X)$ is equivalent to $A_1 \vee \ldots \vee A_l$ where A_i is a conjunction $(\alpha_1, X_1^i) \wedge \ldots \wedge (\alpha_n, X_n^i)$ of signed formulas. Then an *expansion rule* is a tree defined by setting

$$
\begin{array}{c}
(h(\alpha_1,\ldots,\alpha_n), X) \\
\hline
\begin{array}{|c|c|c|c|}
(\alpha_1, X_1^1) & (\alpha_1, X_1^2) & \ldots & (\alpha_1, X_1^l) \\
\ldots & \ldots & \ldots & \ldots \\
(\alpha_n, X_n^1) & (\alpha_n, X_n^2) & \ldots & (\alpha_n, X_n^l)
\end{array}
\end{array}
$$

The set of all tableaux for a given signed formula $(k(\beta_1,\ldots,\beta_m), Z)$ is defined as the set of labelled trees that can be produced by a finite number of applications of the following rules (where a *branch* in a tree is a maximal path):

(*) $(k(\beta_1,\ldots,\beta_m), Z)$ is a tableau for $(k(\beta_1,\ldots,\beta_m), Z)$;

(**) If T is a tableau for $(k(\beta_1,\ldots,\beta_m), Z)$ and $(h(\alpha_1,\ldots,\alpha_n), X)$ a node label from T, then a new tableau T' is constructed by extending all branches of T; containing $(h(\alpha_1,\ldots,\alpha_n), X)$ in accordance with the related expansion rule,

(***) If T is a tableau for $(k(\beta_1,\ldots,\beta_m), Z)$ and (α,X), (α,Y) are two signed formulas in T, then we can extend any branch containing both (α,X) and (α,Y) by the signed formula $(\alpha, X \cap Y)$.

The proof of the following proposition is trivial:

Proposition 8.4. Let (α,X) be a signed formula, T a tableau for it and $\pi_1, \ldots, \pi_t$ the branches in T. Then m is a model of (α,X) if and only if, for a suitable index i, m is a model of the formulas in π_i.

A branch in a tableau is *closed* if it contains a *trivially inconsistent* signed formula $(\alpha,\varnothing)$. A branch is *complete* if either it is closed or no rule application to formulas on the branch produces a formula that was not already present. A tableau is *closed* (*complete*) if all of its branches are closed (*complete*). It is evident that any tableau can be completed.

Definition 8.5. A signed formula (α, X) is *tableau refutable* if a closed tableau for (α, X) exists.

It is possible to prove the following completeness theorem:

Theorem 8.6. *Let $\mathcal{M}$ be a tableaux-manageable fuzzy semantics. Then a signed formula is tableau refutable iff it is a contradiction.*

Proof. Assume that (α,X) is tableaux refutable. Then, since no closed branch has a model, from Proposition 8.4 it follows that (α,X) doesn't have a model. Conversely, assume that (α, X) is a contradiction, let T be a complete tableau for (α,X) and π_j any branch in T. Moreover, assume that π_j is not closed and therefore that, for any propositional variable p_i, the set

$$X(p_i) = \bigcap\{X \in \mathcal{C} : (p_i, X) \text{ occurs in } \pi_j\}$$

is nonempty. Then we can build up a model m by associating any propositional variable p_i with any value $m(p_i)$ in $X(p_i)$. It is obvious that m is a model of all the formulas in π_j and therefore a model of (α, X). This contradicts the hypothesis that no model for (α, X) exists. Thus, π_j is closed. □

It is evident how to give the notion of *tableaux-refutable under a given set of hypotheses*. Observe that complexity is the basic obstacle for the tableaux-refutable approach. Indeed, the exponential character of the tableaux creates serious difficulties in this sense. It is possible to reduce only to binary trees provided that, for every logical connective h and $X \in \mathcal{C}$, the set $\underline{h}^{-1}(X)$ is a union of two rectangles.

CHAPTER 6

CANONICAL EXTENSION OF A CRISP HILBERT LOGIC

1. EXTENDING A CRISP HILBERT DEDUCTION SYSTEM

As observed in Section 6 in Chapter 3, any crisp (abstract) logic can be extended to a fuzzy (abstract) logic in a canonical way. In this chapter we will consider an extension principle which fits the Hilbert systems well (see Gerla [1995]). Namely, given any crisp *H*-system $\mathcal{S} = (\mathbb{A}, \mathbb{R})$ and a crisp rule $r \in \mathbb{R}$, we define the fuzzy rule $r^* = (r', r'')$ by setting

$$r' = r \quad , \quad r''(x_1,\dots,x_n) = x_1 \wedge \dots \wedge x_n,$$

where $\wedge$ denotes the minimum operation. We call r^* the *canonical extension of r*.

Definition 1.1. Let $\mathcal{S} = (\mathbb{A}, \mathbb{R})$ be a crisp *H*-system. Then, *the canonical extension of* $\mathcal{S}$ is the fuzzy *H*-system $\mathcal{S}^* = (\mathbb{A}, \mathbb{R}^*)$ where

$$\mathbb{R}^* = \{r^* : r \in \mathbb{R}\}.$$

The following theorem shows that such a definition is related to the definition of canonical extension of a closure operator.

Theorem 1.2. *Let* $\mathcal{S}$ *be a crisp H-system whose deduction operator is* $\mathcal{D}$ *and whose class of theories is* $\mathcal{T}$. *Then,*

(i) *the class of fuzzy theories of* $\mathcal{S}^*$ *is the canonical extension* $\mathcal{T}^*$ *of* $\mathcal{T}$,

(ii) *the deduction operator of* $\mathcal{S}^*$ *is the canonical extension* $\mathcal{D}^*$ *of* $\mathcal{D}$.

Consequently, for every initial valuation v and $\alpha \in \mathbb{F}$,

$$\mathcal{D}^*(v)(\alpha) = Sup\{Incl(X,v) : X \in \mathcal{P}_f(\mathbb{F}),\ X \vdash \alpha\} \tag{1.1}$$

Proof. In order to prove (i), let τ be an element of $\mathcal{T}^*$, $r^* = (r', r'')$ an *n*-ary fuzzy rule of inference in $\mathcal{S}^*$ and $\alpha_1,\dots,\alpha_n$ formulas. Then, if $\lambda = \tau(\alpha_1) \wedge \dots \wedge \tau(\alpha_n)$, since $C(\tau,\lambda)$ is a theory in $\mathcal{S}$, r' is a rule in $\mathbb{R}$ and $\alpha_1,\dots,\alpha_n \in C(\tau,\lambda)$, we get that $r'(\alpha_1,\dots,\alpha_n) \in C(\tau,\lambda)$. This proves that

$$\tau(r'(\alpha_1,\dots,\alpha_n)) \geq \tau(\alpha_1) \wedge \dots \wedge \tau(\alpha_n) = r''(\tau(\alpha_1),\dots,\tau(\alpha_n)),$$

i.e., that τ is closed under any rule in $\mathcal{S}^*$. As $C(\tau,1)$ is a theory in $\mathcal{S}$, $C(\tau,1) \supseteq \mathbb{A}$ and hence $\tau \supseteq \mathbb{A}$. Thus τ is a theory of $\mathcal{S}^*$.

Conversely, let τ be a theory of $\mathcal{S}^*$. Then,

$$\tau(r'(\alpha_1,\dots,\alpha_n)) \geq \tau(\alpha_1) \wedge \dots \wedge \tau(\alpha_n),$$

and, given any cut $C(\tau,\lambda)$, from $\alpha_1,\dots,\alpha_n \in C(\tau,\lambda)$ it follows that $r'(\alpha_1,\dots,\alpha_n) \in C(\tau,\lambda)$. Since $\tau \supseteq \mathbb{A}$, we have that $C(\tau,\lambda) \supseteq \mathbb{A}$ and this demonstrates that $C(\tau,\lambda)$ is a theory. In conclusion, τ is an element of $\mathcal{T}^*$.

To prove (ii), observe that, by (2.3) of Proposition 2.2 in Chapter 3, $\mathcal{D}^* = Co(\mathcal{T})^* = Co(\mathcal{T}^*)$ and that $Co(\mathcal{T}^*)$ is the deduction operator of $\mathcal{S}^*$. □

(1.1) means that in a canonical extension of an H-system the valuation of a proof π only depends on the assumptions occurring in π and not on the structure of π. This is not true for any fuzzy H-system (see Theorem 8.4).

Corollary 1.3. *Let $\mathcal{S}$ be a crisp H-system. Then:*
(i) $\mathcal{S}^*$ *is not logically compact.*
(ii) $Inc(\mathcal{S}^*) = 1$.
(iii) $Contr(\mathcal{M}) = \varnothing$.

Proof. See Theorem 6.3 in Chapter 3 and Proposition 9.1 in Chapter 2. □

In accordance with the results in Section 5 of Chapter 4, we can get $\mathcal{D}^*(v)$ as a fixed point of an immediate consequence operator. Indeed, consider the immediate consequence operator H of the crisp system $\mathcal{S}$,

$$H(X) = \{r(\alpha_1,\ldots,\alpha_n) : \alpha_1,\ldots,\alpha_n \in X,\ r\in \mathbb{R}\} \cup X \cup \mathbb{A} \tag{1.2}$$

and its canonical extension H^* defined by

$$H^*(v)(\alpha) = Sup\{Incl(\{\alpha_1, \ldots,\alpha_n\},v) : \alpha = r(\alpha_1,\ldots,\alpha_n),\ r\in \mathbb{R}\} \cup v \cup \mathbb{A}\,. \tag{1.3}$$

Then H^* is the immediate consequence operator of the canonical extension $\mathcal{S}^*$ and therefore,

$$\mathcal{D}^*(v) = Sup_{n\in N} H^{*\,n}(v). \tag{1.4}$$

We conclude this section by giving two characterizations of the canonical extensions of crisp H-systems:

Theorem 1.4. *Let $\mathcal{H}$ be a fuzzy closure operator extending a crisp operator. Then $\mathcal{H}$ is the deduction operator of the canonical extension of an H-system iff*
(i) *$\mathcal{H}$ is continuous,*
(ii) *for every $\mu \in U$ and $v \in \mathcal{F}(\mathbb{F})$,*

$$\mathcal{H}(\mathrm{O}(v,\mu)) = \mathrm{O}(\mathcal{H}(v),\mu).$$

Proof. Assume that $\mathcal{H}$ is the deduction operator of the canonical extension of an H-system. Then $\mathcal{H}$ is the canonical extension of an algebraic closure operator. Hence, $\mathcal{H}$ is continuous. Moreover, by Proposition 5.2 in Chapter 3, (ii) holds.

Conversely, let $\mathcal{H}$ be a fuzzy closure operator satisfying (i) and (ii). Then, by Theorem 7.5 in Chapter 3, $\mathcal{H}$ is the canonical extension of an algebraic closure operator J. Now, by a basic result of closure operator theory (see, e.g., Grätzer [1968]), J is the closure operator associated with a suitable algebraic structure with domain $\mathbb{F}$. Since every algebraic structure is an H-system in an obvious way, the theorem is demonstrated. □

The proof of the next theorem is trivial:

Theorem 1.5. *Let $\mathcal{S}$ be a fuzzy H-system extending a crisp operator and denote the related deduction operator by $\mathcal{H}$. Then the following are equivalent.*

(i) *$\mathcal{S}$ is a canonical extension of an H-system.*

(ii) *For every $\mu \in U$ and $v \in \mathcal{F}(\mathbb{F})$.*

$$\mathcal{H}(\mathrm{O}(v,\mu)) = \mathrm{O}(\mathcal{H}(v),\mu),$$

(iii) *For every $v \in \mathcal{F}(\mathbb{F})$ and for every join-preserving function $f: U \to U$ such that $f(1) = 1$,*

$$\mathcal{H}(f \circ v) = f \circ \mathcal{H}(v).$$

2. CONTROLLING THE INCONSISTENCY

In the canonical extension $\mathcal{S}^*$ of a crisp H-system $\mathcal{S}$ there are theories of any given degree of inconsistency. Nevertheless, there is no difficulty to define extensions of $\mathcal{S}$ in which only degrees of inconsistency smaller than a fixed number $\mu \in U$ are admitted. For example, suppose that a contradiction ϕ exists and consider a *μ-collapsing rule* χ^{μ} saying that everywhere we are able to prove ϕ to degree $\lambda > \mu$, then we can also assert ϕ to degree 1. Namely, the fuzzy rule $\chi^{\mu} = (\chi'_{\mu},\chi''_{\mu})$ such that

- the domain of χ'_{μ} is the set of contradictions,
- $\chi'_{\mu}(\phi) = \phi$ for every contradiction ϕ,
- $\chi''_{\mu}(\lambda) = 1$ if $\lambda > \mu$ and $\chi''_{\mu}(\lambda) = 0$ otherwise.

Such a rule can be pictured as follows:

$$\frac{\phi}{\phi} \quad (\phi \text{ contradiction}) \qquad ; \qquad \frac{\lambda}{1} \quad (\lambda > \mu)\,.$$

We indicate by $\mathcal{S}^{\mu}$ the fuzzy H-system achieved by adding the μ-collapsing rule to $\mathcal{S}^*$ and we say that $\mathcal{S}^{\mu}$ is the *μ-canonical extension of $\mathcal{S}$*. Furthermore, we denote the related deduction operator by $\mathcal{D}^{\mu}$. Observe that $\mathcal{S}^1$ coincides with $\mathcal{S}^*$.

Theorem 2.1. *Suppose that $\mathcal{S}$ is an H-system with a contradiction ϕ and let $\mu \in U$. Then the following sentences are equivalent :*

(i) *τ is a consistent theory of $\mathcal{S}^{\mu}$.*

(ii) *τ is a theory of $\mathcal{S}^*$ such that $Inc(\tau) \le \mu$.*

(iii) *τ is a theory of $\mathcal{S}^*$ such that $\tau(\phi) \le \mu$ for every contradiction ϕ.*

Proof. (i) $\Rightarrow$ (ii). Let τ be a consistent theory of $\mathcal{S}^{\mu}$. Then τ is a theory of $\mathcal{S}$ and $\tau(\phi) \le \mu$. Indeed, otherwise,

$$\tau(\phi) = \tau(\chi'_{\mu}(\phi)) \ge \chi''_{\mu}(\tau(\phi)) = 1.$$

This proves that $Inc(\tau) \le \mu$.

Implication (ii) $\Rightarrow$ (iii) is trivial. In order to prove (iii) $\Rightarrow$ (i), let τ be a theory of $\mathcal{S}^*$ such that $\tau(\phi) \le \mu$ for every contradiction ϕ. Then, since $\chi''_{\mu}(\tau(\phi)) = 0$, τ is closed under the μ-collapsing rule. Thus τ is a consistent theory of $\mathcal{S}^{\mu}$. □

In particular, in the case $\mu = 0$, we have that the theories of $\mathcal{S}^{\mu}$ coincide with the totally consistent theories of $\mathcal{S}^{*}$. Note that, differently from $\mathcal{S}^{*}$, a constant map s^{λ} is consistent in $\mathcal{S}^{\mu}$ iff $\lambda \leq \mu$.

Corollary 2.2. *Suppose that* S *is an H-system with a contradiction* ϕ *and that* $\mu \neq 1$. *Then the* μ*-canonical extension* $\mathcal{S}^{\mu}$ *is a logically compact H-system such that*

$$Inc(\mathcal{S}^{\mu}) = \mu.$$

Moreover,

$$v \text{ consistent in } \mathcal{S}^{\mu} \Leftrightarrow C(v,\lambda) \text{ consistent for every } \lambda > \mu.$$

Proof. Let $(\tau_i)_{i\in I}$ be a directed family of consistent theories of $\mathcal{S}^{\mu}$ and ϕ a contradiction. Then, since $\tau_i(\phi) \leq \mu$ for every $i \in I$, it is $(\bigcup_{i\in I}\tau_i)(\phi) \leq \mu$ and therefore $\bigcup_{i\in I} \tau_i$ is consistent. By Proposition 8.2 in Chapter 1 this proves that $\mathcal{S}^{\mu}$ is logically compact. The rest is obvious. □

Corollary 2.3. *Let* $\mu \in U$ *and* $\mathcal{D}^{\mu}$ *the deduction operator of* $\mathcal{S}^{\mu}$. *Then*

$$\mathcal{D}^{\mu}(v) = \begin{cases} \mathcal{D}^{*}(v) & \text{if } Inc(v) \leq \mu, \\ s^{1} & \text{otherwise.} \end{cases}$$

Unlike $\mathcal{S}^{*}$, $\mathcal{S}^{\mu}$ contains maximal theories. Indeed, we have the following proposition:

Proposition 2.4. *Let* τ *be a fuzzy set of formulas, then*
τ *is a maximal theory of* $\mathcal{S}^{\mu}$ $\Leftrightarrow$ $\tau = \mu \vee T$ *where* T *is a maximal theory of* S.

Proof. $\Leftarrow$. Let T be a maximal theory of S and let τ' be a theory of $\mathcal{S}^{\mu}$ containing $\mu \vee T$. Then, if a formula α exists such that $\tau'(\alpha) > (\mu \vee T)(\alpha)$, we have $\alpha \notin T$ and $\tau'(\alpha) > \mu$. Set $\lambda = \tau'(\alpha)$, then $C(\tau',\lambda)$ is a theory containing T and α. Since T is maximal and $\alpha \notin T$, $C(\tau',\lambda)$ is inconsistent and therefore, $Inc(\tau') \geq \tau'(\alpha) > \mu$. This proves that τ' is inconsistent in $\mathcal{S}^{\mu}$. Thus, $\mu \vee T$ is a maximal theory of $\mathcal{S}^{\mu}$.

$\Rightarrow$. Let τ be a maximal theory of $\mathcal{S}^{\mu}$. Then, since for every $\lambda > \mu$, $C(\tau,\lambda)$ is a consistent theory of S, $T = \bigcup_{\lambda>\mu} C(\tau,\lambda)$ is a consistent theory of S. Since $\mu \vee T$ is a theory of $\mathcal{S}^{\mu}$ containing τ, T is maximal and $\mu \vee T = \tau$. □

Proposition 2.4 entails that we cannot consider the class of maximal theories as an adequate semantics for $\mathcal{S}^{\mu}$. Instead, we have the following proposition:

Proposition 2.5. *Let* $\mathcal{M}$ *be a semantics for the crisp system* S. *Then*

$$\mathcal{M}^{\mu} = \{\lambda \vee M : \lambda \leq \mu \text{ and } M \in \mathcal{M}\}$$

is a fuzzy semantics for the fuzzy system $\mathcal{S}^{\mu}$.

3. NECESSITY LOGIC

In this section we will consider the simple but interesting case in which $\mathcal{S}$ is the deduction apparatus of classical propositional logic. We call *generalized necessity logic* the related extension $\mathcal{S}^*$. Such a logic is related to the interesting papers of D. Dubois and H. Prade on possibilistic logic (see also Biacino and Gerla [1970] and Biacino and Gerla [1992]*b*). More specifically, in $\mathcal{S}^*$

- the set F of formulas coincides with the set of formulas of classical propositional calculus,
- the fuzzy subset of logical axioms coincides with (the characteristic function of) the set of classical tautologies
- the only rule is the fuzzy extension of Modus Ponens

$$\frac{\alpha, \alpha \to \beta}{\beta} \quad ; \quad \frac{\lambda, \mu}{\lambda \wedge \mu}.$$

Equivalently, we can assume as logical axioms only a suitable base for the deduction apparatus of classical propositional calculus. It is obvious that the fuzzy theories are the fuzzy subsets τ such that

(i) $\tau(\alpha) = 1$ for every tautology α,
(ii) $\tau(\beta) \geq \tau(\alpha) \wedge \tau(\alpha \to \beta)$.

Proposition 3.1. *A fuzzy subset* τ *of formulas is a theory in* $\mathcal{S}^*$ *iff for every* $\alpha, \beta \in F$,

(j) $\tau(\alpha) = 1$ *for every tautology* α,
(jj) α *logically equivalent to* $\beta \Rightarrow \tau(\alpha) = \tau(\beta)$,
(jjj) $\tau(\alpha \wedge \beta) = \tau(\alpha) \wedge \tau(\beta)$.

Proof. If τ is a theory, then (j) and (jj) are obvius. In order to prove (jjj), observe that $\alpha \wedge \beta$ implies both α and β and therefore $\tau(\alpha \wedge \beta) \leq \tau(\alpha)$ and $\tau(\alpha \wedge \beta) \leq \tau(\beta)$. Hence, $\tau(\alpha \wedge \beta) \leq \tau(\alpha) \wedge \tau(\beta)$. Moreover, by Corollary 6.8 in Chapter 3, since $\alpha, \beta \vdash \alpha \wedge \beta$, we have $\tau(\alpha \wedge \beta) \geq \tau(\alpha) \wedge \tau(\beta)$.

Conversely, assume that (j), (jj) and (jjj) are satisfied. Then (j) means that τ contains the logical axioms and from (jj) and (jjj) it follows that

$$\tau(\alpha) \wedge \tau(\alpha \to \beta) = \tau(\alpha \wedge (\alpha \to \beta)) = \tau(\alpha \wedge \beta) = \tau(\alpha) \wedge \tau(\beta) \leq \tau(\beta),$$

i.e., τ is closed under the extended Modus Ponens. Thus, τ is a theory. □

Proposition 3.2. *Given an initial valuation* v *and* $\alpha \in F$, *either* $\mathcal{D}^*(v)(\alpha) = Inc(v)$ or $\mathcal{D}^*(v)(\neg\alpha) = Inc(v)$. *In particular, either* $\mathcal{D}^*_n(v)(\alpha) = 0$ *or* $\mathcal{D}^*_n(v)(\neg\alpha) = 0$.

Proof. The fact that $\alpha \wedge \neg\alpha$ is a contradiction entails, by Proposition 3.1, that

$$Inc(v) = \mathcal{D}^*(v)(\alpha \wedge \neg\alpha) = \mathcal{D}^*(v)(\alpha) \wedge \mathcal{D}^*(v)(\neg\alpha).$$ □

Assume that α is a tautology and that v is not totally consistent. Then from Proposition 3.2 it follows that $\mathcal{D}^*(v)(\alpha) + \mathcal{D}^*(v)(\neg\alpha) = 1 + Inc(v) > 1$. This means that generalized necessity logic is not a logic with a negation in the sense of Section 5 in Chapter 5.

Suppose that v_1 and v_2 are two fuzzy subsets of formulas expressing valuations of two different experts. Then it is natural to consider the valuation $v = v_1 \cup v_2$ as a combination of these two different sources of information. The following proposition gives an interesting way to compute $\mathcal{D}^*(v_1 \cup v_2)$.

Proposition 3.3. *Given two initial valuations* v_1, v_2,

$$\mathcal{D}^*(v_1 \cup v_2)(\gamma) = Sup\{\mathcal{D}^*(v_1)(\alpha) \wedge \mathcal{D}^*(v_2)(\beta) : \alpha \wedge \beta \vdash \gamma\}. \tag{3.1}$$

Proof. Set $\tau_1 = \mathcal{D}^*(v_1)$ and $\tau_2 = \mathcal{D}^*(v_2)$. Then

$$\begin{aligned}\mathcal{D}^*(v_1 \cup v_2)(\gamma) &= \mathcal{D}^*(\tau_1 \cup \tau_2)(\gamma)\\ &= Sup\{\tau_1(\alpha_1) \wedge ... \wedge \tau_1(\alpha_n) \wedge \tau_2(\beta_1) \wedge ... \wedge \tau_2(\beta_m) : \alpha_1,...,\alpha_n, \beta_1,...,\beta_m \vdash \gamma\}\\ &= Sup\{\tau_1(\alpha_1 \wedge ... \wedge \alpha_n) \wedge \tau_2(\beta_1 \wedge ... \wedge \beta_m) : \alpha_1,...,\alpha_n, \beta_1,...,\beta_m \vdash \gamma\}\\ &= Sup\{\tau_1(\alpha_1 \wedge ... \wedge \alpha_n) \wedge \tau_2(\beta_1 \wedge ... \wedge \beta_m) :\\ &\qquad (\alpha_1 \wedge ... \wedge \alpha_n) \wedge (\beta_1 \wedge ... \wedge \beta_m) \vdash \gamma\}\\ &= Sup\{\tau_1(\alpha) \wedge \tau_2(\beta) : \alpha \wedge \beta \vdash \gamma\}.\end{aligned}$$

□

Now, it is rather probable that the information v_1 is partially conflicting with the information v_2. The next proposition gives a way to compute the degree of inconsistency of $v_1 \cup v_2$, i.e., the degree of inconsistency of v_1 with v_2.

Proposition 3.4. *Given two initial valuations* v_1, v_2,

$$Inc(v_1 \cup v_2) = Sup\{\mathcal{D}^*(v_1)(\alpha) \wedge \mathcal{D}^*(v_2)(\neg\alpha) : \alpha \in F\}. \tag{3.2}$$

As a consequence, v_1 *is totally consistent with* v_2 *if, for every* $\alpha \in F$,

$$\mathcal{D}^*(v_1)(\alpha) \neq 0 \Rightarrow \mathcal{D}^*(v_2)(\neg\alpha) = 0.$$

Proof. Let γ be a contradiction. Then, by Proposition 3.3,

$$\begin{aligned}Inc(v_1 \cup v_2) = \mathcal{D}^*(v_1 \cup v_2)(\gamma) &= Sup\{\mathcal{D}^*(v_1)(\alpha) \wedge \mathcal{D}^*(v_2)(\beta) : \alpha \wedge \beta \vdash \gamma\}\\ &= Sup\{\mathcal{D}^*(v_1)(\alpha) \wedge \mathcal{D}^*(v_2)(\beta) : \beta \vdash \neg\alpha\}.\end{aligned}$$

By recalling that from $\beta \vdash \neg\alpha$ it follows that $\mathcal{D}^*(\beta) \leq \mathcal{D}^*(\neg\alpha)$, we have that

$$Sup\{\mathcal{D}^*(v_1)(\alpha) \wedge \mathcal{D}^*(v_2)(\beta) : \beta \vdash \neg\alpha\} = Sup\{\mathcal{D}^*(v_1)(\alpha) \wedge \mathcal{D}^*(v_2)(\neg\alpha) : \alpha \in F\}.$$

This concludes the proof. □

Formula (3.2) says that the degree of inconsistency of v_1 with v_2 is the multivalued valuation of the claim

"there is a formula α such that while v_1 entails α, v_2 entails $\neg\alpha$".

In accordance with Theorem 6.9 of Chapter 3, the class

$$\mathcal{M} = \{\lambda \vee m : \lambda \in U, \lambda \neq 1, m \text{ is a 0-1-valuation}\}$$

of U-models is a simple fuzzy semantics for generalized necessity logic. Perhaps it is interesting to characterize the models $\lambda \vee m$ in terms of truth-functional valuations. To this purpose note that, for every $\lambda \in U$, $\lambda \neq 1$, the set $\{\lambda,1\}$ is a Boolean algebra isomorphic to the Boolean algebra $\{0,1\}$. As a consequence we may consider the truth-functional valuations of the formulas in $\mathcal{F}$ in $\{\lambda,1\}$.

Definition 3.5. Let $\lambda \neq 0$, and let $m : \mathcal{F} \to U$ be a valuation of the formulas. Then we say that m is a *λ-1-model* if m is defined by assigning values in $\{\lambda,1\}$ to the propositional variables and by setting

$$m(\alpha \wedge \beta) = m(\alpha) \wedge m(\beta) \ ; \ m(\alpha \vee \beta) = m(\alpha) \vee m(\beta) \ ; \ m(\neg\alpha) = \lambda + 1 - m(\alpha).$$

It is easy to verify that m is a λ-1-model iff a classical 0-1-valuation m' exists such that $m = \lambda \vee m'$. So we have the following theorem:

Theorem 3.6. *The class of* λ*-1-models*

$$\mathcal{M} = \{m \in \mathcal{F}(\mathcal{F}) : m \text{ is a } \lambda\text{-1-model}, \lambda \in U, \lambda \neq 1\},$$

is a fuzzy semantics for generalized necessity logic. Further, every λ*-1-model m is a maximal element in the class* $\{\tau : \tau \text{ is a theory such that } Inc(\tau) \leq \lambda\}$.

Proof. The first part is trivial. The second part is a consequence of Proposition 6.10 in Chapter 3. □

Proposition 3.7. *Generalized necessity logic is not with optimal proofs. Moreover, while for every* $\lambda \in U$,

$$O(\mathcal{D}^*(v),\lambda) = \mathcal{D}(O(v,\lambda)) \quad (3.3)$$

we usually have that

$$C(\mathcal{D}^*(v),\lambda) \neq \mathcal{D}(C(v,\lambda)). \quad (3.4)$$

Proof. Let v be a fuzzy subset such that $v(p_i \to p_1) = 1$ and $v(p_i) = 1-1/i$ for every $i \in N$. Then $\mathcal{D}^*(v)(p_1) = 1$ while $Val(\pi,v) < 1$ for every proof π of p_1. This demonstrates the first part of the proposition. Equation (3.3) is equal to Proposition 5.2 in Chapter 3. To prove (3.4), assume that $\lambda \neq 0$, and let v be the initial valuation defined by setting

$$v(\alpha) = \begin{cases} \lambda & \text{if } \alpha \text{ is } p_i \to p_1, \\ \lambda - \lambda/i & \text{if } \alpha \text{ is } p_i, \\ 0 & \text{otherwise.} \end{cases}$$

Then, since

$$\mathcal{D}^*(v)(p_1) = Sup\{v(p_i \to p_1) \wedge v(p_i) : i = 1, 2, \ldots\} = \lambda,$$

p_1 is an element in $C(\mathcal{D}^*(v),\lambda)$ while, since no p_i belongs to $C(v,\lambda)$, p_1 is not an element in $\mathcal{D}(C(v,\lambda))$. □

In order to obtain the extension S^{μ}, we can add the μ-collapsing rule introduced in Section 2. Equivalently, we may add the rule $\chi = (\chi',\chi'')$ defined by setting

$$Dom(\chi) = \{(\alpha,\neg\alpha) : \alpha \in \mathcal{F}\} \ ; \ \chi'(\alpha,\neg\alpha) = \alpha \wedge \neg\alpha$$

and

$$\chi''(\lambda_1,\lambda_2) = \begin{cases} 1 & \text{if } \lambda_1 \wedge \lambda_2 > \mu, \\ 0 & \text{otherwise.} \end{cases}$$

Indeed, a fuzzy subset τ is closed under χ iff, for every formula α,

$$\tau(\alpha) > \mu \text{ and } \tau(\neg\alpha) > \mu \Rightarrow \tau(\alpha \wedge \neg\alpha) = 1.$$

4. A SIMPLE EXAMPLE OF NON-MONOTONE FUZZY LOGIC

Generalized necessity logic enables us to show the role played by the notion of degree of inconsistency. Indeed, let Inf_1 be a consistent set of statements about an unknown world and imagine that we are not completely sure about the validity of these statements. Then, we may express this by assigning to every formula α in Inf_1 the valuation $v(\alpha) = 0.8$ while we set $v(\alpha) = 0$ if $\alpha \notin Inf_1$. By applying (1.1), we obtain that:

$$\mathcal{D}^*(v)(\alpha) = \begin{cases} 1 & \text{if } \alpha \text{ is a tautology,} \\ 0.8 & \text{if } Inf_1 \vdash \alpha \text{ and } \alpha \text{ is not a tautology,} \\ 0 & \text{otherwise.} \end{cases} \tag{4.1}$$

Moreover, since Inf_1 is consistent, $Inc(v) = 0$ and $\mathcal{D}^*(v) = \mathcal{D}^*_n(v)$. Now, imagine that further information about the unknown world is available but that we consider this less sure than the information furnished by Inf_1. If this piece of information is contained in a set Inf_2 of formulas and $Inf_1 \cap Inf_2 = \emptyset$, then we may express such a stratified type of information by setting

$$v(\alpha) = \begin{cases} 0.8 & \text{if } \alpha \in Inf_1, \\ 0.6 & \text{if } \alpha \in Inf_2, \\ 0 & \text{otherwise.} \end{cases} \tag{4.2}$$

Consequently,

$$\mathcal{D}^*(v)(\alpha) = \begin{cases} 1 & \text{if } \alpha \text{ is a tautology,} \\ 0.8 & \text{if } Inf_1 \vdash \alpha \text{ and } \alpha \text{ is not a tautology,} \\ 0.6 & \text{if } Inf_1 \cup Inf_2 \vdash \alpha \text{ but } \alpha \text{ is not a consequence of } Inf_1, \\ 0 & \text{otherwise.} \end{cases} \tag{4.3}$$

Of course, it is also possible that Inf_2 is inconsistent with Inf_1. In this case every formula is a logical consequence of $Inf_1 \cup Inf_2$ and therefore $Inc(v) = 0.6$. Then we rely on α only if α can be proved to a degree greater than 0.6. To consider the normalized deduction operator, we observe that the normalization v_n of v is the function defined by setting $v_n(\alpha) = 0.5$ if $\alpha \in Inf_1$ and $v_n(\alpha) = 0$ if $\alpha \notin Inf_1$. Consequently, by the equality $\mathcal{D}^*_n(v) = \mathcal{D}^*(v_n)$, we have, for every $\alpha \in \mathcal{F}$,

$$\mathcal{D}^*_n(v)(\alpha) = \begin{cases} 1 & \text{if } \alpha \text{ is a tautology,} \\ 0.5 & \text{if } Inf_1 \vdash \alpha \text{ and } \alpha \text{ is not a tautology,} \\ 0 & \text{otherwise.} \end{cases} \tag{4.4}$$

Then, fuzzy logic is non-monotone whenever an inconsistency phenomenon arises. Indeed, assume that α is a consequence of Inf_1. Then, in absence of the information Inf_2 the formula α is valued 0.8. Instead, by adding the new information Inf_2, the valuation of α becomes 0.5. As a matter of fact, any new piece of information contradicting Inf_1 determines a lowering of the degree of belief in Inf_1. In this sense, even inconsistent information may contain useful information.

We can restate the above considerations with reference to a typical example of non-monotone reasoning. Consider a language containing the following predicates with the related interpretations:

$Sc(x)$: x is a Swedish citizen,
$Ip(x)$: x has Italian parents,
$Pr(x)$: x is a Protestant.

Set

$$Inf_1 = \{Sc(b) \wedge Ip(b) \rightarrow \neg Pr(b)\}, \quad Inf_2 = \{Sc(b) \rightarrow Pr(b)\}.$$

The theory $Inf_1 \cup Inf_2$ is consistent. In fact a model is obtained by assuming that $Sc(b)$ and $Pr(b)$ are true and $Ip(b)$ false. We can assume that a corresponding initial valuation v is defined as in (4.2) and therefore $\mathcal{D}^*(v)$ is generated as in (4.3). Now, if we learn that b is a Swedish citizen, that is $Sc(b)$, then this piece of information is consistent with $Inf_1 \cup Inf_2$ and, if we set $v_1 = v \cup \{Sc(b)\}$,

$$\mathcal{D}^*(v_1)(\alpha) = \begin{cases} 1 & \text{if } Sc(b) \vdash \alpha, \\ 0.8 & \text{if } Inf_1, Sc(b) \vdash \alpha \text{ and it is not } Sc(b) \vdash \alpha, \\ 0.6 & \text{if } Inf_1, Inf_2, Sc(b) \vdash \alpha \text{ and it is not } Inf_1, Sc(b) \vdash \alpha, \\ 0 & \text{otherwise.} \end{cases} \tag{4.5}$$

In particular, we are able to prove $Pr(b)$ to degree 0.6. On the other hand, if we learn both that b is a Swedish citizen and has Italian parents, that is $Sc(b)$ and $Ip(b)$, then this information is inconsistent with $Inf_1 \cup Inf_2$ and, if v_2 denotes the new information $v \cup \{Sc(b), Ip(b)\}$,

$$\mathcal{D}^*(v_2)(\alpha) = \begin{cases} 1 & \text{if } Sc(b), Ip(b) \vdash \alpha, \\ 0.8 & \text{if } Inf_1, Sc(b), Ip(b) \vdash \alpha \text{ and not } Sc(b), Ip(b) \vdash \alpha, \\ 0.6 & \text{otherwise.} \end{cases} \tag{4.6}$$

So,

$$\mathcal{D}_n^*(v_2)(\alpha) = \begin{cases} 1 & \text{if } Sc(b), Ip(b) \vdash \alpha, \\ 0.5 & \text{if } Inf_1, Sc(b), Ip(b) \vdash \alpha \text{ and not } Sc(b), Ip(b) \vdash \alpha, \\ 0 & \text{otherwise.} \end{cases} \tag{4.7}$$

Thus, while $v_1 \subseteq v_2$, we have that
- v_1 enables us to prove $Pr(b)$ to degree 0.6 and $\neg Pr(b)$ to degree 0,
- v_2 enables us to prove $Pr(b)$ to degree 0 and $\neg Pr(b)$ to degree 0.5.

5. FUZZY FILTERS AND FUZZY SUBALGEBRAS

In this section we will consider the canonical extensions of two basic algebraic closure systems: the class of all filters in a lattice and the class of all subalgebras of an algebraic structure.

Let $(L, \leq)$ be a pre-ordered set with a least element 0 and a greatest element 1. Then it is easy to prove that the class C of upper subsets of L is an algebraic closure system in the lattice $\mathcal{P}(L)$. The related canonical extension C^* is the class of all fuzzy subsets $s : L \to U$ such that

$$x \leq y \Rightarrow s(x) \leq s(y).$$

Given any subset X of L, we denote the upper subset $Co(C)(X)$ generated by X by $Upp(X)$. Then,

$$Upp(X) = \{x \in L : x \geq y \text{ for a suitable } y \in X\} \cup \{1\}, \tag{5.1}$$

and $Upp(X)$ is proper iff $0 \notin X$. The related canonical extension Upp^* can be calculate by setting, for any fuzzy subset s of L,

$$Upp^*(s)(x) = Sup\{s(x') : x' \leq x\}.$$

Assume that L is a lattice. The class C of filters is an algebraic closure system. Given a subset X of L, we denote the filter $Co(C)(X)$ generated by X by $Fil(X)$. Trivially,

$$Fil(X) = \{x : \exists x_1 \in X, \ldots, \exists x_m \in X, x \geq x_1 \wedge \ldots \wedge x_m\} \cup \{1\}. \tag{5.2}$$

In particular, for every $z \in L$, the set $Fil(\{z\}) = \{x \in L : z \leq x\}$ is the *principal filter generated by z*. Given $u \in L$ define the operator J_u by setting

$$J_u(x) = x \vee u.$$

Then we have

$$J_u = Co(Fil(\{u\})) \text{ and } Cs(J_u) = Fil(\{u\}).$$

Indeed,

$$\begin{aligned} Co(Fil(\{u\}))(x) &= Inf\{y \in Fil(\{u\}) : y \geq x\} \\ &= Inf\{y \in L : y \geq u \text{ and } y \geq x\} = u \vee x = J_u(x), \end{aligned}$$

and

$$Cs(J_u) = \{x \in L : u \vee x = x\} = \{x \in L : x \geq u\} = Fil(\{u\}).$$

Definition 5.1. We call a *fuzzy filter* of a lattice L any element of the canonical extension of the class of filters in L.

Then, we can identify the fuzzy filters with the continuous chains of filters. The proof of the following proposition is trivial.

Proposition 5.2 A fuzzy subset $s : L \to U$ of L is a fuzzy filter iff

$$s(1) = 1, \quad x \geq y \Rightarrow s(x) \geq s(y), \quad s(x \wedge y) \geq s(x) \wedge (y).$$

Another typical example of an algebraic closure system is furnished by the class of subalgebras of a given algebraic structure. Let $\mathcal{A} = (\mathbb{A}, \mathbb{H}, \mathbb{C})$ be an algebraic structure, where $\mathbb{A}$ is the domain, $\mathbb{H} = \{h_1,\dots,h_s\}$ is the set of operations on $\mathbb{A}$ and $\mathbb{C} \subseteq \mathbb{A}$ the set of constants. Assume that C is the class of subalgebras of $\mathcal{A}$, where, if there is no constant, the empty subset is considered as a subalgebra. Then C is an algebraic closure system and, for every subset X of $\mathbb{A}$, $Co(C)(X)$ is the subalgebra of $\mathcal{A}$ generated by X. In accordance with the literature, we write $<X>$ instead of $Co(C)(X)$. Let $Pol(\mathcal{A})$ be the set of polynomial functions of $\mathcal{A}$, then

$$<X> = \{z \in \mathbb{A} : \exists p \in Pol(\mathcal{A}) \text{ and } \exists x_1, \dots, \exists x_n \in X \text{ s. t. } z = p(x_1, \dots, x_n)\}. \quad (5.3)$$

To get $<X>$ by a fixed point technique, we define, for every operation h_i, a classical operator K_i by setting, for every subset X of $\mathbb{A}$,

$$K_i(X) = \{x \in \mathbb{A} : \exists x_1, \dots, \exists x_n \in X \text{ such that } x = h_i(x_1,\dots,x_n)\}, \quad (5.4)$$

i.e., $K_i(X)$ is the image of X by h_i. Moreover, define K by

$$K(X) = K_1(X) \cup \dots \cup K_s(X) \quad (5.5)$$

and H by

$$H(X) = X \cup \mathbb{C} \cup K(X). \quad (5.6)$$

Proposition 5.3. H is a continuous *a-c*-operator whose fixed points coincide with the subalgebras of $\mathcal{A}$ and, for every subset X of $\mathbb{A}$,

$$<X> = \cup_{n \in N} H^n(X). \quad (5.7)$$

Definition 5.4. Given an algebraic structure $\mathcal{A} = (\mathbb{A}, H, \mathbb{C})$, we say that the fuzzy subset $s : \mathbb{A} \to U$ is a *fuzzy subalgebra* of $\mathcal{A}$ if it is an element of the canonical extension of the class of subalgebras of $\mathcal{A}$.

Then, a fuzzy subalgebra is a fuzzy subset of $\mathbb{A}$ whose closed cuts are subalgebras of $\mathcal{A}$. Equivalently, we can identify the fuzzy subalgebras with the continuous chains of subalgebras of $\mathcal{A}$ (Rosenfeld [1971] and Di Nola and Gerla [1987]).

6. NECESSITY MEASURES AS FUZZY THEORIES

The theory of necessity measures is a basic tool in fuzzy set theory (see Dubois and Prade [1988]). In this section we will emphasize that there is a strict connection between necessity theory and the canonical extension of classical propositional calculus. Indeed, given a Boolean algebra $\boldsymbol{B}$, in Biacino and Gerla [1992]*b*, a *generalized necessity* is defined as a map $n : \boldsymbol{B} \to U$ such that,

$$n(1) = 1 \quad ; \quad n(x \wedge y) = n(x) \wedge n(y)$$

for every $x, y \in \boldsymbol{B}$. The name *"generalized necessity"* is justified by the fact that the generalized necessities n for which $n(\mathbf{0}) = 0$ are known in the literature under the name of *necessities*. Equivalently, a fuzzy subset n of $\boldsymbol{B}$ is a generalized necessity if, for every $x, y \in \boldsymbol{B}$,

$$n(1) = 1 \ ; \ x \geq y \Rightarrow n(x) \geq n(y) \ ; \ n(x \wedge y) \geq n(x) \wedge n(y).$$

The proof of the following theorem is immediate:

Theorem 6.1. *The class of generalized necessities is the canonical extension of the class of filters of* ***B*** *and therefore it coincides with the class of fuzzy filters of* ***B***.

Then, given any fuzzy subset s of $\boldsymbol{B}$, $Fil^*(s)$ is the generalized necessity generated by s. Because Fil is an algebraic closure operator, Fil^* is a continuous closure operator and

$$Fil^*(s)(x) = \begin{cases} 1 & \text{if } x = 1, \\ Sup\{s(x_1) \wedge \ldots \wedge s(x_m) : x_1 \wedge \ldots \wedge x_m \leq x\} & \text{otherwise.} \end{cases} \tag{6.1}$$

Notice that if the support of s is consistent, i.e.,

$$y_1,\ldots,y_n \in Supp(s) \Rightarrow y_1 \wedge \ldots \wedge y_n \neq 0,$$

then $Fil^*(s)(0) = 0$ and (6.1) gives the necessity generated by s (Biacino and Gerla [1992]b).

Theorem 6.2. *A fuzzy subset of formulas τ is a theory of the canonical extension of classical propositional calculus iff there exist*

- *a Boolean algebra* ***B***,
- *a Boolean model m,*
- *a generalized necessity $n : \boldsymbol{B} \to U$ such that $\tau = n \circ m$*, i.e.,

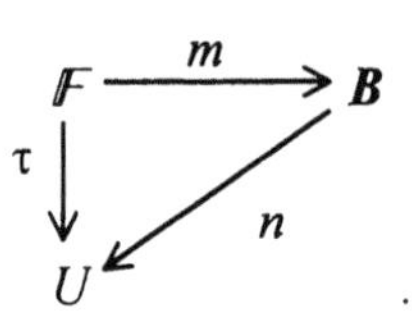

Proof. Consider the Lindenbaum-Tarski algebra of classical propositional calculus, i.e., the Boolean algebra $\boldsymbol{B}$ obtained as a quotient of the set of formulas $\mathcal{F}$ modulo the logical equivalence relation (see, e.g. Sikorski [1960] pag. 142). More specifically, given any formula α, the complete class of equivalence $[\alpha]$ is defined by setting

$$[\alpha] = \{\alpha' \in \mathcal{F} : \alpha' \text{ logically equivalent to } \alpha\}.$$

Moreover, we set

$$\boldsymbol{B} = \{[\alpha] : \alpha \in \mathcal{F}\},$$

and we define the Boolean operations by

$$[\alpha] \wedge [\beta] = [\alpha \wedge \beta] \quad ; \quad [\alpha] \vee [\beta] = [\alpha \vee \beta] \quad ; \quad -[\alpha] = [\neg\alpha].$$

As a consequence, the function $m : \mathcal{F} \to \boldsymbol{B}$ defined by setting $m(\alpha) = [\alpha]$ is a Boolean model (see Section 3 in Chapter 5). Furthermore,

$$[\alpha] \leq [\beta] \Leftrightarrow \beta \text{ is a logical consequence of } \alpha$$

$$\mathbf{1} = \{\alpha : \alpha \text{ is a tautology}\} \quad ; \quad \mathbf{0} = \{\alpha : \alpha \text{ is a contradiction}\}$$

$$[\alpha] \wedge [\beta] = 0 \Leftrightarrow \alpha \text{ is inconsistent with } \beta.$$

Now, assume that τ is a theory and define the map $n : \boldsymbol{B} \to U$ by setting, for every complete class of equivalence $[\alpha] \in \boldsymbol{B}$, $n([\alpha]) = \tau(\alpha)$. Such a definition is well-given since, as we have observed in Proposition 3.1, if two formulas α and α' are logically equivalent, then $\tau(\alpha) = \tau(\alpha')$. Since

$$n(x \wedge y) = n(x) \wedge n(y) \ ; \ \ n(1) = 1,$$

the map n is a generalized necessity measure such that $\tau(\alpha) = n(m(\alpha))$.

Conversely, let $\boldsymbol{B}$ be any Boolean algebra, m a Boolean model and $n : \boldsymbol{B} \to U$ a generalized necessity such that $\tau = n \circ m$. Then a direct inspection shows that τ satisfies (j), (jj) and (jjj) of Proposition 3.1 and therefore that τ is a theory. □

The above theorem yields a translation of all our results for generalized necessity logic into corresponding results for generalized necessities. We can then suggestively regard the canonical extension $\mathcal{S}^*$ of classical propositional calculus as a logic of generalized necessities while the 0-canonical extension $\mathcal{S}^0$ can be regarded as logic of necessity.

7. FUZZY HILBERT SYSTEMS AND FUZZY SUBALGEBRAS

Every algebraic structure defines a crisp H-system in a trivial way.

Definition 7.1. Let $\mathcal{A} = (\mathbb{A}, \mathbb{H}, \mathbb{C})$ be an algebraic structure. Then *the H-system associated with* $\mathcal{A}$ is the Hilbert deduction system $\mathcal{S}(\mathcal{A})$ in which

- $\mathbb{A}$ is the set of formulas,
- $\mathbb{H}$ is the set of rules,
- $\mathbb{C}$ is the set of logical axioms.

The interpretation of an algebraic structure as an H-system is not too artificial. Indeed, consider an algebraic structure $\mathcal{A}$, and let $\mathcal{L}$ be a first order language obtained by adding a monadic predicate name S to a language suitable for $\mathcal{A}$. We identify every element x of $\mathcal{A}$ with the formula $S(x)$ which we interpret as "x is an element of the subalgebra S". In such a way the set $\{S(c) : c \in C\}$ of logical axioms claims that the subalgebra S contains every constant c. Furthermore, given an n-ary operation h, we can interpret it as the rule of inference h saying that from $S(x_1),\ldots, S(x_n)$ we can infer $S(h(x_1,\ldots,x_n))$. This expresses the fact that a subalgebra S is closed under all the operations.

Proposition 7.2. *The class* $\mathcal{T}$ *of theories of* $\mathcal{S}(\mathcal{A})$ *coincides with the class of subalgebras of* $\mathcal{A}$ *and, for every* $X \subseteq \mathbb{A}$, *the theory* $\mathcal{D}(X)$ *is the subalgebra* $<X>$ *generated by* X. *Moreover,*

- *the set of tautologies coincides with the subalgebra* $< \mathbb{C} >$ *generated by the set of constants*
- *an element g of* $\mathcal{A}$ *is a contradiction iff g is a generator of* $\mathcal{A}$.

It is interesting to observe that any proof in $\mathcal{S}(\mathcal{A})$ of a formula x under hypotheses $x_1,\ldots,x_n$ determines a polynomial function p such that $x = p(x_1,\ldots,x_n)$.

Conversely, any polynomial function p such that $x = p(x_1,\ldots,x_n)$ determines a proof in $\mathcal{S}(\mathcal{A})$ of x.

Now, consider the canonical extension $\mathcal{S}(\mathcal{A})^*$ of $\mathcal{S}(\mathcal{A})$, that we call the *fuzzy $\mathcal{A}$-system*. Then, every n-ary operation h defines a fuzzy rule, by

$$\frac{x_1,\ldots,x_n}{h(x_1,\ldots,x_n)} \quad ; \quad \frac{\lambda_1,\ldots,\lambda_n}{\lambda_1 \wedge \ldots \wedge \lambda_n} \; .$$

This rule claims that

IF $x_1,\ldots x_n$ belong to the subalgebra S
(to a degree greater than or equal to $\lambda_1,\ldots,\lambda_n$, respectively)
THEN $h(x_1,\ldots,x_n)$ belongs to S
(to a degree at least equal to $\lambda_1 \wedge \ldots \wedge \lambda_n$).

The fuzzy $\mathcal{A}$-system is strictly related to fuzzy subalgebras theory.

Theorem 7.3. *The class of theories of the fuzzy $\mathcal{A}$-system coincides with the class of fuzzy subalgebras of $\mathcal{A}$. As a consequence,*
(i) *the intersection of a family of fuzzy subalgebras is a fuzzy subalgebra,*
(ii) *for every fuzzy subset s, $\mathcal{D}(s)$ is the fuzzy subalgebra $\langle s \rangle$ generated by s.*
Finally, a fuzzy subset s is a fuzzy subalgebra iff, for every $c \in \mathbb{C}$,

$$s(c) = 1 \tag{7.1}$$

and, for any $h \in H$ and $x_1,\ldots,x_n \in \mathbb{A}$,

$$s(h(x_1,\ldots,x_n)) \geq s(x_1) \wedge \ldots \wedge s(x_n). \tag{7.2}$$

Note that in a fuzzy $\mathcal{A}$-system the inconsistent theory is a meaningful model and hence, intuitively speaking, *"consistent"*. This is not surprising since no negative information is admitted in the related language. On the other hand, this is a usual behavior in logic programming.

The following theorem, whose proof is routine, gives a way to compute the fuzzy subalgebras generated by a given fuzzy subset v.

Theorem 7.4. *Let $Pol(\mathcal{A})$ be the set of polynomial functions of $\mathcal{A}$. Then, for every fuzzy subset s,*

$$\langle s \rangle(z) = \begin{cases} Sup\{s(x_1)\wedge\ldots\wedge s(x_n) : p(x_1,\ldots,x_n) = z \,,\, p \in Pol(\mathcal{A})\} & \text{if } z \notin \langle \mathbb{C} \rangle, \\ 1 & \text{if } z \in \langle \mathbb{C} \rangle. \end{cases} \tag{7.3}$$

To get $\langle s \rangle$ by a fixed point technique, assume that $\mathbb{H} = \{h_1,\ldots,h_s\}$ and, for any operation h_i, let K_i^* be the canonical extension of the operator K_i defined in (5.4), i.e.,

$$K_i^*(s)(x) = Sup\{s(x_1) \wedge \ldots \wedge s(x_n) : x = h_i(x_1,\ldots,x_n)\}. \tag{7.4}$$

Let K^* be the canonical extension of K, i.e.,

$$K^*(s) = \cup_{h\in H} K_i^*(s). \tag{7.5}$$

Then $K^*(s)$ is the fuzzy subset of elements of $\mathcal{A}$ that can be obtained by composing elements in s by an operation in H. If we set

$$H(s) = K^*(s) \cup s \cup \mathbb{A}\,, \tag{7.6}$$

we obtain

$$<s> \ = Sup_{n\in N} H^n(s). \tag{7.7}$$

For instance, let $\mathcal{A} = (\mathbb{A}, \cdot, 1)$ be a semigroup (see, e.g., Lallement [1979]). Then the set of logical axioms of the related fuzzy $\mathcal{A}$-system is $\{1\}$ and the only fuzzy inference rule is

$$\frac{x\,,\,y}{x\cdot y} \quad ; \quad \frac{\lambda_1, \lambda_2}{\lambda_1 \wedge \lambda_2}\,.$$

Moreover, for every fuzzy subset s,

$$<s>(z) = \begin{cases} 1 & \text{if } z = 1, \\ Sup\{s(x_1) \wedge \ldots \wedge s(x_n) : x_1\cdot\ldots\cdot x_n = z\} & \text{if } z \neq 1. \end{cases} \tag{7.8}$$

By setting

$$H(s) = s \cup \{1\} \cup s \cdot s$$

where

$$(s \cdot s)(x) = Sup\{v(x_1) \wedge v(x_2) : x_1 \cdot x_2 = x\},$$

we get that $<s> = \cup_{n\in N} H^n(s)$. If 1 belongs to s, that is $s(1) = 1$, then $H(s) = s \cdot s$ and, more generally, $H^n(s) = s^n$. Then, the semigroup generated by s is given by $<s> = \cup_{n\in N} s^n$.

Theorem 7.5. *Let $\mathcal{A} = (\mathbb{A}, \cdot, 1)$ be a free semigroup* (see, e.g., Lallement [1979]). *Then the fuzzy $\mathcal{A}$-system is with optimal proofs and p-compact. Let $\mathcal{A}$ be the semigroup of the positive real numbers with respect to the product operation. Then the fuzzy $\mathcal{A}$-system is not p-compact and, hence, is not with optimal proofs.*

Proof. The first part of the proposition is a direct consequence of (7.3) Suppose that $\mathcal{A}$ is a free semigroup. Then the fact that each z admits only a finite number of factorizations implies that $x_1,\ldots,x_n$ exist such that $\mathcal{D}(s)(z) = s(x_1) \wedge \ldots \wedge s(x_n)$. The sequence

$$x_1, x_2, \ldots, x_n, x_1 \cdot x_2, x_1 \cdot x_2 \cdot x_3, \ldots, x_1 \cdot \ldots \cdot x_n$$

is a proof π of z such that $\mathcal{D}(s)(z) = Val(\pi, s)$.

Assume that $\mathcal{A}$ is the semigroup $(U, \cdot, 1)$, that s is the identity map and that $z \in (0,1)$. Moreover, observe that, for every $n \in N$, a factoring of z is obtained by setting $x_1 = \ldots = x_n = z^{1/n}$ and that $\lim_{n\to\infty} z^{1/n} = 1$. Then it is easy to prove that

$$\mathcal{D}(s)(z) = Sup\{z^{1/n} : n \in N\} = 1.$$

Moreover, there is no finite fuzzy subset f of U such that $\mathcal{D}(f)(z) = 1$. □

Analogous results can be given for the free, pure, very pure, left unitary, right unitary, unitary fuzzy subsemigroups of a free semigroup $\mathcal{A}$ (see Gerla [1985] and Gerla [1987]*b*).

8. EXTENSIONS BY CONTINUOUS TRIANGULAR NORMS

We have observed earlier in Section 10 of Chapter 3 that (1.1) can be achieved by a valuation in a multivalued logic of the claim:

"*α is a consequence of v if a finite subset X of v exists capable of proving α* "

where the existential quantifier $\exists$ is interpreted by the operator *Sup* and the subsethood relation is defined by the minimum. Now, it is natural to interpret such a claim by different multivalued logics. For example, we can assume that U is any complete lattice and that the logical connective "and" is interpreted by a binary operation $\odot$ different from the minimum. In this section we assume that $\odot$ is associative, commutative, that $x \odot 1 = 1$ and $x \odot 0 = 0$. Moreover, we assume that $\odot$ is join-preserving. As an example, we can consider a *continuous triangular norm* in [0,1] (see Section 2 in Chapter 1). In accordance, the notion of *inclusion degree* $Incl_{\odot}(X,s)$ is defined by setting $Incl_{\odot}(X,s) = 1$ if $X = \varnothing$ and

$$Incl_{\odot}(X,s) = s(x_1) \odot \ldots \odot s(x_n)$$

if $X = \{x_1,\ldots,x_n\}$. The proof of the following properties is evident.

i) $X \subseteq Y \Rightarrow Incl_{\odot}(X,s) \geq Incl_{\odot}(Y,s)$; ii) $s \subseteq s' \Rightarrow Incl_{\odot}(X,s) \leq Incl_{\odot}(X,s')$

iii) $Incl_{\odot}(X \cup Y,s) \geq Incl_{\odot}(X,s) \odot Incl_{\odot}(Y,s)$; iv) $Incl(X,s) \geq Incl_{\odot}(X,s)$.

For example, to prove iv) observe that $x \odot y \leq x \odot 1 = x$ and therefore, since $x \odot y \leq y$, $x \odot y \leq x \wedge y$.

Definition 8.1. Let $J : \mathcal{P}(S) \to \mathcal{P}(S)$ be an operator. Then the *canonical extension of J* by $\odot$ is the operator $\mathcal{J}^{\odot} : U^S \to U^S$ defined by setting

$$\mathcal{J}^{\odot}(s)(x) = Sup\{Incl_{\odot}(X,s) : X \in \mathcal{P}_f(S), x \in J(X)\}. \tag{8.1}$$

If $\odot$ is the minimum, then $\mathcal{J}^{\odot}$ is the usual canonical extension $\mathcal{J}^*$.

Proposition 8.2. *Let J be an algebraic closure operator. Then $\mathcal{J}^{\odot} : U^S \to U^S$ is a continuous fuzzy closure operator extending J. Moreover, $Inc(\mathcal{J}^{\odot}) = 1$ and $\mathcal{J}^{\odot} \leq \mathcal{J}^*$ (while $\mathcal{J}^{\odot} \neq \mathcal{J}^*$, in general).*

Proof. It is obvious that $\mathcal{J}^{\odot}(s) \supseteq s$ and that, since $\odot$ is order-preserving, $\mathcal{J}^{\odot}$ is order-preserving. So, to prove that $\mathcal{J}^{\odot}$ is a closure operator, we need only to prove that, for every $x \in S$, $\mathcal{J}^{\odot}(s)(x) \geq \mathcal{J}^{\odot}(\mathcal{J}^{\odot}(s))(x)$ and therefore that, for every finite set X and $x \in J(X)$,

$$Incl_{\odot}(X, \mathcal{J}^{\odot}(s)) \leq Sup\{Incl_{\odot}(Z, s) : Z \in \mathcal{P}_f(S), x \in J(Z)\}. \tag{8.2}$$

If $X = \varnothing$, (8.2) is evident. Assume that $X = \{x_1, \ldots ,x_n\}$ and set, for $i = 1,\ldots, n$

$$\mathcal{J}^{\odot}(s)(x_i) = Sup\{Incl_{\odot}(Z^i, s) : Z^i \in \mathcal{P}_f(S), x_i \in J(Z^i)\}.$$

Then, since the least upper bounds are preserved by $\odot$, we obtain

$$Incl_{\odot}(X, \mathcal{J}^{\odot}(s)) = \mathcal{J}^{\odot}(s)(x_1) \odot \ldots \odot \mathcal{J}^{\odot}(s)(x_n)$$

$= Sup\{Incl_\odot(Z^1, s)\odot...\odot Incl_\odot(Z^n, s) \ : Z^i \in \mathcal{P}_f(S),\ x_i \in J(Z^i) \text{ for } i = 1,...,n\}$.*
Set $Z = Z^1\cup...\cup Z^n$. Then, since $x_i \in J(Z^i)$ for $i = 1,...,n$ and $x \in J(\{x_1,...,x_n\})$, we have that $x \in J(Z)$, and that

$$Incl_\odot(Z^1, s)\odot...\odot Incl_\odot(Z^n,s) \leq Incl_\odot(Z,s).$$

This yields (8.2).

To prove that $J^\odot$ is an extension of J, let s be a crisp subset, namely the characteristic function of the subset X. Then it is obvious that $J^\odot(s)(x) = 1$ iff either $x \in J(\emptyset)$ or $x_1,...,x_n$ exist in X such that $x \in J(\{x_1,...,x_n\})$. Since J is algebraic, this is equivalent to saying that $x \in J(X)$. Thus, $J^\odot(s)$ is the characteristic function of $J(X)$. To prove that $Inc(J^\odot) = 1$, observe that, for any $\lambda \in U$, s^λ is a theory of $J^\odot$ such that $Inc(s^\lambda) = \lambda$. The remaining part of the proof is trivial. □

Let $(\mathbb{F}, \mathcal{D})$ be an abstract crisp deduction system. Then we can consider the abstract system $(\mathbb{F}, \mathcal{D}^\odot)$ obtained as the canonical extension of $(\mathbb{F}, \mathcal{D})$ by $\odot$. In particular, given a crisp H-system $\mathcal{S} = (\mathbb{A}, \mathbb{R})$ whose deduction operator is $\mathcal{D}$, we can consider the system $(\mathbb{F}, \mathcal{D}^\odot)$. Also, we can proceed as in Section 1, by obtaining a fuzzy H-system whose inference rules are obtained by extending the inference rules in $\mathcal{S}$ by $\odot$.

Definition 8.3. Let $\mathcal{S} = (\mathbb{A}, \mathbb{R})$ be a crisp H-system and $\odot$ a continuous triangular norm. Then *the extension of* $\mathcal{S}$ *by* $\odot$ is the fuzzy H-system $\mathcal{S}^\odot = (\mathbb{A}, \mathbb{R}^\odot)$, where $\mathbb{R}^\odot = \{r^\odot : r \in \mathbb{R}\}$ and, in turn, for any $r \in \mathbb{R}$, $r^\odot = (r',r'')$ is the fuzzy rule such that

$$r' = r \quad\text{and}\quad r''(x_1,...,x_n) = x_1 \odot ... \odot x_n.$$

If $\mathcal{D}$ is the deduction operator of $\mathcal{S} = (\mathbb{A}, \mathbb{R})$, we denote by $\mathcal{D}^{\ominus}$ the deduction operator of $\mathcal{S}^\odot$. Observe that $\mathcal{D}^{\ominus} \neq \mathcal{D}^\odot$. This is so because in (8.1) the formulas $\alpha_1,...,\alpha_n$ are distinct and therefore the valuation of each formula α_i is considered once in the quantity $v(\alpha_1) \odot ... \odot v(\alpha_n)$. On the other hand, if $\alpha_1,...,\alpha_n \vdash \alpha$, then it is possible that no proof π of α exists such that each α_i occurs once.

Theorem 8.4. *Let $\mathcal{S} = (\mathbb{A}, \mathbb{R})$ be a crisp H-system whose deduction operator is $\mathcal{D}$. Then, the extension $\mathcal{S}^\odot$ of $\mathcal{S}$ by $\odot$ is a fuzzy H-system such that $Inc(\mathcal{S}^\odot) = 1$. Moreover, $\mathcal{D}^{\ominus} \leq \mathcal{D}^\odot$ while, in general, $\mathcal{D}^{\ominus} \neq \mathcal{D}^\odot$.*

Proof. Equality $Inc(\mathcal{S}^\odot) = 1$ follows from Theorem 4.2 in Chapter 4. Equivalently, observe that s^λ is a fixed point of $\mathcal{D}^{\ominus}$ and therefore that $Inc(s^\lambda) = \lambda$.

To prove that $\mathcal{D}^{\ominus} \leq \mathcal{D}^\odot$, let α be a formula and $\pi = \alpha_1,...,\alpha_m$ any normalized proof of α whose hypotheses are $\alpha_1,...,\alpha_h$. Then, $n(1), ..,n(h)$ exist such that

$$Val(\pi,v) = v(\alpha_1)^{n(1)} \odot ... \odot v(\alpha_h)^{n(h)}.$$

By observing that $\alpha_1,...,\alpha_h \vdash \alpha$ and that

$$v(\alpha_1)^{n(1)} \odot ... \odot v(\alpha_h)^{n(h)} \leq v(\alpha_1) \odot ... \odot v(\alpha_h),$$

we can conclude that $Val(\pi,v) \leq \mathcal{D}^\odot(v)(\alpha)$. Thus $\mathcal{D}^{\ominus}(v)(\alpha) \leq \mathcal{D}^\odot(v)(\alpha)$.

To give an example of an H-system such that $\mathcal{D}^{\circledcirc} \neq \mathcal{D}^{\odot}$, suppose that $\mathcal{A}$ is the free monoid generated by an element a, let $\mathcal{S}$ be the related crisp H-system and let $\odot$ be the usual product in U. Then the theories of $\mathcal{S}^{\odot}$ are the fuzzy subsets τ such that

$$\tau(1) = 1 \text{ and } \tau(x \cdot y) \geq \tau(x) \cdot \tau(y). \tag{8.3}$$

Let v be the initial valuation defined by setting $v(a) = 0.9$ and $v(x) = 0$ if $x \neq a$ and consider the fuzzy subset τ defined by setting $\tau(1) = 1$, $\tau(a^n) = (0.9)^n$. Then $\tau = H(v)$, i.e., τ is the fuzzy theory generated by v. In fact, τ satisfies (8.3) and therefore τ is a theory. Let s be a fuzzy theory containing v. We claim that for any integer n, $s(a^n) \geq \tau(a^n)$ and therefore, that $s \supseteq \tau$. Indeed, in the case $n = 0$,

$$s(a^0) = s(1) = 1 = \tau(1) = \tau(a^0).$$

In the case $n = 1$ we have that $s(a^1) \geq v(a) = \tau(a)$. Finally, if $n > 1$,

$$s(a^n) = s(a^{n-1} \cdot a) \geq s(a^{n-1}) \cdot s(a) \geq \tau(a^{n-1}) \cdot \tau(a) = \tau(a^n).$$

On the other hand, since for every $n \in N$ it is $a \vdash a^n$, we can conclude that $\mathcal{D}^{\odot}(v)(a^n) = 0.9$ and therefore, that $\mathcal{D}^{\circledcirc}(v)(a^n) \neq \mathcal{D}^{\odot}(v)(a^n)$ for every $n > 1$. ◻

The next theorem shows that the presence among the connectives of an implication for which a deduction theorem holds, enables us to prove that $\mathcal{D}^{\circledcirc} = \mathcal{D}^{\odot}$.

Theorem 8.5. *Let $\mathcal{S} = (\mathcal{A}, \mathcal{R})$ be either the classical propositional or a first order calculus and let $\mathcal{D}$ be its deduction operator. Then, $\mathcal{D}^{\circledcirc} = \mathcal{D}^{\odot}$.*

Proof. Observe that in the first order calculus a weak form of Deduction Theorem holds, i.e.,

$$\alpha_1,\dots,\alpha_n \vdash \alpha \text{ iff } Q(\alpha_1) \to (\dots (Q(\alpha_n) \to \alpha)) \text{ is logically true}$$

where $Q(\alpha)$ is the universal closure of the formula α. Assume that $\alpha_1,\dots,\alpha_n \vdash \alpha$. Then a (classical) proof π of $Q(\alpha_1) \to (\dots (Q(\alpha_n) \to \alpha))$ exists whose valuation is 1. Also, recall that the generalization rule does not change the truth values. Then, the sequence

$$\begin{array}{lll}
 & \alpha_1, & v(\alpha_1), \\
 & \dots, & \dots \\
 & \alpha_n, & v(\alpha_n), \\
 & Q(\alpha_1), & v(\alpha_1), \\
 & \dots & \dots \\
 & Q(\alpha_n), & v(\alpha_n), \\
\pi & \left\{ \begin{array}{l} \dots \\ Q(\alpha_1) \to (\dots (Q(\alpha_n) \to \alpha)), \end{array} \right. & \begin{array}{l} \\ 1, \end{array} \\
 & Q(\alpha_2) \to (\dots (Q(\alpha_n) \to \alpha)), & v(\alpha_1), \\
 & \dots & \dots \\
 & Q(\alpha_n) \to \alpha & v(\alpha_1) \odot \dots \odot v(\alpha_{n-1}), \\
 & \alpha & v(\alpha_1) \odot \dots \odot v(\alpha_n).
\end{array}$$

gives a proof π of α and the related valuation. Thus, since $Val(\pi,v) = v(\alpha_1) \odot \dots \odot v(\alpha_n)$, we have that $\mathcal{D}^{\circledcirc} \geq \mathcal{D}^{\odot}$ and therefore $\mathcal{D}^{\circledcirc} = \mathcal{D}^{\odot}$.

In the case of propositional calculus one proceeds in a similar way. □

Remark 1 (the interpretation of ∃). The interpretation of the existential quantifier by the operator *Sup* : $\mathcal{P}(U) \to U$ is on the basis of Definition 8.1. This is rather questionable everywhere the logical connective "and" is interpreted by a triangular norm different from the minimum. In fact the operator used to interpret ∃ must extend to the infinitary case the interpretation of the binary connective "or", i.e., the co-norm ⊕ associated with ⊙. Now *Sup* satisfies such a condition only in the case that ⊙ is the minimum and therefore ⊕ is the maximum. A natural candidate for the general case is the operator $\bigoplus : \mathcal{P}(U) \to U$ defined by setting, for any subset X of U,

$$\bigoplus(X) = Sup\{x_1 \oplus \ldots \oplus x_n : x_1, \ldots, x_n \in X\}.$$

In accordance, it should be interesting to examine a fuzzy logic whose deduction operator is defined by

$$\mathcal{D}(v)(\alpha) = \bigoplus(\{Incl_{\odot}(X, v) : X \in \mathcal{P}_f(\mathcal{F}) \text{ and } X \vdash \alpha\}). \quad (8.4)$$

Such a proposal requires further investigation. For example, it is not clear whether $\mathcal{D}$ is a closure operator or not.

Remark 2 (the good triangular norm). We conclude this chapter with the following basic question:

which triangular norm should be chosen for fuzzy logic ?

My feeling is that in fuzzy logic several kinds of triangular norms (and therefore several kinds of conjunctions, disjunctions and implications) must be employed at the same time. As an example, assume that we have to build an expert system for suggesting tourist mountain paths, and that p_1, p_2, ... are successive segments of a passage through the mountains. Let's say that q_n is the path obtained by connecting $p_1, \ldots, p_n$. Moreover, assume that each segment p_i is a light challenge, that is, very short and not physically demanding. Assume also that p_i is completely safe from any hazards. Then, by considering a language with two predicate constants *LIGHT* and *SAFE*, we can assume the axioms:

$LIGHT(q_1)$ (to degree 1),
$LIGHT(q_i) \to LIGHT(q_{i+1})$ (at least to degree λ)
$SAFE(q_1)$ (to degree 1),
$SAFE(q_i) \to SAFE(q_{i+1})$ (at least to degree μ)

with λ and μ different from 0 and 1. By proceeding as in the heap paradox, we can conclude that $LIGHT(q_n)$ and $SAFE(q_n)$ can be proved to degree λ^{n-1} and μ^{n-1}, respectively. Now, the lightness of a path depends strongly on its length. Therefore, our expectation is that an integer m exists such that, for any $n \geq m$, $\lambda^n = 0$. On the other hand, the safety of a path does not depend on its length. Therefore, if we express the fact that every segment p_i is sufficiently safe by the axiom

$SAFE(p_i)$ (at least to degree 0.8),

we expect that $\mu^n \geq 0.8$ for any integer n. We can obtain both of these conditions only by admitting that the triangular norm in the power λ^n is different from the triangular norm in μ^n. As an example, we can suppose that λ^n refers to ukasiewicz

Łukasiewicz conjunction and μ^{n} refers to the minimum. Accordingly, we need two *MP* rules whose valuation parts coincide with Łukasiewicz conjunction and the minimum, respectively. To do this, we need two implications $\rightarrow$ and $\hookrightarrow$ in our language enabling us to express the two kinds of axioms by

$$LIGHT(q_i) \rightarrow LIGHT(q_{i+1}) \quad \text{and} \quad SAFE(q_i) \hookrightarrow SAFE(q_{i+1})$$

More generally, we think that is not possible to propose a finite set of logical connectives as a definitive base for multivalued logic. Indeed, in classical logic this is possible only since a functional completeness theorem holds, i.e., any n-ary function from $\{0,1\}$ to $\{0,1\}$ can be defined by $\wedge$, $\vee$ and $\neg$. In particular any possible connective can be defined by these connectives and this makes any new connectives useless. A cardinality argument shows that a similar theorem is impossible for functions defined in $[0,1]$.

CHAPTER 7

GRADED CONSEQUENCE RELATIONS

1. GRADED INFORMATION WITH GRADED DEDUCTIVE TOOLS

Formula (6.1) in Chapter 3 for the canonical extension of a closure operator enables us to apply a crisp deduction apparatus to fuzzy information, i.e., information *"stratified"* at several levels of validity. Now, it is also possible to apply a *"stratified"* deduction apparatus (i.e., various deductive instruments each with a related degree of validity) to a crisp information. We can represent such a state of affairs assuming that, for every $\lambda \in U$, a crisp deduction operator $\mathcal{D}_\lambda$ is defined. Given a set X of formulas, we interpret $\mathcal{D}_\lambda(X)$ as the set of formulas that we can derive from X by using arguments which are *"reliable"* to degree λ.

More generally, it is possible that the available information and the deduction apparatus are both stratified. In this case, given a fuzzy information v, wherever $\alpha \in \mathcal{D}_\lambda(\mathrm{C}(v,\lambda))$ for a suitable $\lambda \in U$, we say that *α is a consequence of v to degree λ*. Obviously, we must consider the lower-constraint for the truth degree of α which is the best we are able to get. Then, it is natural to consider the number

$$\mathcal{D}(v)(\alpha) = Sup\{\lambda \in U : \alpha \in \mathcal{D}_\lambda(\mathrm{C}(v,\lambda))\} \tag{1.1}$$

as the best lower-constraint for this truth degree. This suggests a generalization of the formula for the canonical extension of a classical closure operator we will consider in the next section. Such a generalization will be useful to investigate the interesting notion of a graded consequence relation proposed in Chakraborty [1988].

2. STRATIFIED FUZZY CLOSURE OPERATORS

The following definition enables us to associate any family of crisp operators with a fuzzy operator.

Definition 2.1. Let $(J_\lambda)_{\lambda\in U}$ be a family of operators in a set S and let J be the fuzzy operator defined by setting, for every $s \in \mathcal{F}(S)$ and $x \in S$,

$$J(s)(x) = Sup\{\lambda \in U : x \in J_\lambda(\mathrm{C}(s,\lambda))\}. \tag{2.1}$$

Then we say that J is the *fuzzy operator associated with* $(J_\lambda)_{\lambda\in U}$.

We say that a family $(J_\lambda)_{\lambda\in U}$ of operators is a *chain* provided that $(J_\lambda(X))_{\lambda\in U}$ is a chain for every subset X, i.e.,

(i) J_0 is the map constantly equal to S,

(ii) $(J_\lambda)_{\lambda\in U}$ is order-reversing.

We say that $(J_\lambda)_{\lambda\in U}$ is a *continuous chain* provided that $(J_\lambda(X))_{\lambda\in U}$ is a continuous chain for every subset X, i.e.,

(j) J_0 is the map constantly equal to S,
(jj) $J_\mu(X) = \bigcap_{\lambda<\mu} J_\lambda(X)$ for every subset X and $\mu \in U$.

Definition 2.1 extends the notion of canonical extension. Indeed, we have the following proposition:

Proposition 2.2. *Let J be associated with a chain $(J_\lambda)_{\lambda\in U}$ and H a crisp operator. Then the following are equivalent:*
(a) *J is an extension of H.*
(b) *$J_\lambda = H$ for every $0 < \lambda < 1$.*
(c) *J is the canonical extension of H.*

Proof. Implications (b) $\Rightarrow$ (c) and (c) $\Rightarrow$ (a) are evident, so we only have to prove (a) $\Rightarrow$ (b). Assume that J extends H, i.e., for every subset X of S, $x \in H(X)$ iff $J(X)(x)=1$. Then, first we prove that $H(X) \subseteq J_\mu(X)$ for any $\mu \neq 0$ and $\mu \neq 1$. Indeed otherwise an element x of $H(X)$ exists such that $x \notin J_\mu(X)$. Then $1 = Sup\{\lambda \in U : x \in J_\lambda(X)\} \leq \mu < 1$, a contradiction. On the other hand, $J_\mu(X) \subseteq H(X)$. Indeed, $x \in J_\mu(X)$ entails $J(X)(x) \neq 0$ and hence $J(X)(x) = 1$. □

Of course, we are interested in families of closure operators.

Theorem 2.3. *Let $(J_\lambda)_{\lambda\in U}$ be a family of closure operators and let J be the associated operator. Then J is a fuzzy a-c-operator which is not, in general, a closure operator. Assume that $(J_\lambda)_{\lambda\in U}$ is a chain, then J is a fuzzy closure operator.*

Proof. Trivially, J is order-preserving. In order to prove the inclusion property, observe that, since $C(s,\lambda) \subseteq J_\lambda(C(s,\lambda))$, we have

$$s(x) = Sup\{\lambda \in U : x \in C(s,\lambda)\} \leq Sup\{\lambda \in U : x \in J_\lambda(C(s,\lambda))\} = J(s)(x).$$

To show an example in which J is not a closure operator, let $\boldsymbol{R}$ be the real line and, for every subset X of $\boldsymbol{R}$, denote the topological closure of X by $c(X)$ and the smallest closed convex subset containing X by $<X>$. Then we can consider the order-preserving family $(J_\lambda)_{\lambda\in U}$ defined by setting $J_\lambda(X) = c(X)$ for any $\lambda < 0.5$ and $J_\lambda(X) = <X>$ for $\lambda \geq 0.5$. We claim that the associated operator J is not a closure operator. Indeed, consider the fuzzy subset s defined by setting $s(x) = |x|$ if $-0.5 < x < 0.5$ and, otherwise, by setting $s(x) = 0$. Then, for every $\lambda \neq 0$, $C(s,\lambda) = (-0.5,\lambda] \cup [\lambda,0.5)$ if $\lambda < 0.5$ and $C(s,\lambda) = \varnothing$ if $\lambda \geq 0.5$. Consequently, since

$$J(s)(x) = Sup\{\lambda \in U : x \in J_\lambda(C(s,\lambda))\} = Sup\{\lambda < 0.5 : x \in c(C(s,\lambda))\},$$

it is $J(s)(x) = |x|$ if $-0.5 \leq x \leq 0.5$ and, otherwise, $J(s)(x) = 0$. On the other hand, since $J_{0.5}(C(J(s),0.5)) = <\{-0.5, +0.5\}> = [-0.5,+0.5]$, we have that $J(J(s))(x) = 0.5$ for every $x \in [-0.5,+0.5]$. This proves that $J(J(s)) \neq J(s)$.

Finally, assume that $(J_\lambda)_{\lambda\in U}$ is a chain. Then, to prove that $J(J(s)) = J(s)$, it is sufficient to prove that every cut $C(J(s),\lambda)$ is a fixed point for J_λ. Indeed, in this case

$$J(J(s))(x) = Sup\{\lambda \in U : x \in J_\lambda(C(J(s),\lambda))\}$$

$= Sup\{\lambda \in U : x \in C(J(s),\lambda)\} = J(s)(x).$

Now, observe that $(J_\lambda(C(s,\lambda)))_{\lambda \in U}$ is a chain of subsets of S. In fact, $J_0(C(s,0)) = S$ and if $\lambda \leq \lambda'$, then $J_\lambda(C(s,\lambda)) \supseteq J_\lambda(C(s,\lambda')) \supseteq J_{\lambda'}(C(s,\lambda'))$. Furthermore, observe that if $\mu \leq \lambda$, then every fixed point for J_μ is a fixed point for J_λ. In particular, $J_\mu(C(s,\mu))$ is a fixed point for J_λ. By recalling that the intersection of a class of fixed points is a fixed point and that, by Proposition 4.7 in Chapter 2,

$$C(J(s),\lambda) = \bigcap_{\mu<\lambda} J_\mu(C(s,\mu)),$$

we conclude that $C(J(s),\lambda)$ is a fixed point for J_λ. □

Observe that from Proposition 4.7 in Chapter 2 it follows that if $(J_\lambda)_{\lambda \in U}$ is a chain of closure operators and J the associated operator, then, for every $\mu \in U$,

$$O(J(s),\mu) = \bigcup_{\lambda>\mu} J_\lambda(C(s,\lambda)) \subseteq J_\mu(C(s,\mu)) \subseteq \bigcap_{\lambda<\mu} J_\lambda(C(s,\lambda)) = C(J(s),\mu).$$

The following proposition shows that every fuzzy closure operator obtained by a chain of closure operators can be obtained by a continuous chain of closure operators.

Proposition 2.4. *Let $(J_\lambda)_{\lambda \in U}$ be any chain of closure operators and set, for every $\lambda \in U$ and $X \subseteq S$,*

$$J''_\lambda(X) = \bigcap_{\mu<\lambda} J_\mu(X). \tag{2.3}$$

Then $(J''_\lambda)_{\lambda \in U}$ is a continuous chain of closure operators. Moreover, the fuzzy closure operator associated with $(J_\lambda)_{\lambda \in U}$ coincides with the operator associated with $(J''_\lambda)_{\lambda \in U}$.

Proof. J''_λ is a closure operator since it is an intersection of closure operators. The remaining part of the proposition is a direct consequence of the results given in Chapter 2. □

Definition 2.5. Let $(J_\lambda)_{\lambda \in U}$ be a family of closure operators and let J' be the associated fuzzy operator. We define the *fuzzy closure operator associated with* $(J_\lambda)_{\lambda \in U}$ as the closure operator J generated by J'. In this case we say that J is *stratified*. If $(J_\lambda)_{\lambda \in U}$ is a chain, we say that J is *well-stratified*.

An example of a fuzzy operator which is stratified but not well-stratified is given at the end of Section 4.

Example. In Castro, Trillas, Cubillo [1994] any fuzzy relation $Imp : S \times S \to U$ in S such that, for any x, y, z in S,

(a) $Imp(x,x) = 1$ (reflexivity),

(b) $Imp(x,y) \wedge Imp(y,z) \leq Imp(x,z)$ (transitivity)

is called a *fuzzy implication.* Given a fuzzy implication Imp the fuzzy operator J *associated with Imp* is defined by setting

$$J(s)(z) = Sup\{s(x) \wedge Imp(x,z) : x \in S\}. \tag{2.4}$$

We claim that J is well-stratified. Indeed, set, for any $\lambda \in U$,

$$J_\lambda(X) = \{x \in S : \exists x' \in X \text{ such that } Imp(x',x) \geq \lambda\}. \tag{2.5}$$

It is easy to demonstrate that $(J_\lambda)_{\lambda \in U}$ is a chain of closure operators and that J is the fuzzy closure operator J associated with this chain. Note that if *Imp* is not crisp, J is not an extension of a crisp operator and therefore, by Proposition 2.2, J is not a canonical extension of a crisp operator.

3. STRATIFIED CLOSURE SYSTEMS

Now, we define a notion of stratified closure system which is well related to the notion of stratified closure operator.

Definition 3.1. Let $(C_\lambda)_{\lambda \in U}$ be a family whose elements are classes of subsets of a given set S. Then the class of fuzzy subsets of S

$$C = \{s \in \mathcal{F}(S) : \mathrm{C}(s,\lambda) \in C_\lambda \text{ for every } \lambda \neq 0\} \tag{3.1}$$

is said to be *the class of fuzzy subsets associated with* $(C_\lambda)_{\lambda \in U}$.

Proposition 3.2. *Let C be the class of fuzzy subsets associated with the family $(C_\lambda)_{\lambda \in U}$. Then X is a crisp element of C iff, for any $\lambda \neq 0$, X is a crisp element of C_λ. In other words, $\bigcap_{\lambda \neq 0} C_\lambda$ is the class of crisp elements of C.*

Proof. If s is a crisp fuzzy set and $X = Supp(s)$, then $s \in C$ iff $X = \mathrm{C}(s,\lambda) \in C_\lambda$ for every $\lambda \neq 0$. □

Obviously, we are interested in families of closure systems, obviously. We say that a family of closure systems $(C_\lambda)_{\lambda \in U}$ is a *chain* if

(i) $C_0 = \{S\}$,

(ii) $(C_\lambda)_{\lambda \in U}$ is order-preserving.

Such a family is called a *continuous chain* if, for every $\lambda \in U$,

(j) $C_0 = \{S\}$,

(jj) $C_\lambda = Sup\{C_\mu : \mu < \lambda\}$.

Here the operator *Sup* is the join in the lattice of closure systems and hence (jj) means that C_λ is the closure system generated by $\bigcup_{\mu < \lambda} C_\mu$. This notion is well related to the notion of continuous chain for closure operators.

(3.1) generalizes the formula for the canonical extension of a class of subsets. More precisely:

Proposition 3.3. *Let C be the fuzzy system associated with a chain $(C_\lambda)_{\lambda \in U}$ of classes of subsets. Then C is the canonical extension of a crisp class $\mathcal{H}$ iff every C_λ, with $\lambda \neq 0$, coincides with $\mathcal{H}$.*

Proof. Suppose $C = \mathcal{H}^*$ where $\mathcal{H}$ is a class of subsets. Then, by Proposition 3.2, $\mathcal{H} = \bigcap_{\lambda \neq 0} C_\lambda$. Assume that $\lambda \neq 0$ exists such that $C_\lambda \neq \mathcal{H}$ and let X be an element of C_λ such that $X \notin \mathcal{H}$. Now, if $s = \lambda \vee X$, then $\mathrm{C}(s,\mu) = S$ if $\mu < \lambda$, otherwise $\mathrm{C}(s,\mu) = X$. As a consequence, since $(C_\lambda)_{\lambda \in U}$ is a chain, $\mathrm{C}(s,\mu) \in C_\mu$ for every $\mu \in$

U. This demonstrates that $s \in C$ while it is obvious that $s \notin \mathcal{H}^*$. This contradiction proves that $C_\lambda = \mathcal{H}$. The converse implication is trivial. □

Theorem 3.4. *Let $(C_\lambda)_{\lambda \in U}$ be a family of closure systems. Then the class C associated with $(C_\lambda)_{\lambda \in U}$ is a fuzzy closure system.*

Proof. Let $(s_i)_{i \in I}$ be a family of elements of C. Then, since for every $\lambda \in U$,

$$\mathrm{C}(\bigcap_{i \in I} s_i, \lambda) = \bigcap_{i \in I} \mathrm{C}(s_i, \lambda) \in C_\lambda,$$

we have that $\bigcap_{i \in I} s_i \in C$ and this proves that C is a closure system. □

Note that, differently from the closure operators, a fuzzy closure system C associated with a family of closure systems $(C_\lambda)_{\lambda \in U}$ is always an extension of a classical closure system, namely the system $\bigcap_{\lambda \neq 0} C_\lambda$.

Definition 3.5. Let C be the fuzzy closure system associated with a family $(C_\lambda)_{\lambda \in U}$ of closure systems, then C is said to be *stratified*. If $(C_\lambda)_{\lambda \in U}$ is a chain, then we say that C is *well-stratified*.

Notice that the class of stratified closure systems is closed under finite and infinite intersections and therefore is a closure system.

Any family $(J_\lambda)_{\lambda \in U}$ of closure operators defines a corresponding family $(Cs(J_\lambda))_{\lambda \in U}$ of closure systems and any family $(C_\lambda)_{\lambda \in U}$ of closure systems defines a corresponding family $(Co(C_\lambda))_{\lambda \in U}$ of closure operators. Indeed, we have the following natural equivalences:

Proposition 3.6. *Let $(J_\lambda)_{\lambda \in U}$ be a family of closure operators. Then*

$$(J_\lambda)_{\lambda \in U} \text{ is a chain} \Leftrightarrow (Cs(J_\lambda))_{\lambda \in U} \text{ is a chain}$$

$$(J_\lambda)_{\lambda \in U} \text{ is a continuous chain} \Leftrightarrow (Cs(J_\lambda))_{\lambda \in U} \text{ is a continuous chain.}$$

Proof. The first equivalence is trivial. Assume that $(J_\lambda)_{\lambda \in U}$ is a continuous chain. We claim that $Cs(J_\mu)$ is the closure system generated by $\bigcup_{\lambda<\mu} Cs(J_\lambda)$. In fact, it is obvious that $Cs(J_\mu)$ is a closure system containing $\bigcup_{\lambda<\mu} Cs(J_\lambda)$. Let C be a closure system containing the class $\bigcup_{\lambda<\mu} Cs(J_\lambda)$ and let X be an element of $Cs(J_\mu)$. Then, since $X = J_\mu(X) = \bigcap_{\lambda<\mu} J_\lambda(X)$ and $J_\lambda(X)$ is an element of $Cs(J_\lambda)$, X is an intersection of elements of $\bigcup_{\lambda<\mu} Cs(J_\lambda)$. Hence $X \in C$ and this proves that $C \supseteq Cs(J_\mu)$.

Suppose $(Cs(J_\lambda))_{\lambda \in U}$ is a continuous chain. Then $(J_\lambda)_{\lambda \in U}$ is a chain and $Cs(J_\mu)$ is the closure system generated by $\bigcup_{\lambda<\mu} Cs(J_\lambda)$. Consequently, since $J_\mu(X)$ is an element of $Cs(J_\mu)$, a family $(X_{\lambda(i)})_{i \in I}$ of subsets of S exists such that $\lambda(i) < \mu$, $X_{\lambda(i)}$ is a fixed point of $J_{\lambda(i)}$ and $J_\mu(X) = \bigcap_{i \in I} X_{\lambda(i)}$. Furthermore, since $X_{\lambda(i)} \supseteq X$ entails $X_{\lambda(i)} \supseteq J_{\lambda(i)}(X)$, we have

$$J_\mu(X) = \bigcap_{i \in I} J_{\lambda(i)}(X) \supseteq \bigcap_{\lambda<\mu} J_\lambda(X).$$

Now, $(J_\lambda)_{\lambda\in U}$ is order-reversing and therefore $J_\mu(X) \subseteq \bigcap_{\lambda<\mu} J_\lambda(X)$. Thus, $J_\mu(X) = \bigcap_{\lambda<\mu} J_\lambda(X)$ and this proves the continuity of $(J_\lambda)_{\lambda\in U}$. □

The next theorem says that Definitions 2.1 and 3.1 are related in a natural way:

Theorem 3.7. *Let J be the closure operator associated with a family $(J_\lambda)_{\lambda\in U}$ of closure operators and C the closure system associated with the family $(Cs(J_\lambda))_{\lambda\in U}$ of closure systems. Then, $J - Co(C)$, that is*

$$\begin{array}{ccc} (J_\lambda)_{\lambda\in U} & \longrightarrow & (Cs(J_\lambda))_{\lambda\in U} \\ \downarrow & & \downarrow \\ J & \longleftarrow & C. \end{array}$$

Proof. Let J' be the operator associated with $(J_\lambda)_{\lambda\in U}$. In order to prove $J = Co(C)$ we can prove that $Cs(J) = C$, i.e., that $C = Cs(J')$. Let s be an element of C, then every cut $C(s,\lambda)$ belongs to $Cs(J_\lambda)$ and therefore it is a fixed point for J_λ. Then

$$J'(s)(x) = Sup\{\lambda \in U : x \in J_\lambda(C(s,\lambda))\} = Sup\{\lambda \in U : x \in C(s,\lambda)\} = s(x)$$

and this proves that $s \in Cs(J')$. Conversely, if $J'(s) = s$, then, for every $x \in S$,

$$Sup\{\lambda \in U : x \in J_\lambda(C(s,\lambda))\} = s(x).$$

In other words, $x \in J_\lambda(C(s,\lambda))$ implies $\lambda \le s(x)$ and this proves that $x \in C(s,\lambda)$. Then, since $J_\lambda(C(s,\lambda))$ is contained in $C(s,\lambda)$, $C(s,\lambda)$ is a fixed point for J_λ. Thus s is an element of C. □

In a similar way one demonstrates the next theorem:

Theorem 3.8. *Let C be the fuzzy closure system associated with a family $(C_\lambda)_{\lambda\in U}$ of closure systems and J the closure operator associated with the family $(Co(C_\lambda))_{\lambda\in U}$ of closure operators. Then, $C = Cs(J)$, that is*

$$\begin{array}{ccc} (C_\lambda)_{\lambda\in U} & \longrightarrow & (Co(C_\lambda))_{\lambda\in U} \\ \downarrow & & \downarrow \\ C & \longleftarrow & J. \end{array}$$

Corollary 3.9. *If J is a closure operator, then*

J stratified $\Leftrightarrow$ *Cs(J) stratified,*

J well-stratified $\Leftrightarrow$ *Cs(J) well-stratified.*

If C is a fuzzy closure system, then

C stratified $\Leftrightarrow$ *Co(C) stratified,*

C well-stratified $\Leftrightarrow$ *Co(C) well-stratified.*

Proposition 3.10. *Assume that J_1 and J_2 are two (well) stratified closure operators. Then the closure operator generated by the product $J_1 \circ J_2$ is (well) stratified.*

Proof. Observe that $Cs(J_1)$ and $Cs(J_2)$ are both (well) stratified and therefore $Cs(J_1) \cap Cs(J_2)$ is also (well) stratified. Moreover, by Theorem 9.1 in Chapter 1, $Cs(J_1) \cap Cs(J_2)$ is the set of fixed points of $J_1 \circ J_2$. This completes the proof. □

4. A CHARACTERIZATION OF STRATIFIED CLOSURE SYSTEMS

Observe that every fuzzy closure system determines a family of closure systems in a natural way.

Definition 4.1. Let C be a class of fuzzy subsets and set, for every $\lambda \in U$

$$\mathcal{H}(C,\lambda) = \{\mathrm{C}(s,\lambda) : s \in C\}. \tag{4.1}$$

Then we say that $(\mathcal{H}(C,\lambda))_{\lambda \in U}$ is *the family of classes associated with C.*

Proposition 4.2. *For every fuzzy closure system C, and $\lambda \in U$, the class $\mathcal{H}(C,\lambda)$ is a closure system. Moreover, if C is associated with a family $(C_\lambda)_{\lambda\in U}$ of closure systems then, for every $\lambda \in U$,*

$$\mathcal{H}(C,\lambda) \subseteq C_\lambda.$$

Finally, in the case that $(C_\lambda)_{\lambda\in U}$ is a continuous chain,

$$\mathcal{H}(C,\lambda) = C_\lambda.$$

Proof. Let $(X_i)_{i\in I}$ be a family of elements of $\mathcal{H}(C,\lambda)$. Then a family $(s_i)_{i\in I}$ of elements of C exists such that $X_i = \mathrm{C}(s_i,\lambda)$. Since $\bigcap_{i\in I}X_i = \mathrm{C}(\bigcap_{i\in I}s_i,\lambda)$ and $\bigcap_{i\in I}s_i$ belongs to C, the set $\bigcap_{i\in I}X_i$ belongs to $\mathcal{H}(C,\lambda)$. This demonstrates that $\mathcal{H}(C,\lambda)$ is a closure system.

Assume that C is associated with a family $(C_\lambda)_{\lambda\in U}$ of closure systems and let $X \in \mathcal{H}(C,\lambda)$. Then an element s_λ in C exists such that $\mathrm{C}(s_\lambda,\lambda) = X$. The fact that every λ-cut of an element in C belongs to C_λ enables us to conclude that $X \in C_\lambda$.

Assume that $(C_\lambda)_{\lambda\in U}$ is a continuous chain and that $X \in C_\lambda$. Then, since C_λ is the closure operator generated by $\bigcup_{\mu<\lambda} C_\mu$, $X = \bigcap_{i\in I}X_i$ where each X_i is an element of a suitable $C_{\mu(i)}$, for $\mu(i) < \lambda$. Consider the fuzzy subset $s_i = X_i \vee \mu(i)$. We claim that $s_i \in C$. Indeed, since S belongs to any closure system, in the case $t \leq \mu(i)$ we have $\mathrm{C}(s_i,t) = S \in C_t$. In the case $t > \mu(i)$, since $\mathrm{C}(s,t) = X_i$ and $X_i \in C_{\mu(i)}$, from the inclusion $C_{\mu(i)} \subseteq C_t$, it follows that $\mathrm{C}(s,t) \in C_t$. From $s_i \in C$ it follows that $X_i = \mathrm{C}(s_i,\lambda) \in \mathcal{H}(C,\lambda)$. Since $\mathcal{H}(C,\lambda)$ is a closure system, this demonstrates that $X = \bigcap_{i\in I}X_i \in \mathcal{H}(C,\lambda)$. □

Definition 4.3. Given a fuzzy closure system C, we denote by C^* the fuzzy closure system associated with the family $(\mathcal{H}(C,\lambda))_{\lambda\in U}$ and we say that C^* is *the stratified closure system associated with C.*

In other words, we set

$$C^* = \{s \in \mathcal{F}(S) : \mathrm{C}(s,\lambda) \in \mathcal{H}(C,\lambda) \text{ for every } \lambda \in U\}. \tag{4.2}$$

The following proposition, whose proof is trivial, shows that such a notation is in accordance with the notation for the canonical extension of a classical closure system given in Chapter 3.

Proposition 4.4. *If C is a classical closure system, then the stratified closure system associated with C by* (4.2) *coincides with the canonical extension of C.*

The following proposition shows that the map associating any fuzzy closure system C with the fuzzy closure system C^* defines a closure operator in the lattice of fuzzy closure systems. Furthermore, it characterizes the stratified fuzzy closure systems as the fixed points of such an operator.

Theorem 4.5. *Assume that C, C_1 and C_2 are fuzzy closure systems. Then*

(i) $C \subseteq C^*$,

(ii) $C_1 \subseteq C_2 \Rightarrow C_1^* \subseteq C_2^*$,

(iii) $(C^*)^* = C^*$.

Moreover,

$$C = C^* \Leftrightarrow C \text{ is a stratified closure system.}$$

Proof. Properties (i), (ii) and (iii) are obvious. It is self-evident that if $C = C^*$, then C is stratified. Assume that C is the stratified closure system associated with a family $(C_\lambda)_{\lambda\in U}$ of closure systems. Then from the inclusion $\mathcal{H}(C,\lambda) \subseteq C_\lambda$ we have that $C^* \subseteq C$ and therefore that $C^* = C$.

The converse implication is trivial. □

Let $\mathcal{T}$ be a class of fuzzy subsets. Then we define $Q(\mathcal{T})$ by setting

$$Q(\mathcal{T}) = \{\lambda \vee \mathrm{C}(s,\lambda) : s \in \mathcal{T}, \lambda \in U\}. \tag{4.3}$$

Proposition 4.6. For every fuzzy closure system C,

$$C \subseteq C^* \subseteq c(Q(C))$$

where $c(Q(C))$ is the fuzzy closure system generated by $Q(C)$. Consequently,

$$C \supseteq Q(C) \Rightarrow C = C^* = c(Q(C)).$$

Furthermore, if $(\mathcal{H}(C,\lambda))_{\lambda\in U}$ is a chain,

$$C^* = c(Q(C))$$

and therefore,

$$C = c(Q(C)) \Leftrightarrow C = C^*.$$

Proof. Let $s \in C^*$ and observe that

$C^* = \{s \in \mathcal{F}(S) :$ for every $\lambda \in U$, $\mathrm{C}(s,\lambda) = \mathrm{C}(s_\lambda,\lambda)$ for a suitable $s_\lambda \in C\}$.

Then,

$$s = \bigcap_{\lambda\in U} \lambda \vee \mathrm{C}(s,\lambda) = \bigcap_{\lambda\in U} \lambda \vee \mathrm{C}(s_\lambda,\lambda),$$

and therefore that $s \in c(Q(C))$. This proves that $C^* \subseteq c(Q(C))$.

Suppose that $(\mathcal{H}(C,\lambda))_{\lambda\in U}$ is a chain, and let $s \in C$. Then, since $\mathrm{C}(s,\lambda) \in \mathcal{H}(C,\lambda) \subseteq \mathcal{H}(C,\mu)$ for every $\mu > \lambda$ and

$$C(\lambda \vee C(s,\lambda), \mu) = \begin{cases} S & \text{if } \mu \leq \lambda, \\ C(s,\lambda) & \text{if } \mu > \lambda, \end{cases}$$

we may conclude that $Q(C) \subseteq C^*$ and therefore, that $c(Q(C)) \subseteq C^*$. □

Examples. Let E be a finite dimensional Euclidean space, CS the class of closed subsets of E and CCS the class of closed convex subsets of S. Then we obtain a continuous chain $(C_\lambda)_{\lambda \in U}$ by setting $C_0 = \{S\}$ and

$$C_\lambda = \begin{cases} CCS & \text{if } 0 < \lambda \leq 0.5, \\ CS & \text{otherwise.} \end{cases}$$

Denote the well-stratified closure system associated with this family by C. Then

$s \in C \Leftrightarrow C(s,\lambda)$ is closed for $\lambda > 0.5$ and closed and convex for $\lambda \leq 0.5$.

Also, we have

$$C = C^* = c(Q(C)) \text{ and } \mathcal{H}(C,\lambda) = C_\lambda$$

and the class of crisp elements of C coincides with CCS.

A counterexample can be achieved by exchanging CCS with CS in defining $(C_\lambda)_{\lambda \in U}$, i.e., by setting

$$C_\lambda = \begin{cases} CS & \text{if } 0 < \lambda \leq 0.5, \\ CCS & \text{otherwise.} \end{cases}$$

Then the obtained family is order-reversing. Moreover, if C is the fuzzy closure system associated with $(C_\lambda)_{\lambda \in U}$,

$s \in C \Leftrightarrow C(s,\lambda)$ is closed for $\lambda \leq 0.5$ and closed and convex for $\lambda > 0.5$.

Trivially, since C is stratified, $C = C^*$. We claim that

$$Q(C) \not\subseteq C. \tag{4.4}$$

Indeed, let X and Y be two disjoint closed subsets such that $X \cup Y$ is not convex and Y is convex and define s by

$$s(x) = \begin{cases} 1 & \text{if } x \in Y, \\ 0.5 & \text{if } x \in X, \\ 0 & \text{otherwise.} \end{cases}$$

We have that $s \in C$ and that $\lambda \vee C(s,\lambda) = \lambda \vee (X \cup Y)$ for every $\lambda \leq 0.5$. Now, since $C(\lambda \vee X \cup Y,\mu) = X \cup Y$ for every $\mu > \lambda$, we have that $C(\lambda \vee X \cup Y,\mu) \notin C_\mu$ for every $\mu > 0.5$. Thus, $\lambda \vee C(s,\lambda) \notin C$ and this proves that $Q(C)$ is not contained in C. As a consequence of (4.4) we have

$$C^* \neq c(Q(C)). \tag{4.5}$$

Another interesting property is that, for any $\lambda \in U$,

$$C_\lambda = \mathcal{H}(C,\lambda). \tag{4.6}$$

In fact, let X be an element of C_λ. Then in the case $\lambda > 0.5$ the set X is closed and convex. Hence $X \in C$ and $X \in \mathcal{H}(C,\lambda)$. If $\lambda \leq 0.5$, then $\lambda \wedge X$ is an element of C such that $X = \mathrm{C}(\lambda \wedge X,\lambda)$. This proves that $X \in \mathcal{H}(C,\lambda)$.

Finally, note that C is not well-stratified. Indeed, otherwise, let $(Q_\lambda)_{\lambda \in U}$ be a continuous chain of closure systems whose associate system is C. Then by Proposition 4.2, we have that $Q_\lambda = \mathcal{H}(C,\lambda) = C_\lambda$. This is absurd. In fact, $(C_\lambda)_{\lambda \in U}$, is not order-preserving.

$Co(C)$ is an example of fuzzy operator which is stratified and not well-stratified.

5. A CHARACTERIZATION OF STRATIFIED OPERATORS

Given a fuzzy closure operator J, we may define a family $(\mathcal{K}(J,\lambda))_{\lambda \in U}$ of classical operators by setting

$$\mathcal{K}(J,\lambda)(X) = \mathrm{C}(J(\lambda \wedge X),\lambda) \tag{5.1}$$

for every $\lambda \in U$. If J is a deduction operator of a fuzzy logic, then we can interpret $\mathcal{K}(J,\lambda)(X)$ as the set of formulas which are consequences at least to degree λ of the formulas in X assumed at least to degree λ.

The following proposition shows that $\mathcal{K}$ and $\mathcal{H}$ are related in accordance with the diagrams:

$$\begin{array}{ccc} J & \longrightarrow & Cs(J) \\ \downarrow & & \downarrow \\ K(J,\lambda) & \longleftarrow & \mathcal{H}(Cs(J),\lambda) \end{array} \qquad\qquad \begin{array}{ccc} C & \longrightarrow & Co(C) \\ \downarrow & & \downarrow \\ \mathcal{H}(C,\lambda) & \longleftarrow & K(Co(C),\lambda). \end{array}$$

Proposition 5.1 (Castro [1993]) *Given a fuzzy closure operator J, $(\mathcal{K}(J,\lambda))_{\lambda \in U}$ is a family of closure operators. More specifically, we have*

$$\mathcal{K}(J,\lambda) = Co(\mathcal{H}(Cs(J),\lambda)). \tag{5.2}$$

Given a fuzzy closure system C, we have

$$\mathcal{H}(C,\lambda) = Cs(\mathcal{K}(Co(C),\lambda)). \tag{5.3}$$

Proof. Let X be a subset of S and suppose $x \in Co(\mathcal{H}(Cs(J),\lambda))(X)$. Then, since $\mathcal{H}(Cs(J),\lambda)$ is the class of the λ-cuts of $Cs(J)$, $x \in \mathrm{C}(s,\lambda)$ for every $s \in Cs(J)$ such that $X \subseteq \mathrm{C}(s,\lambda)$. Taking $s = J(\lambda \wedge X)$, since $s \in Cs(J)$ and

$$X = \mathrm{C}(\lambda \wedge X,\lambda) \subseteq \mathrm{C}(J(\lambda \wedge X),\lambda) = \mathrm{C}(s,\lambda),$$

we have $x \in \mathrm{C}(J(\lambda \wedge X),\lambda) = \mathcal{K}(J,\lambda)(X)$.

Conversely, suppose $x \in \mathcal{K}(J,\lambda)(X)$. Then $J(\lambda \wedge X)(x) \geq \lambda$ and hence, for any $s \in Cs(J)$ such that $s \supseteq \lambda \wedge X$, we have $x \in \mathrm{C}(s,\lambda)$. Thus, since $s \supseteq \lambda \wedge X$ iff $\mathrm{C}(s,\lambda) \supseteq X$, for every $s \in Cs(J)$ such that $\mathrm{C}(s,\lambda) \supseteq X$, we have $x \in \mathrm{C}(s,\lambda)$. This proves that $x \in Co(\mathcal{H}(Cs(J),\lambda))(X)$.

In order to prove (5.3), we apply (5.2) to the fuzzy closure operator $Co(C)$ by obtaining

$$\mathcal{K}(Co(C),\lambda) = Co(\mathcal{H}(C,\lambda)).$$

This equation is equivalent to (5.3). □

Definition 5.2. Given a fuzzy closure operator J we denote by J^* the fuzzy closure operator associated with the family $(\mathcal{K}(J,\lambda))_{\lambda \in U}$ of closure operators and we say that J^* is *the stratified operator associated with J.*

The above notation is in accordance with the notation for the canonical extension of a classical closure operator. Indeed, given an operator J, denote by J' the fuzzy operator defined by setting, for every fuzzy subset s,

$$J'(s) = J(Supp(s)).$$

Then, the following proposition holds:

Proposition 5.3. *Let J be a classical closure operator. Then the stratified operator associated with J′ coincides with the canonical extension of J.*

Proof. Observe that

$$Sup\{\lambda \in U : x \in J(C(s,\lambda))\} = Sup\{\lambda \in U : x \in J(Supp(\lambda \wedge C(s,\lambda),\lambda))\}$$
$$= Sup\{\lambda \in U : J'(\lambda \wedge C(s,\lambda))(x) \geq \lambda\}.$$

This demonstrates both that the *a-c*-closure operator associated with the family $(\mathcal{K}(J',\lambda))_{\lambda \in U}$ of closure operators is a closure operator and that this operator coincides with the canonical extension as defined in Chapter 3. □

The next proposition shows that the notion of stratified closure system C^* associated with a fuzzy closure system C is strictly related to the notion of stratified fuzzy operator J^* associated with a fuzzy closure operator J. Indeed, the following diagrams commute:

$$\begin{array}{ccc} C & \longrightarrow & Co(C) \\ \downarrow & & \downarrow \\ C^* & \longleftarrow & Co(C)^* \end{array} \qquad\qquad \begin{array}{ccc} J & \longrightarrow & Cs(J) \\ \downarrow & & \downarrow \\ J^* & \longleftarrow & Cs(J)^* \end{array}$$

Proposition 5.4. *Let C and J be a fuzzy closure system and a fuzzy closure operator, respectively. Then*

$$C^* = Cs(Co(C)^*) \ ; \ J^* = Co(Cs(J)^*). \tag{5.4}$$

Proof. If J is a fuzzy closure operator, then by Proposition 3.6, $J^* = Co(C)$ where C is the closure system associated with the family $(Cs(\mathcal{K}(J,\lambda)))_{\lambda \in U}$. Moreover, since by Proposition 5.1 $\mathcal{K}(J,\lambda) = Co(\mathcal{H}(Cs(J),\lambda))$, we have that $Cs(\mathcal{K}(J,\lambda)) = \mathcal{H}(Cs(J),\lambda)$. This proves the first part of the proposition.

Let C be a fuzzy closure system and set $J = Co(C)$. Then from the proved equality we obtain that $(Co(C))^* = Co(Cs(Co(C))^*) = Co(C^*)$ and therefore $Cs(Co(C)^*) = Cs(Co(C^*)) = C^*$. □

The following Theorem shows that Definition 5.2 gives an interior operator in the lattice of fuzzy closure operators. Furthermore, it characterizes the stratified closure operators as the fixed points of such an operator.

Theorem 5.5. *For every fuzzy closure operator J,*

$$J^* \leq J \quad ; \quad J_1 \leq J_2 \Rightarrow J_1^* \leq J_2^* \quad ; \quad J^{**} = J^*.$$

Moreover,

$$J = J^* \Leftrightarrow J \textit{ is a stratified closure operator.}$$

Proof. Inclusion $Cs(J)^* \supseteq Cs(J)$ entails that $J^* = Co(Cs(J)^*) \leq Co(Cs(J)) = J$. Likewise, since $Cs(J^*) = Cs(J)^*$,

$$J^{**} = Co(Cs(J^*)^*) = Co(Cs(J)^*) = J^*.$$

Moreover, from $J_1 \leq J_2$ it follows that $C_1 \supseteq C_2$ and, hence, that $C_1^* \supseteq C_2^*$. So, $J_1^* = Co((Cs(J_1))^*) \leq Co(Cs(J_2)^*) = J_2^*$. Finally, $J = J^*$ iff $J = Co((Cs(J))^*)$ iff $Cs(J) = Cs(J)^*$ iff a family $(C_\lambda)_{\lambda \in U}$ exists such that $Cs(J)$ is associated with it. In turn, this happens iff J is associated with $(Co(C_\lambda))_{\lambda \in U}$. □

6. STRATIFIED DEDUCTION SYSTEMS

We can interpret the definitions and the results in the previous sections in terms of deduction systems. Indeed, we interpret a family $((\mathcal{F},\mathcal{D}_\lambda))_{\lambda \in U}$ of deduction systems as a deduction apparatus stratified in accordance with the reliability of the deductive instruments used. More precisely, given a set X of formulas, we interpret $\mathcal{D}_\lambda(X)$ as the set of formulas we can derive from X to degree λ.

Definition 6.1. Let $((\mathcal{F},\mathcal{D}_\lambda))_{\lambda \in U}$ be a family of crisp deduction systems and $\mathcal{D}$ the closure operator associated with $(\mathcal{D}_\lambda)_{\lambda \in U}$. Then $(\mathcal{F},\mathcal{D})$ is called, *the fuzzy deduction system associated with* $((\mathcal{F},\mathcal{D}_\lambda))_{\lambda \in U}$. In this case, we say that $(\mathcal{F},\mathcal{D})$ is *stratified.* If $((\mathcal{F},\mathcal{D}_\lambda))_{\lambda \in U}$ is a chain, we say that $(\mathcal{F},\mathcal{D})$ is *well-stratified.*

It is rather natural to admit that $(\mathcal{D}_\lambda)_{\lambda \in U}$ is a continuous chain. Indeed, given a set X of formulas, condition $\mathcal{D}_0(X) = \mathcal{F}$ means that every formula can be considered as a consequence of X (at least) to degree zero. The inclusion $J_\mu(X) \subseteq \bigcap_{\lambda<\mu} J_\lambda(X)$, means that, for every $\lambda < \mu$, if x is a consequence of X (at least) to degree μ, then x is a consequence of X (at least) to degree λ. Condition $J_\mu(X) \supseteq \bigcap_{\lambda<\mu} J_\lambda(X)$ says that if x is a consequence of X (at least) to degree λ for any $\lambda < \mu$, then x is a consequence of X (at least) to degree μ.

If $(\mathcal{D}_\lambda)_{\lambda \in U}$ is a chain, $\mathcal{D}$ can be defined in a more expressive way by the relation $\vdash_\lambda$ defined by setting $v \vdash_\lambda \alpha$ everywhere $\alpha \in \mathcal{D}_\lambda(C(v,\lambda))$. Indeed,

$$\mathcal{D}(v)(\alpha) = Sup\{\lambda \in U : v \vdash_\lambda \alpha\}. \tag{6.1}$$

Trivially, if Tau is the fuzzy subset of tautologies of $\mathcal{D}$ and Tau_λ the set of tautologies of $\mathcal{D}_\lambda$, $Tau(\alpha) = Sup\{\lambda \in U : \alpha \in Tau_\lambda\}$ or, equivalently,

$$Tau = \bigcup_{\lambda \in U}(\lambda \wedge Tau_\lambda). \tag{6.2}$$

In particular, α is a tautology of $(\mathcal{F},\mathcal{D})$ iff α is a tautology of all the deduction systems $(\mathcal{F},\mathcal{D}_\lambda)$.

Theorem 6.2. *Let $(\mathcal{F},\mathcal{D})$ be a stratified fuzzy deduction system. Then, for every set X of formulas and $\mu \in U$,*

$$\mathcal{D}(\mu \wedge X) = (\mu \wedge \mathcal{D}(X)) \cup Tau. \tag{6.3}$$

Moreover,

$$Inc(\mathcal{D}) = 1. \tag{6.4}$$

Proof. To prove (6.3) assume that $\mathcal{D}$ is associated with a family of $(\mathcal{D}_\lambda)_{\lambda\in U}$ of closure operators. Since

$$C(\mu \wedge X, \lambda) = \begin{cases} X & \text{if } \lambda \leq \mu \\ \varnothing & \text{if } \lambda > \mu, \end{cases}$$

we have that

$$\mathcal{D}(\mu \wedge X)(x) = Sup\{\lambda \in U : x \in \mathcal{D}_\lambda(X),\ \lambda \leq \mu\} \vee Sup\{\lambda \in U : x \in \mathcal{D}_\lambda(\varnothing),\ \lambda > \mu\}.$$

Assume that $x \in \mathcal{D}_\lambda(\varnothing)$ for a suitable $\lambda > \mu$. Then, since

$$Sup\{\lambda \in U : x \in \mathcal{D}_\lambda(\varnothing),\ \lambda > \mu\} > \mu \geq Sup\{\lambda \in U : x \in \mathcal{D}_\lambda(X),\ \lambda \leq \mu\}$$

it is

$$\mu \wedge \mathcal{D}(X)(x) = Sup\{\lambda \in U : x \in \mathcal{D}_\lambda(X),\ \lambda \leq \mu\}$$

Because $\mathcal{D}$ is order-preserving,

$$\mathcal{D}(\varnothing)(x) \leq \mathcal{D}(\mu \wedge X)(x) = Sup\{\lambda \in U : x \in \mathcal{D}_\lambda(\varnothing),\ \lambda > \mu\} \leq \mathcal{D}(\varnothing)(x)$$

and $\mathcal{D}(\mu \wedge X)(x) = \mathcal{D}(\varnothing)(x)$. Then

$$\mathcal{D}(\varnothing)(x) = ((\mu \wedge \mathcal{D}(X)(x)) \vee \mathcal{D}(\varnothing))(x).$$

Assume that, for every $\lambda > \mu$, $x \notin \mathcal{D}_\lambda(\varnothing)$. Then $Sup\{\lambda \in U : x \in \mathcal{D}_\lambda(\varnothing),\ \lambda > \mu\} = 0$ and $\mathcal{D}(\varnothing)(x) \leq \mu$. As a consequence

$$\mathcal{D}(\mu \wedge X)(x) = Sup\{\lambda \in U : x \in \mathcal{D}_\lambda(X),\ \lambda \leq \mu\} = \mu \wedge \mathcal{D}(X)(x)$$
$$= ((\mu \wedge \mathcal{D}(X)) \vee \mathcal{D}(\varnothing))(x).$$

To prove (6.4), set X equal to the whole set of formulas $\mathcal{F}$. Then (6.3) becomes

$$\mathcal{D}(s^\mu) = s^\mu \cup Tau.$$

Consequently, $Inc(s^\mu) \geq \mu$ and therefore, $Inc(\mathcal{D}) \geq Sup\{Inc(s^\mu) : \mu \in U\} = 1$. □

The proof of the following theorem is trivial.

Theorem 6.3. *Let $(\mathcal{F},\mathcal{D})$ be a fuzzy deduction system. Then $(\mathcal{F},\mathcal{D})$ is stratified iff the class of its theories is a stratified closure system. Moreover, if $(\mathcal{F},\mathcal{D})$ is associated with the family $((\mathcal{F},\mathcal{D}_\lambda))_{\lambda\in U}$ of deduction systems, then*

τ is a theory of $(\mathcal{F},\mathcal{D})$ $\Leftrightarrow$ every cut $C(\tau,\lambda)$ of τ is a theory of $(\mathcal{F},\mathcal{D}_\lambda)$.

Finally,

X is a crisp theory of $(\mathcal{F},\mathcal{D})$ $\Leftrightarrow$ X is a theory of any $(\mathcal{F},\mathcal{D}_\lambda)$.

Proposition 4.6 and Theorem 4.5 entail the next theorem:

Theorem 6.4. *Assume that, for every* $\lambda \in U$,

$$\tau \text{ theory of } (\mathcal{F},\mathcal{D}) \Rightarrow \lambda \vee C(\tau,\lambda) \text{ theory of } (\mathcal{F},\mathcal{D}).$$

Then $(\mathcal{F},\mathcal{D})$ *is stratified.*

By recalling that the deduction operator of a canonical similarity logic is generated by the product of two stratified operators, from Proposition 3.10 we obtain:

Theorem 6.5. *Any canonical similarity logic is well-stratified.*

7. SEQUENTS AND CONSEQUENCE RELATIONS

We call *sequent* any element of the set $SEQ = \mathcal{P}(\mathcal{F}) \times \mathcal{F}$, i.e., any pair (X,α) where X is a set of formulas and α a formula. A sequent (X,α) represents the metalogical claim that α is a consequence of the set X of formulas. In this section we begin by giving the basic notions of the theory of crisp consequence relations. We call a *conclusion relation* any set of sequents, i.e., any binary relation $\vdash$ from $\mathcal{P}(\mathcal{F})$ to $\mathcal{F}$. Given $X \in \mathcal{P}(\mathcal{F})$ and $\alpha \in \mathcal{F}$, we write $X \vdash \alpha$ to denote that $(X,\alpha) \in \vdash$. Given $Z \in \mathcal{P}(\mathcal{F})$, we write $X \vdash Z$ to denote that $X \vdash \alpha$ for any formula α in Z.

Definition 7.1. A conclusion relation $\vdash$ is a *consequence relation* if
(i) $X \vdash \alpha$ whenever $\alpha \in X$,
(ii) $X \vdash \alpha \Rightarrow X \cup Y \vdash \alpha$,
(iii) $X \vdash Z$ and $X \cup Z \vdash \alpha \Rightarrow X \vdash \alpha$.

If $\vdash$ is a consequence relation and $X \vdash \alpha$, then we say that *α is a consequence of X*. The meanings of the above conditions are apparent. Condition (i) says that every formula in X is a consequence of X, condition (ii) that the logic under consideration is monotone, (iii) that if the set of formulas Z follows from X and we are able to prove α from $X \cup Z$, then we may prove α directly from X.

There is a strict connection between the operators and the conclusion relations.

Definition 7.2. Given an operator J, we define $\vdash_J$ by setting

$$X \vdash_J \alpha \Leftrightarrow \alpha \in J(X). \tag{7.1}$$

Given a conclusion relation $\vdash$ we denote by $J_\vdash$ the operator defined by

$$J_\vdash(X) = \{\alpha \in \mathcal{F} : X \vdash \alpha\}. \tag{7.2}$$

These definitions enable us to define a bijective correspondence between the class of operators and the class of conclusion relations.

Proposition 7.3. *Let J be an operator and let $\vdash$ be its associated conclusion relation. Then*

$$J_\vdash = J. \tag{7.3}$$

Let $\vdash$ be a conclusion relation and denote its associated operator by J. Then

$$\vdash_J = \vdash. \tag{7.4}$$

Proof. For every $X \subseteq \mathbb{F}$,

$$J_{\vdash}(X) = \{\alpha \in \mathbb{F} : X \vdash_J \alpha\} = \{\alpha \in \mathbb{F} : \alpha \in J(X)\} = J(X).$$

Likewise, for every $X \subseteq \mathbb{F}$ and $\alpha \in \mathbb{F}$

$$X \vdash_J \alpha \Leftrightarrow \alpha \in J_{\vdash}(X) \Leftrightarrow X \vdash \alpha .$$ □

Definitions (7.1) and (7.2) establish also a one-one correspondence between the crisp consequence relations and the closure operators.

Theorem 7.4. *Let $\vdash$ be a conclusion relation. Then*

$$\vdash \text{ is a consequence relation } \Leftrightarrow J_{\vdash} \text{ is a closure operator.}$$

Let $J : \mathcal{P}(\mathbb{F}) \to \mathcal{P}(\mathbb{F})$ be an operator. Then

$$J \text{ is a closure operator } \Leftrightarrow \vdash_J \text{ is a consequence relation.}$$

Proof. Assume that $\vdash$ is a consequence relation. Then, from (i) it follows that $J_{\vdash}(X) \supseteq X$ and from (ii) that $X \supseteq Y$ implies $J_{\vdash}(X) \supseteq J_{\vdash}(Y)$. In order to prove that $J_{\vdash}(J_{\vdash}(X)) = J_{\vdash}(X)$, observe that, since $X \vdash \beta$ for every $\beta \in J_{\vdash}(X)$, by (iii),

$$J_{\vdash}(X) \vdash \alpha \Rightarrow X \vdash \alpha.$$

Thus, $J_{\vdash}(J_{\vdash}(X)) = \{\alpha \in \mathbb{F} : J_{\vdash}(X) \vdash \alpha\} \subseteq J_{\vdash}(X)$ and $J_{\vdash}(J_{\vdash}(X)) = J_{\vdash}(X)$.

Assume that J is a closure operator. Then, the inclusion property $J(X) \supseteq X$ entails that $X \vdash_J \alpha$ for every $\alpha \in X$, and this demonstrates (i). In order to prove (ii), suppose $X \vdash_J \alpha$ and $Y \supseteq X$. Then, since $\alpha \in J(X)$ and J is order-preserving, $\alpha \in J(Y)$, i.e., $Y \vdash_J \alpha$. In order to prove (iii), assume that $X \vdash_J \beta$ for every $\beta \in Z$ and that $X \cup Z \vdash_J \alpha$. Then, $Z \subseteq J(X)$ and $\alpha \in J(X \cup Z)$. Consequently, since $X \cup Z \subseteq J(X)$ and $\alpha \in J(X \cup Z) \subseteq J(J(X)) = J(X)$, we have that that X $\vdash_J \alpha$. Thus $\vdash_J$ is a consequence relation.

The remaining part of the proposition follows from Proposition 7.3. □

For instance if J is the identity map, then the corresponding relation $\vdash_J$ is the usual membership relation. As a matter of fact, by (i) of Definition 7.1, this relation is the least crisp consequence relation.

The next theorem shows that the theory of consequence relations coincides with the theory of the abstract deduction systems.

Theorem 7.5. *A conclusion relation $\vdash$ is a consequence relation iff a deduction system $(\mathbb{F},\mathcal{D})$ exists such that*

$$X \vdash \alpha \Leftrightarrow \alpha \in \mathcal{D}(X). \tag{7.5}$$

Proof. Let $\vdash$ be a consequence relation and let $\mathcal{D} = J_{\vdash}$. Then $\mathcal{D}$ is a closure operator satisfying (7.5). Conversely, if $(\mathbb{F},\mathcal{D})$ is a deduction system satisfying (7.5), then $\vdash$ coincides with $\vdash_{\mathcal{D}}$. So, $\vdash$ is a consequence relation. □

8. GRADED CONSEQUENCES AND SEQUENT CALCULUS

The concept of graded consequence relation was proposed in Chakraborty [1988] as a graded extension of the crisp abstract concept of consequence relation $\vdash$. We call a *graded conclusion relation* any fuzzy subset of sequents, i.e., any fuzzy relation $g : SEQ \to U$ from $\mathcal{P}(\mathbb{F})$ to $\mathbb{F}$. If X is a set of formulas and α a formula, we write $g(X \vdash \alpha)$ instead of $g(X,\alpha)$. Moreover, given another set Z of formulas, we set

$$g(X \vdash Z) = Inf\{g(X \vdash z) : z \in Z\}. \quad (8.1)$$

Definition 8.1. We say that a graded conclusion relation g is a *graded consequence relation* if, for every $X, Y, Z \in \mathcal{P}(\mathbb{F})$ and $\alpha \in \mathbb{F}$,

(i) $g(X \vdash \alpha) = 1$ for every $\alpha \in X$,
(ii) $g(X \cup Y \vdash \alpha) \geq g(X \vdash \alpha)$,
(iii) $g(X \vdash \alpha) \geq g(X \vdash Z) \wedge g(X \cup Z \vdash \alpha)$.

If $\lambda = g(X \vdash \alpha)$ we say that *α is a consequence of X at least to degree λ*. The following proposition, whose proof is trivial, summarizes some elementary properties of the graded consequences.

Proposition 8.2. *Let g be a graded consequence. Then, for X, Y, Z, X_1, X_2, Y_1, Y_2 subsets of $\mathbb{F}$ and $(Y_i)_{i\in I}$ family of subsets of $\mathbb{F}$,*

(i) $g(X \vdash X) = 1$,
(ii) $g(X \cup Y \vdash Z) \geq g'(X \vdash Z)$,
(iii) $g(X \vdash Y) \geq g(X \vdash Z) \wedge g(X \cup Z \vdash Y)$,
(iv) $g(X \vdash \cup_{i\in I} Y_i) = Inf\{g(X \vdash Y_i) : i \in I\}$,
(v) $X_1 \subseteq X_2 \Rightarrow g(X_1 \vdash Y) \leq g(X_2 \vdash Y)$,
(vi) $Y_1 \subseteq Y_2 \Rightarrow g(X \vdash Y_1) \geq g(X \vdash Y_2)$,
(vii) $X \supseteq Y \Rightarrow g(X \vdash Y) = 1$,
(viii) $g(X \vdash Y) \geq g(X \vdash Z) \wedge g(Z \vdash Y)$.

It is possible to interpret the crisp consequence relations as theories of a suitable H-system with an infinitary inference rule. In fact, first we consider a crisp H-system $\mathcal{S} = (\mathbb{A},\mathbb{R})$ which we call *minimal sequent calculus*, such that

- the set of formulas is the set SEQ of sequents,
- the set $\mathbb{A}$ of logical axioms is $\{(X,x) : x \in X\}$,
- there is a finitary rule:

$$\frac{(X,\alpha)}{(X \cup Y, \alpha)}$$

and an infinitary rule:

$$\frac{\{(X,\beta) : \beta \in Z\}\ ,\ (X \cup Z, \alpha)}{(X, \alpha)}\ .$$

Proposition 8.3. *The class of theories of the minimal sequent calculus coincides with the class of crisp consequence relations.*

Proof. Indeed, with reference to Definition 7.1, a set $\vdash$ of sequents contains the set of logical axioms iff it satisfies (i). $\vdash$ is closed under the finitary rule iff it satisfies (ii), $\vdash$ is closed under the infinitary rule iff it satisfies (iii). □

The next proposition shows that the class of graded consequence relations is the canonical extension of the class of crisp consequence relations.

Proposition 8.4. *The following are equivalent:*
(a) *$g : SEQ \to U$ is a graded consequence relation.*
(b) *every cut* $C(g,\lambda)$ *is a consequence relation.*
In other words, the class of graded consequence relations is the canonical extension of the class of consequence relations.

Proof. (a) $\Rightarrow$ (b) It is self-evident that $C(g,\lambda)$ satisfies (i) and (ii) of Definition 7.1. In order to prove (iii), suppose $(X \cup Z,\alpha) \in C(g,\lambda)$ and $(X,z) \in C(g,\lambda)$ for every $z \in Z$. Then $g(X \cup Z \vdash \alpha) \geq \lambda$ and $g(X \vdash z) \geq \lambda$ for every $z \in Z$. Consequently, $g(X \vdash Z) \geq \lambda$ and by (iii) of Definition 8.1 this implies that $g(X \vdash \alpha) \geq \lambda$. Hence, $(X,\alpha) \in C(g,\lambda)$.
(b) $\Rightarrow$ (a) Let X be a set of formulas and $x \in X$. Then, the fact that $C(g,1)$ is a consequence relation entails that $(X,x) \in C(g,1)$, i.e., $g(X \vdash x) = 1$. Let Y be a set of formulas containing X, and $\lambda = g(X \vdash \alpha)$. Then, since $(X,\alpha) \in C(g,\lambda)$ and $Y \supseteq X$, we have $(Y,\alpha) \in C(g,\lambda)$, i.e., $g(Y \vdash \alpha) \geq \lambda = g(X \vdash \alpha)$. Finally, given any set Z of formulas, set

$$\lambda = Inf(\{g(X \vdash z) : z \in Z\}) \wedge g(X \cup Z \vdash \alpha).$$

Then, since $C(g,\lambda)$ is a consequence relation, $(X,z) \in C(g,\lambda)$ for every $z \in Z$ and $(X \cup Z, \alpha) \in C(g,\lambda)$, we may conclude that $(X,\alpha) \in C(g,\lambda)$. Thus

$$g(X \vdash \alpha) \geq Inf(\{g(X \vdash z) : z \in Z\}) \wedge g(X \cup Z \vdash \alpha).$$ □

We indicate the canonical extension of the minimal sequent calculus by $\mathcal{S}^*$, where how to extend an infinitary rule in a fuzzy infinitary rule is evident. Then

- the fuzzy set of logical axioms is

$$\{((X,x) : x \in X\},$$

- we have a finitary rule

$$\frac{(X,\alpha)}{(X \cup Y)} \quad ; \quad \frac{\lambda}{\lambda}$$

- we have an infinitary rule

$$\frac{\{(X,\beta):\beta\in Z\},\ (X\cup Z,\alpha)}{(X,\ \alpha)}\quad ;\quad \frac{S,\ \lambda}{Inf(S)\wedge\lambda}$$

The following theorem holds:

Theorem 8.5. *The class of graded consequence relations coincides with the class of theories of the canonical extension S^* of the minimal sequent calculus S.*

Proof. This is trivial. □

In particular, the class of graded consequence relations is a closure system and any fuzzy conclusion relation can be extended to a consequence relation (see Castro, Trillas, Cubillo [1994]).

9. FINITE SEQUENT CALCULUS AND COMPACT GRADED CONSEQUENCES

It is possible to avoid the infinitary inference rules provided that we only consider compact graded consequences. In the following, if X is a set we denote the class of finite subsets of X by $\mathcal{P}_f(X)$. A conclusion relation $\vdash$ is *compact*, if

$$X\vdash\alpha \Leftrightarrow \text{there exists } X_f\in\mathcal{P}_f(X) \text{ such that } X_f\vdash\alpha.$$

Trivially, if $\vdash$ is a conclusion relation, then

$$\vdash \text{ is compact } \Leftrightarrow J_\vdash \text{ is compact,}$$

if J is an operator, then

$$J \text{ is compact } \Leftrightarrow \vdash_J \text{ is compact.}$$

A graded conclusion relation g is said to be *compact* if

$$g(X,\alpha) = Sup\{g(X_f\vdash\alpha) : X_f\in\mathcal{P}_f(X)\}. \tag{9.1}$$

Proposition 9.1. *A graded conclusion g is a compact graded consequence iff it satisfies* (9.1) *and*

(i) $g(X\vdash\alpha) = 1$ *for every* $\alpha\in X$,

(ii) $g(X\cup Y\vdash\alpha)\ge g(X\vdash\alpha)$,

(iii) $g(X\vdash\alpha)\ge g(X\vdash z)\wedge g(X\cup\{z\}\vdash\alpha)$.

Proof. Let g be is a compact graded consequence. Then it is evident that (9.1), (i), (ii) and (iii) are satisfied. Conversely, assume these conditions are satisfied. Then, first we demonstrate that, for every finite set $Z_f=\{z_1,\dots,z_n\}$,

$$g(X\vdash\alpha)\ge(Inf\{g(X\vdash z) : z\in Z_f\})\wedge g(X\cup Z_f\vdash\alpha).$$

Indeed, such an inequality coincides with (iii) for $n = 1$. Moreover, by induction hypothesis,

$$\begin{aligned} g(X\vdash\alpha) &\ge (Inf\{g(X\vdash z) : z\in\{z_1,\dots,z_{n-1}\}\})\wedge g(X\cup\{z_1,\dots,z_{n-1}\}\vdash\alpha)\\ &\ge (Inf\{g(X\vdash z) : z\in\{z_1,\dots,z_{n-1}\}\}))\wedge g(X\cup\{z_1,\dots,z_{n-1}\}\vdash z_n)\\ &\quad\wedge g(X\cup\{z_1,\dots,z_{n-1},z_n\}\vdash\alpha)\\ &\ge (Inf\{g(X\vdash z) : z\in\{z_1,\dots,z_{n-1}\}\}))\wedge g(X\vdash z_n)\wedge g(X\cup Z_f\vdash\alpha)\\ &= (Inf\{g(X\vdash z) : z\in Z_f\})\wedge g(X\cup Z_f\vdash\alpha). \end{aligned}$$

Let Z be any set, then,

$$(Inf\{g(X \vdash z) : z \in Z\}) \wedge g(X \cup Z \vdash \alpha)$$
$$= (Inf\{g(X \vdash z) : z \in Z\}) \wedge (Sup\{g(X \cup Z_f \vdash \alpha) : Z_f \in \mathcal{P}_f(Z)\}$$
$$= Sup\{(Inf\{g(X \vdash z) : z \in Z\}) \wedge g(X \cup Z_f \vdash \alpha) : Z_f \in \mathcal{P}_f(Z)\}$$
$$\leq Sup\{Inf(g(X \vdash z) : z \in Z_f\}) \wedge g(X \cup Z_f \vdash \alpha) : Z_f \in \mathcal{P}_f(Z)\} \leq g(X \vdash \alpha).$$ □

We call *finite sequent* any sequent (X,α) in which X is finite and we denote the set of finite sequents by SEQ_f. If g is compact, then g is completely defined by its restriction to SEQ_f. Conversely, let h be a fuzzy subset of SEQ_f and set

$$g(X,\alpha) = Sup\{h(X_f,\alpha) : X_f \text{ is a finite subset of } X\}.$$

Then g is a compact graded conclusion relation. If h satisfies (ii), then g is an extension of h we indicate as *the compact extension of h*.

Proposition 9.2. *g is a compact graded consequence relation iff g is the compact extension of a fuzzy relation* $h : \mathcal{P}_f(\mathbb{F}) \times \mathbb{F} \to U$ *satisfying*

(j) $h(X,\alpha) = 1$ *for every* $\alpha \in X$,

(jj) $h(X \cup Y, \alpha) \geq h(X,\alpha)$,

(jjj) $h(X,\alpha) \geq h(X,\beta) \wedge h(X \cup \{\beta\},\alpha)$.

Proof. If g is a compact conclusion relation, then it is obvious that its restriction h to SEQ_f satisfies (j), (jj) and (jjj). Conversely, let g be the compact extension of a fuzzy relation h satisfying (j), (jj) and (jjj). Then, by using Proposition 9.1 we can prove g is a graded consequence relation by proving g satisfies (i) and (ii) and

$$g(X,\alpha) \geq g(X,\beta) \wedge g(X \cup \{\beta\}, \alpha). \tag{9.3}$$

Now, (i) and (ii) are trivial. In order to prove (9.3) observe that

$$g(X,\alpha) = Sup\{h(X_f,\alpha) : X_f \in \mathcal{P}_f(X)\} \geq Sup\{h(X_f,\beta) \wedge h(X_f \cup \{\beta\},\alpha) : X_f \in \mathcal{P}_f(X)\}.$$

On the other hand,

$$g(X,\beta) \wedge g(X \cup \{\beta\},\alpha)$$
$$= (Sup\{h(X_1,\beta)) : X_1 \in \mathcal{P}_f(X)\}) \wedge (Sup\{h(X_2 \cup \{\beta\},\alpha) : X_2 \in \mathcal{P}_f(X)\})$$
$$= Sup\{h(X_1,\beta) \wedge h(X_2 \cup \{\beta\},\alpha) : X_1, X_2 \in \mathcal{P}_f(X)\}.$$

Now, observe that, by setting $X_f = X_1 \cup X_2$,

$$h(X_f,B) \wedge h(X_f \cup \beta,\alpha) \geq h(X_1,\beta) \wedge h(X_2 \cup \{\beta\},\alpha).$$

Then, we can conclude that $g(X,\alpha) \geq g(X,\beta) \wedge g(X \cup \{\beta\},\alpha)$. □

Proposition 9.2 enables us to relate the compact graded consequence relations with the theories of the canonical extension of a suitable sequent calculus. In fact, let $S_f = (\mathbb{A},\mathbb{R})$ be the H-system such that

- SEQ_f is the set of formulas,
- the set $\mathbb{A}$ of logical axioms is $\{(X,x) \in SEQ_f : x \in X\}$,
- there are the following rules:

$$\frac{(Y,\alpha)}{(X \cup Y,\ \alpha)} \quad ; \quad \frac{(X,\beta),\ (X \cup \{\beta\},\ \alpha)}{(X,\ \alpha)} \quad .$$

We call *minimal finite-sequent calculus* such a system. Then, we have the following theorem:

Theorem 9.3. *Let S_f^* be the canonical extension of S_f. Then, g is a compact graded consequence relation iff g is (the compact extension of) a theory of S_f^*.*

Proof. It is evident that $h : SEQ_f \to U$ satisfies (j), (jj) and (jjj) of Proposition 9.2 iff h is a theory of S_f^*. □

10. GRADED CONSEQUENCES AND STRATIFIED OPERATORS

The following theorem shows that we can identify the graded consequence relations with the continuous chains of consequence relations:

Theorem 10.1. *A conclusion relation g is a graded consequence relation iff a continuous family $(\vdash_\lambda)_{\lambda \in U}$ of consequence relations exists such that*

$$g(X \vdash \alpha) = Sup\{\lambda \in U : X \vdash_\lambda \alpha\}. \tag{10.1}$$

Proof. Given $\lambda \in U$, denote the conclusion relation $C(g,\lambda)$ by $\vdash_\lambda$. Then

$$g(X \vdash \alpha) = Sup\{\lambda \in U : (X,\alpha) \in C(g,\lambda)\} = Sup\{\lambda \in U : X \vdash_\lambda \alpha\}.$$

So, the proof follows from Proposition 8.4. □

In Theorem 7.5 we observed that a conclusion relation is a consequence relation iff a closure operator J exists such that

$$X \vdash \alpha \Leftrightarrow \alpha \in J(X).$$

The question arises whether such a connection holds also for the graded consequence relations and the fuzzy closure operators.

Proposition 10.2. *Let J be a fuzzy closure operator and define the graded conclusion relation g by setting*

$$g(X \vdash \alpha) = J(X)(\alpha) \tag{10.2}$$

for every $X \subseteq \mathcal{F}$ and $\alpha \in \mathcal{F}$. Then, in general, *g satisfies* (i) *and* (ii) *but not* (iii).

Proof. A straightforward verification proves the first part of the proposition. In the following example (iii) is not satisfied (M. K. Chakraborty, personal communication). Let $\mathcal{F} = \{\alpha_1, \alpha_2, \alpha_3, \alpha_4\}$ and let s_1 and s_2 be the two fuzzy subsets of $\mathcal{F}$ defined by setting

$$s_1(\alpha_1) = s_1(\alpha_3) = 1 \text{ , } s_1(\alpha_2) = 0.7 \text{ , } s_1(\alpha_4) = 0.8$$

and

$$s_2(\alpha_1) = s_2(\alpha_3) = s_2(\alpha_4) = 1 \text{ , } s_2(\alpha_2) = 0.9 \text{ .}$$

Then, the class $C = \{s_1, s_2\}$ defines a fuzzy closure operator J. Namely, for every fuzzy subset s and $\alpha \in \mathcal{F}$,

$$J(s)(\alpha) = Inf\{s_i(\alpha) : s_i \supseteq s\}.$$

Take $X = \{\alpha_1, \alpha_3\}$ and $Z = \{\alpha_4\}$. Then, a simple calculation gives

$$J(X)(\alpha_2) = 0.7 \text{ , } J(X \cup Z)(\alpha_2) = 0.9 \text{ , } J(X)(\alpha_4) = 0.8.$$

So, if g is the conclusion relation associated with J,

$$g(X \vdash \alpha_2) = 0.7,\ Inf\{g(X \vdash z) : z \in Z\} = 0.8, \text{ and } g(X \cup Z \vdash \alpha_2) = 0.9.$$

Hence,

$$g(X \vdash \alpha_2) < (Inf\{g(X \vdash z) : z \in Z\}) \wedge g(X \cup Z \vdash \alpha_2).$$

This demonstrates that (iii) is not satisfied. □

The following theorem shows that we can extend Theorem 7.5 to the graded consequences provided we confine ourselves to the well-stratified deduction systems (Gerla [1996]).

Theorem 10.3. *A fuzzy conclusion relation* $g : \mathcal{P}(\mathcal{F}) \times \mathcal{F} \to U$ *is a graded consequence relation iff a well-stratified deduction system* $(\mathcal{F},\mathcal{D})$ *exists such that*

$$g(X \vdash \alpha) = \mathcal{D}(X)(\alpha) \tag{10.3}$$

for every X *subset of* $\mathcal{F}$ *and* $\alpha \in \mathcal{F}$.

Proof. Let g be a graded consequence relation and, for every $\lambda \in U$, let $\mathcal{D}_\lambda$ be the deduction operator associated with the consequence relation $C(g,\lambda)$, that is

$$\mathcal{D}_\lambda(X) = \{x \in \mathcal{F} : g(X \vdash x) \geq \lambda\}.$$

Moreover, denote by $\mathcal{D}$ the closure operator associated with $(\mathcal{D}_\lambda)_{\lambda \in U}$. Then

$$g(X \vdash \alpha) = Sup\{\lambda \in U : g(X \vdash \alpha) \geq \lambda\} = Sup\{\lambda \in U : \alpha \in \mathcal{D}_\lambda(X)\} = \mathcal{D}(X)(\alpha).$$

So, we must prove only that $(\mathcal{D}_\lambda)_{\lambda \in U}$ is a continuous chain. Let X be a set of formulas. Then, trivially, $\mathcal{D}_0(X) = \mathcal{F}$. Furthermore, if $\mu \in U$, then

$$x \in \mathcal{D}_\mu(X) \Leftrightarrow g(X \vdash x) \geq \mu \Leftrightarrow g(X \vdash x) \geq \lambda \text{ for every } \lambda < \mu$$
$$\Leftrightarrow x \in \cap_{\lambda<\mu}\mathcal{D}_\lambda(X).$$

Conversely, let $(\mathcal{F},\mathcal{D})$ be the fuzzy deduction system associated with a given continuous chain $(\mathcal{D}_\lambda)_{\lambda \in U}$ of deduction systems and, for every $\lambda \in U$, denote by $\vdash_\lambda$ the consequence relation associated with $\mathcal{D}_\lambda$, that is $\vdash_\lambda = \{(X,x) : x \in \mathcal{D}_\lambda(X)\}$. We claim that $(\vdash_\lambda)_{\lambda \in U}$ is a continuous family. Indeed, $\vdash_0 = SEQ$ and

$$(X,x) \in \vdash_\mu \Leftrightarrow x \in \mathcal{D}_\mu(X) \Leftrightarrow x \in \mathcal{D}_\lambda(X) \text{ for every } \lambda < \mu$$
$$\Leftrightarrow (X,x) \in \vdash_\lambda \text{ for every } \lambda < \mu.$$

Thus, by Theorem 10.1 the conclusion relation g defined by (10.3) is a graded consequence relation. □

Theorem 10.3 enables us to find examples of graded consequence in a simple way. For instance, let $\mathcal{S}_1$ and $\mathcal{S}_2$ be two different deductive systems on the same set $\mathcal{F}$ of formulas and let $\mathcal{D}_1$ and $\mathcal{D}_2$ be the related deduction operators. Moreover, assume that $\mathcal{S}_2$ is more powerful than $\mathcal{S}_1$, that is $\mathcal{D}_1(X) \subseteq \mathcal{D}_2(X)$ for every set X of formulas but that, at the same time, $\mathcal{S}_2$ is less reliable than $\mathcal{S}_1$. Then, a continuous family of closure operators is achieved by setting $J_\lambda = \mathcal{D}_2$ for $\lambda \leq 0.5$ and $J_\lambda = \mathcal{D}_1$ for $\lambda > 0.5$. We obtain the corresponding graded consequence relation g by

$$g(X \vdash x) = \begin{cases} 1 & \text{if } x \in \mathcal{D}_1(X), \\ 0.5 & \text{if } x \in \mathcal{D}_2(X) - \mathcal{D}_1(X), \\ 0 & \text{otherwise.} \end{cases} \tag{10.4}$$

Further examples of graded consequences are furnished by the canonical similarity logics. In fact, the deduction operators of these logics are well-stratified.

Theorem 10.4. *Let Con* : $\mathcal{P}(\mathbb{F}) \times \mathbb{F} \rightarrow U$ *the fuzzy relation associated with a canonical similarity logic* (*see Definition* 9.1 *in Chapter* 3). *Then, Con is a graded consequence relation.*

Proof. See Theorem 6.5. □

Remark. Theorem 10.3 suggests a natural way to extend a graded consequence relation g in a fuzzy relation g^e from the lattice $\mathcal{F}(\mathbb{F})$ to $\mathbb{F}$. Indeed, it is sufficient to consider the stratified deduction operator $\mathcal{D}$ associated with g and to set

$$g^e(s \vdash \alpha) = \mathcal{D}(s)(\alpha). \tag{10.5}$$

Equivalently, we can set

$$g^e(s \vdash \alpha) = Sup\{\lambda \in U : g(C(s,\lambda) \vdash \alpha) \geq \lambda\} \tag{10.6}$$

for every $s \in \mathcal{F}(\mathbb{F})$ and $\alpha \in \mathbb{F}$. This suggests to examine the possibility of extending the definition of graded consequence by calling *fuzzy sequent* any element of the set $\underline{SEQ} = \mathcal{F}(\mathbb{F}) \times \mathbb{F}$ and defining the graded consequence relations as suitable fuzzy subsets $Con : \underline{SEQ} \rightarrow U$ of $\underline{SEQ}$.

CHAPTER 8

TRUTH-FUNCTIONAL LOGIC AND FUZZY LOGIC

1. TRUTH-FUNCTIONAL FUZZY SEMANTICS

Generally speaking, all fuzzy logics considered so far arise from the fuzzification of notions having a metalogic character (see Section 10 in Chapter 3). By contrast, in this chapter we will investigate several fuzzy logics directly arising from *"fuzzy worlds"*, i.e., worlds whose properties can be vague and therefore whose truth-value assignments can be graded. This leads to truth-functional semantics, i.e., semantics in which the truth value of a composite sentence is determined by the truth-values of its constituents. More precisely, we consider a language by starting from

- an infinite set $VAR = \{p_1, \ldots, p_n, \ldots\}$ of *propositional variables*,
- a set of logical connectives containing the usual connectives, namely $\wedge$, $\vee$, $\neg$.

We define the set $\mathbb{F}$ of formulas assuming that:

- propositional variables are formulas,
- if h is an n-ary connective and $\alpha_1,\ldots,\alpha_n$ are formulas, then $h(\alpha_1,\ldots,\alpha_n)$ is a formula.

Moreover, we assume that every n-ary connective h is interpreted by a suitable n-ary operation $\underline{h}$ in the lattice U in such a way that:

(i) each operation $\underline{h}$ is either order-preserving or order-reversing with respect to each variable,

(ii) $\wedge$, $\vee$ and $\neg$ extend the related classical interpretations,

(iii) the interpretation $\sim : U \to U$ of $\neg$ is order-reversing,

(iv) the interpretations of $\wedge$ and $\vee$ are order-preserving,

(v) if $\to$ occurs, we interpret it by a function which is order preserving in the second variable and order-reversing in the first variable.

The set U equipped with the operations $\underline{h}$ used to interpret the logical connectives is called the *valuation structure* for the semantics under consideration. As usual we denote by $\wedge$ and $\vee$ both the logical connectives and the lattice operations.

Definition 1.1. A *truth-functional valuation* is any fuzzy subset m of $\mathbb{F}$ such that

$$m(h(\alpha_1,\ldots,\alpha_n)) = \underline{h}(m(\alpha_1),\ldots,m(\alpha_n)) \tag{1.1}$$

for every n-ary logical connective h and $\alpha_1,\ldots,\alpha_n$ in $\mathbb{F}$.

The class $\mathcal{M}$ of all truth-functional valuations is a fuzzy semantics because $s^1 \notin \mathcal{M}$. Indeed, otherwise, by condition (i),

$$1 = s^1(p_1 \wedge \neg p_1) = s^1(p_1) \wedge (\sim s^1(p_1)) = Min\{1,0\} = 0.$$

Definition 1.2. Let $\mathcal{M}$ be the class of truth-functional valuations. Then $\mathcal{M}$ is called a *truth-functional* (fuzzy) semantics. In the case $U = [0,1]$, $\mathcal{M}$ is called *continuous* if all the connectives are interpreted by continuous operations in $[0,1]$.

As usual, any truth-functional semantics is associated with a logical consequence operator $Lc : \mathcal{F}(\mathbb{F}) \to \mathcal{F}(\mathbb{F})$. In this chapter, in accordance with the point of view of fuzzy logic, we will investigate about the existence of a (fuzzy) deduction apparatus to calculate Lc. More precisely, given a truth-functional semantics $\mathcal{M}$, we examine the following question:

is $\mathcal{M}$ axiomatizable by a suitable deduction apparatus?

Of course the answer depends on the meaning we assign to the expression *"deduction apparatus"*. For example, we can either consider the Hilbert systems as defined in Chapter 4 or we can impose to these systems more constructive conditions (see Chapter 11). Also, we can define deduction apparatus for fuzzy logic which are different from the Hilbert systems (tableaux methods, sequents calculi and so on). In any case, as we will see in the following, many semantics exist for which the answer is negative.

Now, an initial step in this direction was made in Pavelka [1979]*c*, where the author demonstrates that of all the residuated lattice carried on by $[0,1]$, only those whose operations are continuous serve as a basis for an axiomatizable truth-functional semantics. This suggests continuity as a candidate for a characterization of the axiomatizable truth-functional semantics.

We emphasize that traditional research in truth-functional logic is sharply different from the approach to approximate reasoning we consider in this book. Indeed, in the classical approach all deductive machinery is crisp, and it is devoted to produce a crisp set of formulas that are considered *"valid"* for some reason. Some considerations and results about the connection between the traditional approach in multi-valued logic and fuzzy logic will be given in Section 8.

2. THE MAIN PROPERTIES

Let $v : VAR \to U$ be an assignment of truth values to the propositional variables. Then, there exists exactly one truth-functional valuation m_v such that $m_v(p_i) = v(p_i)$ for every $i \in N$. In this case we say that m_v is *the truth-functional extension of* v. It is interesting to observe that if $v : \mathbb{F} \to U$ is a fuzzy subset of formulas whose support is the set of propositional variables, then, in general, the theory $Lc(v)$ generated by v is different from the truth-functional extension m_v of v. This is so because if v is assumed as an initial valuation, then $v(p_i)$ expresses the information

"the actual truth value of p_i is at least $v(p_i)$".

Instead, if v is considered as an assignment of truth values to the propositional variables, the information conveyed by $v(p_i)$ is that

"the actual truth value of p_i is exactly $v(p_i)$".

For instance, let v be the initial valuation constantly equal to zero. Then, since a model m' of v is obtained by setting $m'(p_i) = 1$ for every $i \in N$, and $m'(\neg p_i) = \sim(1) = 0$, we have that $Lc(v)(\neg p_i) = 0$ while

$$m_v(\neg p_i) = \sim(m_v(p_i)) = \sim(0) = 1.$$

The following proposition gives the correct way to relate *"truth-functional extensions"* with *"fuzzy theories"*:

Proposition 2.1. *Assume that the interpretation* $\sim$ *of the negation is injective and let* $v : VAR \to U$ *be an assignment of truth values to the propositional variables. Moreover, denote by* v' *the extension of* v *to the whole set* $\mathbb{F}$ *of formulas obtained by setting*

$$v'(x) = \begin{cases} v(p_i) & \text{if } x = p_i, \\ \sim(v(p_i)) & \text{if } x = \neg p_i, \\ 0 & \text{otherwise.} \end{cases}$$

Then

$$Lc(v') = m_v$$

and v' *is a categorical initial valuation whose only model is* m_v.

Proof. Trivially, m_v is a model of v'. Moreover, if m is a model of v', then $m(p_i) \geq v'(p_i)$. Also, since

$$\sim(m(p_i)) = m(\neg p_i) \geq \sim(v'(p_i))$$

and $\sim$ is injective, we get that $m(p_i) \leq v'(p_i)$ and $m(p_i) = v'(p_i) = m_v(p_i)$. This proves that $m = m_v$. □

The following theorem shows that the degree of inconsistency in a truth-functional logic is different from 1.

Theorem 2.2. *For every truth-functional semantics* $\mathcal{M}$

$$Inc(\mathcal{M}) \neq 1. \tag{2.1}$$

Moreover, for any $b \in U$, $b \neq 1$, *a multi-valued semantics* $\mathcal{M}$ *exists such that*

$$Inc(\mathcal{M}) = b. \tag{2.2}$$

Proof. Observe that

$$Inc(\mathcal{M}) \leq Inf\{\lambda \in U : \lambda \geq \sim\lambda\}. \tag{2.3}$$

Indeed, to prove this inequality it is sufficient to prove that, for every $\lambda \in U$,

$$\lambda \geq \sim(\lambda) \Rightarrow \lambda \geq Inc(\mathcal{M})$$

or, equivalently,

$$\lambda < Inc(\mathcal{M}) \Rightarrow \sim(\lambda) > \lambda. \tag{2.4}$$

Now, since

$$Inc(\mathcal{M}) = Sup\{Inc(m) : m \in \mathcal{M}\},$$

if $\lambda < Inc(\mathcal{M})$ a model $m_\lambda \in \mathcal{M}$ exists such that $Inc(m_\lambda) > \lambda$. This entails that, for every formula α,

$$m_\lambda(\alpha) > \lambda \text{ and } \sim(m_\lambda(\alpha)) = m_\lambda(\neg\alpha) > \lambda.$$

Hence, because $\sim$ is order-reversing,

$$\sim(\lambda) \geq \sim(m_\lambda(\alpha)) > \lambda.$$

In order to prove (2.1), assume $Inc(\mathcal{M}) = 1$. We claim that $\sim(\lambda) = 1$ if $\lambda \neq 1$ and $\sim(\lambda) = 0$ if $\lambda = 1$. In fact, assume that $\lambda \neq 1$ and that $\sim(\lambda) \neq 1$. Then, by (2.3), $\lambda < \sim(\lambda)$ and $\sim(\lambda) < \sim(\sim(\lambda))$. This is a contradiction because from $\lambda < \sim(\lambda)$ it follows that $\sim(\lambda) \geq \sim(\sim(\lambda))$. Since $\sim(\lambda) \in \{0,1\}$ for every $\lambda \in U$, for every model m,

$$m(\neg(\neg\alpha) \wedge \neg\alpha) = \sim(\sim(m(\alpha))) \wedge \sim(m(\alpha)) = 0.$$

This contradicts the hypothesis $Inc(\mathcal{M}) = 1$.

Finally, let b be any element in the real interval U different from 1, and consider an order-reversing map $\sim : U \rightarrow U$ such that

$$\sim(b) = b, \sim(0) = 1, \sim(1) = 0.$$

By interpreting the logical connective $\wedge$ and $\vee$ by the minimum and the maximum, respectively, we obtain a truth-functional semantics $\mathcal{M}$. Let m be a model such that $m(p_i) = b$ for every propositional variable p_i. Then, m is constantly equal to b and therefore, $Inc(\mathcal{M}) \geq Inc(m) = b$. Since (2.3) gives $Inc(\mathcal{M}) \leq b$, we can conclude that $Inc(\mathcal{M}) = b$. □

The next proposition gives some information about the maximal models in $\mathcal{M}$.

Proposition 2.3. *If $\sim$ is injective, then all the elements of $\mathcal{M}$ are maximal in $\mathcal{M}$.*

Proof. Let m be any element of $\mathcal{M}$ and suppose $m' \supseteq m$, $m' \in \mathcal{M}$. Then, for every formula α, $m'(\alpha) \geq m(\alpha)$ and therefore, since $\sim$ is order-reversing, we have that $\sim(m'(\alpha)) \leq \sim(m(\alpha))$. Consequently, by observing that $\sim(m'(\alpha)) = m'(\neg\alpha) \geq m(\neg\alpha) = \sim(m(\alpha))$, we have that $\sim(m'(\alpha)) = \sim(m(\alpha))$. Because $\sim$ is injective, we can conclude that $m'(\alpha) = m(\alpha)$ for every formula α and therefore that m is maximal. □

Sometimes it is useful to consider a truth-functional semantics as a class of homomorphisms. Namely, consider $\mathcal{F}$ equipped with the logical connectives as a (free) algebraic structure and denote by $\mathcal{U}$ the valuation structure. Then, it is self-evident that $\mathcal{M}$ is the class of all homomorphisms from $\mathcal{F}$ to $\mathcal{U}$.

Proposition 2.4. *The following are equivalent:*

(a) *$f: U \rightarrow U$ is a homomorphism of the structure $\mathcal{U}$.*

(b) *$f: U \rightarrow U$ is a homomorphism of $\mathcal{M}$, i.e.,*

$$m \in \mathcal{M} \Rightarrow f \circ m \in \mathcal{M}.$$

Proof. (a) $\Rightarrow$ (b). It is sufficient to observe that the product of a homomorphism m from $\mathcal{F}$ to $\mathcal{U}$ and a homomorphism f from $\mathcal{U}$ into $\mathcal{U}$ is a homomorphism from $\mathcal{F}$ into $\mathcal{U}$.

(b) $\Rightarrow$ (a). We have to prove that for any logical connective h, and $\lambda_1,\ldots,\lambda_n$ in U,

$$f(\underline{h}(\lambda_1,\ldots,\lambda_n)) = \underline{h}(f(\lambda_1),\ldots,f(\lambda_n)).$$

Let m be a model such that $m(p_i) = \lambda_i$. Then, since $f \circ m$ is a model,

$$f(\underline{h}(\lambda_1,\ldots,\lambda_n)) = f(m(h(p_1,\ldots,p_n))) = \underline{h}(f(m(p_1)),\ldots,f(m(p_n)))) = \underline{h}(f(\lambda_1),\ldots,f(\lambda_n)).$$ □

If the semantics is crisp, such a proposition is not useful. In fact, the only homomorphism from {0,1} into {0,1} is the identity map. In the next section we will give some application of such a proposition. Finally, the following proposition shows that, in the case of lower constraints, the only possible tautologies are the classical tautologies.

Proposition 2.5. *Let Taut be the fuzzy subset of tautologies defined with respect to the lower constraint frame. Then,* $Taut(\alpha) = 0$ *whenever* α *is not a classical tautology and* $Contr(\alpha) = 0$ *whenever* α *is not a classical contradiction.*

Proof. Recall that the interpretations of $\wedge$, $\vee$ and $\neg$ extend the corresponding classical interpretations and assume that α is not a classical tautology. Then there is a {0,1}-valuation v of the propositional variables assigning the value 0 to α (with respect to the classical logic). This valuation can be extended to a truth-functional valuation $m \in \mathcal{M}$ such that $m(\alpha) = 0$. Thus $Taut(\alpha) = 0$. Likewise we demonstrate that $Contr(\alpha) = 0$ whenever α is not a classical contradiction. □

The next theorem emphasizes that the fuzzy logic defined by a truth-functional semantics is totally different from the logics discussed in the previous chapters.

Theorem 2.6. *A truth-functional logic is neither stratified nor a canonical extension of a crisp logic by a triangular norm.*

Proof. It is sufficient to observe that the degree of inconsistency of either a stratified logic or a canonical extension is equal to 1. □

The proof of this theorem fails for the case μ-canonical extensions of a crisp system $\mathcal{S}$, (for $\mu \neq 1$). The question arises whether a truth-functional semantics can be axiomatized by such a kind of extensions. The following theorem gives a partial answer:

Theorem 2.7. *There is no truth-functional semantics for a μ-canonical extension of classical propositional calculus.*

Proof. Let $\mathcal{M}$ be a truth-functional semantics and m a non-crisp element of $\mathcal{M}$. We claim that m cannot be a theory for the canonical extension of classical propositional calculus. Indeed, otherwise:

- $m(\alpha) = 1$ for every classical tautology α,
- if α is logically equivalent to β, then $m(\alpha) = m(\beta)$,
- $m(\alpha \wedge \beta) = m(\alpha) \wedge m(\beta)$.

Let p be any propositional variable and observe that $p \wedge \neg p$ is logically equivalent to $\neg(\neg(p \wedge \neg p))$. Then, $m(p \wedge \neg p) = m(\neg(\neg(p \wedge \neg p))) = {\sim}({\sim}(m(p \wedge \neg p)))$. Since $\neg(p \wedge \neg p)$ is a classical tautology, ${\sim}(m(p \wedge \neg p)) = m(\neg (p \wedge \neg p)) = 1$. Hence,

$$m(p) \wedge {\sim}(m(p)) = m(p) \wedge m(\neg p) = m(p \wedge \neg p) = {\sim}({\sim}(m(p \wedge \neg p))) = {\sim}(1) = 0.$$

Thus, either $m(p) = 0$ or ${\sim}(m(p)) = 0$, i.e., either $m(p) = 0$ or $m(p) = 1$. □

3. TWO DISCONTINUOUS TRUTH-FUNCTIONAL SEMANTICS

In this section we consider two simple examples of truth-functional semantics which are not continuous. The set $\mathcal{F}$ of formulas is the same as for the classical two-valued propositional calculus. This means that the logical connectives are $\wedge$, $\vee$ and $\neg$. As usual, $\wedge$ and $\vee$ also denote the operations defined by

$$x \wedge y = Min\{x,y\} \quad ; \quad x \vee y = Max\{x,y\}.$$

The first example is defined as follows:

Definition 3.1. Denote by $\mathcal{U}_1 = (U, \wedge, \vee, \sim_1)$ the valuation structure in which $\sim_1$ is defined by

$$\sim_1(\lambda) = \begin{cases} 1 & \text{if } \lambda = 0, \\ 0 & \text{otherwise.} \end{cases}$$

Then we denote the related truth-functional semantics by $\mathcal{M}_1$. Such a semantics is not continuous. The following proposition holds:

Proposition 3.2. *A map $f: U \to U$ is a homomorphism of $\mathcal{U}_1$ iff*
(i) *f is order-preserving,*
(ii) $f(\lambda) = 0 \Leftrightarrow \lambda = 0$,
(iii) $f(1) = 1$.

Proof. Assume that (i), (ii) and (iii) are satisfied. Then, because f is order-preserving, the operations $\wedge$ and $\vee$ are preserved. Moreover, if $\lambda = 0$, then

$$f(\sim_1(\lambda)) = f(1) = 1 = \sim_1(f(\lambda)).$$

If $\lambda \neq 0$, then

$$f(\sim_1(\lambda)) = f(0) = 0 = \sim_1(f(\lambda)).$$

Conversely, assume that f is a homomorphism. Then it is obvious that f is order-preserving. Furthermore, suppose $f(0) \neq 0$. Then

$$0 = \sim_1(f(0)) = f(\sim_1(0)) = f(1)$$

and, hence, $f(1) < f(0)$, which is absurd. This proves that $f(0) = 0$ and, hence, that $f(1) = f(\sim_1(0)) = \sim_1(f(0)) = 1$. Finally, $f(\lambda) = 0$ entails that $1 = \sim_1(f(\lambda)) = f(\sim_1(\lambda))$. Thus, since $\sim_1(\lambda) \neq 0$, we have that $\sim_1(\lambda) = 1$ and $\lambda = 0$. □

Proposition 3.3. *The fuzzy semantics $\mathcal{M}_1$ satisfies the following properties:*

(a) *For every $m \in \mathcal{M}_1$ the fuzzy subset m' defined by setting*

$$m'(\alpha) = \begin{cases} 0 & \text{if } m(\alpha) = 0, \\ 1 & \text{otherwise} \end{cases}$$

is a maximal crisp model containing m.

(b) *The maximal elements of* $\mathcal{M}_1$ *coincide with the classical models.*
(c) *Every classical contradiction is a contradiction in* $\mathcal{M}_1$.
(d) *Every negation of a classical contradiction is a tautology in* $\mathcal{M}_1$.
(e) $Inc(\mathcal{M}_1) = 0$.

Proof. In order to prove (a) let f be the map defined by setting $f(\lambda) = 0$ if $\lambda = 0$ and otherwise $f(\lambda) = 1$. Then f is a homomorphism in $\mathcal{U}_1$. This demonstrates that m' is a crisp model containing m. Assume that m'' is a model containing m'. Then $m''' = f \circ m''$ is a crisp model containing m'' and hence m'. Since a crisp model cannot contain another crisp model, $m''' = m'$ and this demonstrates that $m'' = m'$ and this demonstrates that m' is maximal. Property (b) is a direct consequence of (a). In order to prove (c), observe that if χ is a classical contradiction, then $m(\chi) \leq m'(\chi) = 0$. Properties (d) and (e) are direct consequences of (c). □

Note that (d) does not entail that every classical tautology is a tautology in $\mathcal{M}_1$. For instance, $Taut(p_i \vee \neg p_i) = 0$. In fact, let $\lambda \in U$, $\lambda \neq 0$ and $\lambda \neq 1$, and let m be a model such that $m(p_i) = \lambda$. Then, by observing that $m(p_i \vee \neg p_i) = m(p_i) \vee {\sim_1} m(p_i) = \lambda$, the inequality $Taut(p_i \vee \neg p_i) \leq \lambda$ holds. Thus $Taut(p_i \vee \neg p_i) = 0$. From (b) it follows that the elements in $\mathcal{M}_1$ are not necessarily maximal in $\mathcal{M}_1$. This shows that in Proposition 2.3 the assumption $\sim$ injective is essential.

We now consider a second example.

Definition 3.4. Set $\mathcal{U}_2 = (U, \wedge, \vee, \sim_2)$ where

$$\sim_2(\lambda) = \begin{cases} 0 & \text{if } \lambda = 1, \\ 1 & \text{otherwise.} \end{cases}$$

Then we denote the related multi-valued semantics by $\mathcal{M}_2$. The following propositions dualize Proposition 3.2 and 3.3, respectively.

Proposition 3.5. *A map* $f : U \to U$ *is a homomorphism in* $\mathcal{U}_2$ *iff*
(i) f *is order-preserving,*
(ii) $f(1) = 1 \Leftrightarrow \lambda = 1$,
(iii) $f(0) = 0$.

Proposition 3.6. *The following properties hold :*
(a) *For every* $m \in \mathcal{M}_2$ *the fuzzy subset* m' *defined by setting*

$$m'(\alpha) = \begin{cases} 1 & \text{if } m(\alpha) = 1, \\ 0 & \text{otherwise} \end{cases}$$

is a minimal classical model contained in m.
(b) *The minimal elements of* $\mathcal{M}_2$ *coincide with the classical models.*
(c) *Every classical tautology is a tautology in* $\mathcal{M}_2$.

(d) *Every negation of a classical tautology is a contradiction in* $\mathcal{M}_2$.
(e) $Inc(\mathcal{M}_2) = 0$.

Note that in this case a classical contradiction is not necessarily a contradiction in $\mathcal{M}_2$, for instance $Contr(p_i \wedge \neg p_i) = 0$. Indeed,

$$\begin{aligned} Contr(p_i \wedge \neg p_i) &= 1 - Sup\{\lambda \wedge {\sim_2}(\lambda) : \lambda \in U\} \\ &\leq 1 - Sup\{\lambda \wedge {\sim_2}(\lambda) : \lambda \in U, \lambda \neq 1\} \\ &= 1 - Sup\{\lambda \in U : \lambda \neq 1\} = 0. \end{aligned}$$

4. ANY CONTINUOUS TRUTH-FUNCTIONAL SEMANTICS IS AXIOMATIZABLE

This section is devoted to demonstrating that any continuous truth-functional semantics $\mathcal{M}$ is axiomatizable. First, observe that any formula α whose propositional variables are among $p_1,\ldots,p_n$ is associated with a (polynomial) function $\underline{\alpha} : U^n \to U$ defined by induction on the complexity of α. Indeed, we set, for any $\lambda_1,\ldots,\lambda_n$ in U,

$$\underline{\alpha}(\lambda_1,\ldots,\lambda_n) = \lambda_i \qquad \text{if } \alpha = p_i,$$
$$\underline{\alpha}(\lambda_1,\ldots,\lambda_n) = \underline{h}(\underline{\alpha_1}(\lambda_1,\ldots,\lambda_n),\ldots,\underline{\alpha_t}(\lambda_1,\ldots,\lambda_n)) \quad \text{if } \alpha = h(\alpha_1,\ldots,\alpha_t).$$

To simplify our treatment, we assume that in the language under consideration there are only a finite number of propositional variables $p_1,\ldots,p_k$. As a consequence, the elements in a truth-functional semantics $\mathcal{M}$ coincide with the points $(\lambda_1,\ldots,\lambda_k)$ in the space U^k.

Definition 4.1. Let $\gamma_1, \gamma_2, \ldots$ be a complete enumeration of all the formulas. Then for every initial valuation v we denote by $M(v)$ the set of solutions $(\lambda_1,\ldots,\lambda_k) \in U^k$ of the system

$$\begin{aligned} &\underline{\gamma_1}(\lambda_1,\ldots,\lambda_k) \geq v(\gamma_1) \\ &\underline{\gamma_2}(\lambda_1,\ldots,\lambda_k) \geq v(\gamma_2) \\ &\ldots \end{aligned} \tag{4.1}$$

Proposition 4.2. *The models of an initial valuation v coincide with the points in M(v) and, if v is satisfiable, for every formula* α,

$$Lc(v)(\alpha) = Inf\{\underline{\alpha}(\lambda_1,\ldots,\lambda_k) : (\lambda_1,\ldots,\lambda_k) \in M(v)\} \tag{4.2}$$

In the case $\mathcal{M}$ *balanced,*

$$Lc(v)^{\perp}(\alpha) = Sup\{\underline{\alpha}(\lambda_1,\ldots,\lambda_k) : (\lambda_1,\ldots,\lambda_k) \in M(v)\}. \tag{4.3}$$

Moreover,

$$M(v) = \bigcap\{M(v_f) : v_f \textit{ is finite and } v_f \ll v\}. \tag{4.4}$$

Proof. Equality (4.2) is trivial. In order to prove (4.3) observe that $Lc(v)^{\perp}(\alpha) = 1 - Lc(v)(\neg\alpha)$ and $\underline{\neg\alpha}(\lambda_1,\ldots,\lambda_k) = 1 - \underline{\alpha}(\lambda_1,\ldots,\lambda_k)$. To prove (4.4), observe that $M(v) \subseteq M(v_f)$ for every $v_f \ll v$ and therefore that $M(v)$ is contained in $\bigcap\{M(v_f) : v_f$ *is finite and* $v_f \ll v\}$. To prove the inverse inclusion, suppose $(\lambda_1,\ldots,\lambda_k) \in M(v_f)$, i.e.,

$$\underline{\gamma_1}(\lambda_1,\ldots,\lambda_k) \geq v_f(\gamma_1)$$

$$\underline{\gamma_2}(\lambda_1,\ldots,\lambda_k) \geq v_f(\gamma_2)$$
$$\ldots$$

for every v_f finite such that $v_f \ll v$. Then,

$$\underline{\gamma_1}(\lambda_1,\ldots,\lambda_k) \geq Sup\{v_f(\gamma_1) : v_f \text{ is finite and } v_f \ll v\} = v(\gamma_1)$$
$$\underline{\gamma_2}(\lambda_1,\ldots,\lambda_k) \geq Sup\{v_f(\gamma_2) : v_f \text{ is finite and } v_f \ll v\} = v(\gamma_2)$$
$$\ldots$$

and therefore $(\lambda_1,\ldots,\lambda_k) \in M(v)$. □

A similar procedure holds also in the case that the constraints are closed intervals. For instance, consider an interval constraint (v_1,v_2). In this case we denote by $M(v_1,v_2)$ the set of points $(\lambda_1,\ldots,\lambda_k) \in U^k$ satisfying conditions like

$$v_2(\gamma_1) \geq \underline{\gamma_1}(\lambda_1,\ldots,\lambda_k) \geq v_1(\gamma_1)$$
$$v_2(\gamma_2) \geq \underline{\gamma_2}(\lambda_1,\ldots,\lambda_k) \geq v_1(\gamma_2)$$
$$\ldots$$

In accordance, the logical consequence operator is defined by setting

$$Lc(v_1,v_2)(\alpha) = [J(v_1,v_2)(\alpha)\ ,\ J'(v_1,v_2)(\alpha)]$$

where

$$J(v_1,v_2)(\alpha) = Inf\{\underline{\alpha}(\lambda_1,\ldots,\lambda_k) : (\lambda_1,\ldots,\lambda_k) \in M(v_1,v_2)\}$$

and

$$J'(v_1,v_2)(\alpha) = Sup\{\underline{\alpha}(\lambda_1,\ldots,\lambda_k) : (\lambda_1,\ldots,\lambda_k) \in M(v_1,v_2)\}.$$

In the next theorem, whose proof is obvious, we denote the minimum of a function f in a set X by $Min(f, X)$.

Theorem 4.3. *Let $\mathcal{M}$ be a continuous truth-functional semantics. Then*
- for every formula α the polynomial $\underline{\alpha}$ is continuous
- for every initial valuation v, the set $M(v)$ of models of v is compact.
Consequently,

$$Lc(v)(\alpha) = Min(\underline{\alpha}, M(v)). \tag{4.5}$$

Proposition 4.4. *Let $\mathcal{M}$ be a continuous truth functional semantics. Then*
(a) *$\mathcal{M}$ is closed under ultraproducts.*
(b) *$\mathcal{M}$ is logically compact.*
(c) *Every element of $\mathcal{M}$ is contained in a maximal element of $\mathcal{M}$.*
(d) *The logical consequence operator Lc : $\mathcal{F}(F) \to \mathcal{F}(F)$* associated with *$\mathcal{M}$ is continuous and logically compact.*

Proof. To prove (a), recall that $\mathcal{M}$ is the class of fuzzy subsets m satisfying

$$m(h(\alpha_1,\ldots,\alpha_k)) = \underline{h}(m(\alpha_1),\ldots,m(\alpha_k))$$

for every logical connective h and $\alpha_1,\ldots,\alpha_k$. Then the closure with respect to the ultraproducts follows from Proposition 8.2 in Chapter 2 provided that we set p_i equal to the i-projection for $i = 1,\ldots,k$ and

$$\mathcal{R} = \{(\lambda_0,\ldots,\lambda_k) : \lambda_0 = \underline{h}(\lambda_1,\ldots,\lambda_k)\},$$
$$p_0(\alpha_1,\ldots,\alpha_k) = h(\alpha_1,\ldots,\alpha_k).$$

Propositions (b) and (c) are trivial. To prove (d) observe that, since $\mathcal{M}$ is logically compact, Lc is also logically compact. A more direct proof can be achieved by recalling that, by Proposition 7.1 in Chapter 2, Lc is logically compact provided that

$$v \text{ consistent} \Leftrightarrow \text{every } v_f \text{ finite such that } v_f \ll v \text{ is consistent}$$

and, hence, provided that

$$M(v) \neq \varnothing \Leftrightarrow M(v_f) \neq \varnothing \text{ for every } v_f \text{ finite such that } v_f \ll v.$$

Such an equivalence is a direct consequence of (4.4), the compactness of the sets $M(v_f)$ and the fact that the family $\{M(v_f) : v_f \text{ is finite and } v_f \ll v\}$ is directed.

Finally, by Proposition 6.3 in Chapter 2, to prove that Lc is continuous, we have to prove that

$$Lc(v)(\alpha) = Sup\{Lc(v_f)(\alpha) : v_f \text{ is finite and } v_f \ll v\},$$

that is,

$$Min(\underline{\alpha}, M(v)) = Sup\{Min(\underline{\alpha}, M(v_f)) : \text{where } v_f \text{ is finite and } v_f \ll v\}.$$

To prove such an equality, set

$$M = \{Min(\underline{\alpha}, M(v_f)) : \text{where } v_f \text{ is finite and } v_f \ll v\}.$$

Then, it is obvious that $Min(\underline{\alpha}, M(v))$ is an upper bound of M. Let l be any upper bound of M and, for any v_f finite such that $v_f \ll v$, set

$$T(v_f) = \{(\lambda_1,\dots,\lambda_k) \in U^k : 0 \leq \underline{\alpha}(\lambda_1,\dots,\lambda_k) \leq l\} \cap M(v_f).$$

Because the class $\{v_f : v_f \text{ finite and } v_f \ll v\}$ is upward directed, $C = \{T(v_f) : \text{where } v_f$ is finite and $v_f \ll v\}$ is a downward-directed class of subsets of U^k. Obviously, the elements in this class are compact. Moreover, since at least the points of $M(v_f)$ in which $\underline{\alpha}$ assumes the value $Min(\underline{\alpha}, M(v_f))$ belong to $T(v_f)$, $T(v_f) \neq \varnothing$. Then

$$\cap C = \{(\lambda_1,\dots,\lambda_k) \in U^k : 0 \leq \underline{\alpha}(\lambda_1,\dots,\lambda_k) \leq l\} \cap M(v) \neq \varnothing.$$

Let $(\delta_1,\dots,\delta_k)$ be an element in $M(v)$ such that $\underline{\alpha}(\delta_1,\dots,\delta_k) \leq l$. Then, $Min(\underline{\alpha}, M(v)) \leq \underline{\alpha}(\delta_1,\dots,\delta_k) \leq l$. Thus, $Min(\underline{\alpha}, M(v))$ is the least upper bound of T. □

We can conclude with the main theorem in this section.

Theorem 4.5. *Any continuous truth-functional semantics is axiomatizable.*

5. ANY AXIOMATIZABLE TRUTH-FUNCTIONAL SEMANTICS IS CONTINUOUS

In this section we will try to demonstrate the converse part of Theorem 4.5.

Proposition 5.1. *Let* $\mathcal{M}$ *be a logically compact truth-functional semantics and assume that* $\mathcal{M}$ *is axiomatizable. Then, given any formula* β *such that the associated function* $\underline{\beta}(x_1,\dots,x_n)$ *is order-reversing with respect to each variable* x_i, ,

$$Inf_{a\in A}\, \underline{\beta}(\lambda_1,\dots,\lambda_{t-1}, a, \lambda_{t+1},\dots,\lambda_n) = \underline{\beta}(\lambda_1,\dots\lambda_{t-1}, Sup(A), \lambda_{t+1},\dots,\lambda_n) \quad (5.1)$$

for any $t \in \{1,\dots,n\}$, $\lambda_1, \dots ,\lambda_{t-1},\lambda_{t+1}, \dots ,\lambda_n$ *elements in* U *and* A *nonempty subset of* U. *In particular, for any* $A \subseteq U$

$$Inf(\sim(A)) = \sim(Sup(A)). \quad (5.2)$$

Proof. Let f be the function defined by setting

$$f(x) = \underline{\beta}(\lambda_1,\dots,\lambda_{t-1}, x, \lambda_{t+1},\dots,\lambda_n).$$

Then, the fact that f is order-reversing entails that, for every $a \in A$, $f(a) \geq f(Sup(A))$. Hence, $Inf_{a \in A} f(a) \geq f(Sup(A))$. Also, for every $\lambda \in A$, let v_λ be the initial valuation defined by setting, for any $x \in F$

$$v_\lambda(x) = \begin{cases} \lambda & \text{if } x = p_t, \\ \lambda_i & \text{if } x = p_i \text{ and } i \neq t, \\ Inf_{a \in A} f(a) & \text{if } x = \beta, \\ 0 & \text{otherwise.} \end{cases}$$

Such a valuation is satisfiable. In fact, a model of v_λ can be obtained by considering any truth-functional valuation m such that $m(p_t) = \lambda$ and, for $i \neq t$, $m(p_i) = \lambda_i$. Indeed, since $\lambda \in A$, we have that $f(\lambda) \geq Inf_{a \in A} f(a)$ and therefore,

$$m(\beta) = \underline{\beta}(m(p_1),..., m(p_{t-1}), m(p_t), m(p_{t+1}),...,m(p_n))$$
$$= \underline{\beta}(\lambda_1,...\lambda_{t-1}, \lambda, \lambda_{t+1},..., \lambda_n) = f(\lambda) \geq Inf_{a \in A} f(a) = v_\lambda(\beta).$$

It is obvious that $(v_\lambda)_{\lambda \in A}$ is directed. Consider the corresponding limit v,

$$v(x) = \begin{cases} Sup(A) & \text{if } x = p_t \\ \lambda_i & \text{if } x = p_i,\ i \neq t, \\ Inf_{a \in A} f(a) & \text{if } x = \beta, \\ 0 & \text{otherwise.} \end{cases}$$

Because $\mathcal{M}$ is logically compact and each v_λ is satisfiable, v is satisfiable and a model $m \in \mathcal{M}$ exists such that $m \supseteq v$. Then, since $m(p_t) \geq v(p_t) = Sup(A)$ and $m(p_i) \geq \lambda_i$ for $i \neq t$, we have

$$f(Sup(A)) \geq f(m(p_t)) = \underline{\beta}(\lambda_1,...,\lambda_{t-1}, m(p_t), \lambda_{t+1},..., \lambda_n)$$
$$\geq \underline{\beta}(m(p_1),..., m(p_{t-1}), m(p_t), m(p_{t+1}),..., m(p_n))$$
$$= m(\beta) \geq v(\beta) = Inf_{a \in A} f(a).$$

Thus $f(Sup(A)) = Inf_{a \in A} f(a)$ and this proves (5.1).

We get (5.2) by setting β equal to $\neg p_1$. □

Now, we are able to give the first example of multi-valued semantics which is not axiomatizable.

Theorem 5.2. *The multi-valued semantics $\mathcal{M}_2$ is not axiomatizable.*

Proof. Suppose $\mathcal{M}_2$ is axiomatizable. Then the existence of a contradiction entails, by Proposition 5.3 of Chapter 4, that $\mathcal{M}_2$ is logically compact. On the other hand, (5.2) is not satisfied since $Inf(\sim_2([0,1))) = 1$ while $\sim_2(Sup(A)) = \sim_2(1) = 0$. □

Proposition 5.3. *Let $\mathcal{M}$ be an axiomatizable logically compact truth-functional semantics such that $\sim$ is injective and β a formula such that the associated function $\underline{\beta}(x_1,...,x_n)$ is order-preserving with respect to each variable x_i. Then,*

$$Sup_{a\in A}\,\underline{\beta}(\lambda_1,\ldots,\lambda_{t-1},\, a,\lambda_{t+1},\ldots,\lambda_n) = \underline{\beta}(\lambda_1,\ldots\lambda_{t-1},\, Sup(A),\lambda_{t+1},\ldots,\lambda_n) \qquad (5.3)$$

for any $t\in\{1,\ldots,n\}$, *for any nonempty subset* A *of* U *and* $\lambda_1,\ldots,\lambda_{t-1},\lambda_{t+1},\ldots,\lambda_n$ *elements in* U.

Proof. Consider the negation $\neg\,\beta$ of β. Since $\sim$ is order-reversing and $\underline{\neg\beta}(x) = \sim(\underline{\beta}(x))$, we have that $\underline{\neg\beta}$ is order-reversing. Hence, by (5.1),

$$Inf_{a\in A}\,\underline{\neg\beta}(\lambda_1,\ldots\lambda_{t-1},\, a,\lambda_{t+1},\ldots,\lambda_n) = \underline{\neg\beta}(\lambda_1,\ldots\lambda_{t-1},\, Sup(A),\lambda_{t+1},\ldots,\lambda_n).$$

Moreover, since

$$\underline{\neg\beta}(\lambda_1,\ldots\lambda_{t-1},\, a,\lambda_{t+1},\ldots,\lambda_n) = \sim\underline{\beta}(\lambda_1,\ldots\lambda_{t-1},\, a,\lambda_{t+1},\ldots,\lambda_n),$$

and

$$\underline{\neg\beta}(\lambda_1,\ldots\lambda_{t-1},\, Sup(A),\lambda_{t+1},\ldots,\lambda_n) = \sim\underline{\beta}(\lambda_1,\ldots\lambda_{t-1},\, Sup(A),\lambda_{t+1},\ldots,\lambda_n),$$

by (5.2),

$$Inf_{a\in A}\sim\underline{\beta}(\lambda_1,\ldots\lambda_{t-1},\, a,\lambda_{t+1},\ldots,\lambda_n) = \sim Sup_{a\in A}(\underline{\beta}(\lambda_1,\ldots\lambda_{t-1},a,\lambda_{t+1},\ldots,\lambda_n)),$$

we get

$$\sim Sup_{a\in A}(\underline{\beta}(\lambda_1,\ldots\lambda_{t-1},a,\lambda_{t+1},\ldots,\lambda_n)) = \sim\underline{\beta}(\lambda_1,\ldots\lambda_{t-1},\, Sup(A),\lambda_{t+1},\ldots,\lambda_n).$$

Since $\sim$ is injective this proves (5.3). □

Proposition 5.4. *Let* $\mathcal{M}$ *be a logically compact truth-functional semantics such that* $\sim$ *is a bijection and* β *a formula such that, given* x_i, *the associated function* $\underline{\beta}(x_1,\ldots,x_n)$ *is either order-preserving or order-reversing with respect to* x_i. *Then*

$$\mathcal{M} \text{ axiomatizable} \Rightarrow \underline{\beta} \text{ continuous.}$$

Proof. Assume that $\underline{\beta}$ is order-preserving with respect to x_t. We claim that for every $\mu_1,\ldots\mu_{t-1},\mu_{t+1},\ldots,\mu_n$ in U and B nonempty subset of U

$$Inf_{b\in B}\,\underline{\beta}(\mu_1,\ldots\mu_{t-1},\, b,\mu_{t+1},\ldots,\mu_n) = \underline{\beta}(\mu_1,\ldots\mu_{t-1},\, Inf(B),\mu_{t+1},\ldots,\mu_n). \qquad (5.4)$$

Indeed, let q_i be equal to $\neg\, p_i$ if $\underline{\beta}(x_1,\ldots,x_n)$ is order-preserving with respect to x_i, and let q_i be equal to p_i if $\underline{\beta}(x_1,\ldots,x_n)$ is order-reversing with respect to x_i. Denote by β^* the formula obtained from β by substituting each p_i with q_i. Then, the function $\underline{\beta}^*$ is order-reversing with respect to any variable and, by (5.1), for every $\lambda_1,\ldots\lambda_{t-1}$, $\lambda_{t+1},\ldots,\lambda_n$ in U and every nonempty subset A of U

$$Inf_{a\in A}\,\underline{\beta}^*(\lambda_1,\ldots,\lambda_{t-1},\, a,\lambda_{t+1},\ldots,\lambda_n) = \underline{\beta}^*(\lambda_1,\ldots,\lambda_{t-1},\, Sup(A),\lambda_{t+1},\ldots,\lambda_n)$$

Since $\sim$ is surjective, we can assume that the values λ_i are such that $\sim\lambda_i = \mu_i$ if $\underline{\beta}(x_1,\ldots,x_n)$ is order-preserving with respect to x_i and otherwise, that $\lambda_i = \mu_i$. Consequently, for any $x\in U$,

$$\underline{\beta}^*(\lambda_1,\ldots,\lambda_{t-1},\, x,\lambda_{t+1},\ldots,\lambda_n) = \underline{\beta}(\mu_1,\ldots,\mu_{t-1},\, \sim x,\mu_{t+1},\ldots,\mu_n).$$

Assume that A is a subset of U such that $\sim(A) = B$ and recall that $\sim(Sup(A)) = Inf(\sim(A))$. Thus

$$\begin{aligned}Inf_{b\in B}\,\underline{\beta}(\mu_1,\ldots\mu_{t-1},\, b,\mu_{t+1},\ldots,\mu_n) &= Inf_{a\in A}\,\underline{\beta}(\mu_1,\ldots\mu_{t-1},\, \sim a,\mu_{t+1},\ldots,\mu_n)\\ &= Inf_{a\in A}\,\underline{\beta}^*(\lambda_1,\ldots,\lambda_{t-1},\, a,\lambda_{t+1},\ldots,\lambda_n)\\ &= \underline{\beta}^*(\lambda_1,\ldots,\lambda_{t-1},\, Sup(A),\lambda_{t+1},\ldots,\lambda_n)\\ &= \underline{\beta}(\mu_1,\ldots,\mu_{t-1},\, \sim Sup(A),\mu_{t+1},\ldots,\mu_n)\\ &= \underline{\beta}(\mu_1,\ldots,\mu_{t-1},\, Inf(\sim(A)),\mu_{t+1},\ldots,\mu_n)\\ &= \underline{\beta}(\mu_1,\ldots,\mu_{t-1},\, Inf(B),\mu_{t+1},\ldots,\mu_n)\end{aligned}$$

which proves (5.4). Equation (5.4) together with (5.3) proves that $\underline{\beta}$ is continuous.

Assume that $\underline{\beta}$ is order-reversing with respect to x_i. Then we claim that for every $\mu_1,\ldots\mu_{i-1},\mu_{i+1},\ldots,\mu_n$ in U and $B \subseteq U$

$$Sup_{b\in B}\,\underline{\beta}(\mu_1,\ldots\mu_{i-1}, b,\mu_{i+1},\ldots,\mu_n) = \underline{\beta}(\mu_1,\ldots\mu_{i-1}, Inf(B),\mu_{i+1},\ldots,\mu_n). \tag{5.5}$$

Indeed, set q_i equal to p_i if $\underline{\beta}(x_1,\ldots,x_n)$ is order-preserving with respect to x_i and equal to $\neg p_i$ if $\underline{\beta}(x_1,\ldots,x_n)$ is order-reversing with respect to x_i. Again, denote by β^* the formula obtained from β by substituting each p_i with q_i. Then, the function $\underline{\beta}^*$ is order-preserving with respect to any variable and, taking (5.3) into account, for every $\lambda_1,\ldots\lambda_{i-1},\lambda_{i+1},\ldots,\lambda_n$ in U and every nonempty subset A of U,

$$Sup_{a\in A}\,\underline{\beta}^*(\lambda_1,\ldots,\lambda_{i-1}, a,\lambda_{i+1},\ldots,\lambda_n) = \underline{\beta}^*(\lambda_1,\ldots,\lambda_{i-1}, Sup(A),\lambda_{i+1},\ldots,\lambda_n)$$

Assume that $\sim(A) = B$ and that the values λ_i are such that $\lambda_i = \mu_i$ if $\underline{\beta}(x_1,\ldots,x_n)$ is order-preserving with respect to x_i and otherwise, $\sim\lambda_i = \mu_i$. Then, for any $x \in U$,

$$\underline{\beta}^*(\lambda_1,\ldots,\lambda_{i-1}, x,\lambda_{i+1},\ldots,\lambda_n) = \underline{\beta}(\mu_1,\ldots,\mu_{i-1}, \sim x,\mu_{i+1},\ldots,\mu_n)$$

and

$$\begin{aligned}Sup_{b\in B}\,\underline{\beta}(\mu_1,\ldots\mu_{i-1}, b,\mu_{i+1},\ldots,\mu_n) &= Sup_{a\in A}\,\underline{\beta}(\mu_1,\ldots\mu_{i-1}, \sim a,\mu_{i+1},\ldots,\mu_n)\\ &= Sup_{a\in A}\,\underline{\beta}^*(\lambda_1,\ldots,\lambda_{i-1}, a,\lambda_{i+1},\ldots,\lambda_n)\\ &= \underline{\beta}^*(\lambda_1,\ldots,\lambda_{i-1}, Sup(A),\lambda_{i+1},\ldots,\lambda_n)\\ &= \underline{\beta}(\mu_1,\ldots,\mu_{i-1}, \sim Sup(A),\mu_{i+1},\ldots,\mu_n)\\ &= \underline{\beta}(\mu_1,\ldots,\mu_{i-1}, Inf(\sim(A)),\mu_{i+1},\ldots,\mu_n)\\ &= \underline{\beta}(\mu_1,\ldots,\mu_{i-1}, Inf(B),\mu_{i+1},\ldots,\mu_n).\end{aligned}$$

From (5.1) and (5.5) it follows that $\underline{\beta}$ is continuous. □

Theorem 5.5. *Let $\mathcal{M}$ be a logically compact truth-functional semantics and assume that $\sim$ is a bijection. Then*

$$\mathcal{M} \text{ axiomatizable } \Rightarrow \mathcal{M} \text{ continuous.}$$

6. ZADEH (CONTINUOUS) LOGIC.

In Section 3 of Chapter 5 Zadeh logic was defined as the truth-functional logic whose continuous semantics $\mathcal{M}_Z$ is defined by the valuation structure $\mathcal{U}_Z = (U, \wedge, \vee, \sim_Z)$ where $\wedge$ and $\vee$ are the minimum and the maximum and $\sim_Z(\lambda) = 1-\lambda$.

Proposition 6.1. *If we consider interval constraints, then*

$$Taut(\alpha) = \begin{cases} [1/2, 1] & \text{if } \alpha \text{ is a classical tautology,}\\ [0, 1/2] & \text{if } \alpha \text{ is a classical contradiction,}\\ [0, 1] & \text{otherwise.}\end{cases}$$

Consequently, if we consider lower constraints, then

$$Taut(\alpha) = \begin{cases} 0.5 & \textit{if } \alpha \textit{ is a classical tautology,}\\ 0 & \textit{otherwise.}\end{cases}$$

Proof. Set $X = \{m(\alpha) : m \in \mathcal{M}\}$. Then the continuity of $\mathcal{M}_Z$ entails that X is a closed interval. Moreover, by assigning the value 0.5 to the propositional variables, we get a model m such that $m(\alpha) = 0.5$. This proves that X contains 0.5.

Assume that α is neither a tautology nor a contradiction. Then X is an interval containing both 0 and 1 and therefore $X = [0,1]$.

Assume that α is a tautology, then $1 \in X$ and therefore X contains $[0.5,1]$. We claim that for any $m \in \mathcal{M}_Z$ $m(\alpha) \geq 0.5$ and therefore that $X = [0.5,1]$. Indeed, remember that a *literal* is either a propositional variable p_i or its negation $\neg p_i$ and that a *clause* is a disjunction of literals. Assume that α is a clause. Then, since α is a tautology, in α there are two opposite literals like p_j and $\neg p_j$. Hence, for any $m \in \mathcal{M}_Z$, we have

$$m(\alpha) \geq m(p_j \vee \neg p_j) \geq 1/2.$$

Assume that α is not a clause and observe that, due to the properties of a de Morgan algebra, the conjunctive normal form theorem holds. Then, α is equivalent to a conjunction of clauses $\alpha_1,...,\alpha_h$. Since α is a tautology, each α_i is a tautology and $m(\alpha) = m(\alpha_1) \wedge ... \wedge m(\alpha_h) \geq 0.5$.

Assume that α is a classical contradiction, then $\neg\alpha$ is a classical tautology. Then, because

$$X = \{m(\alpha) : m \in \mathcal{M}\} = \{1-m(\neg \alpha) : m \in \mathcal{M}\} = \{1-x : x \in [0.5,1]\} = [0,0.5],$$

the proof is complete. □

The next proposition, whose proof is trivial, gives a simple characterization of the homomorphisms in $\mathcal{U}_Z$.

Proposition 6.2. *All the homomorphisms in $\mathcal{U}_Z$ can be achieved by considering an order-preserving function $f : [0, 1/2] \to [0,1/2]$ such that $f(1/2) = 1/2$ extended to the whole interval $[0,1]$ by setting*

$$f(\lambda) = 1-f(1-\lambda)$$

for every $\lambda \in [1/2,1]$.

Proposition 6.3. *For every model m the mapping $m' : \mathcal{F} \to U$ defined by setting*

$$m'(\alpha) = \begin{cases} 0 & \text{if } m(\alpha) < 0.5, \\ 1 & \text{if } m(\alpha) > 0.5, \\ 0.5 & \text{if } m(\alpha) = 0.5, \end{cases}$$

is a model we call the "crisp version" of m.

Proof. Observe that the map f defined by setting $f(x) = 0$ if $x < 0.5$, $f(0.5) = 0.5$ and $f(x) = 1$ if $x > 0.5$ is a homomorphism. □

The continuity of $\mathcal{M}_Z$ enables us to apply the algorithm in Theorem 4.3. The following proposition enables us to simplify the computation of $Lc(v)$ and $Lc(v)^{\perp}$:

Proposition 6.4. *For every initial valuation v and $\alpha,\beta \in \mathcal{F}$*

$$Lc(v)(\alpha \wedge \beta) = Lc(v)(\alpha) \wedge Lc(v)(\beta). \tag{6.1}$$

Moreover, neither negation nor disjunction are preserved by $Lc(v)$.

Proof. In order to prove (6.1), let v be any initial valuation. Then,

$$\begin{aligned} Lc(v)(\alpha \wedge \beta) &= Inf\{m(\alpha \wedge \beta) : m \in \mathcal{M}_Z, m \supseteq v\} \\ &= Inf\{m(\alpha) \wedge m(\beta) : m \in \mathcal{M}_Z, m \supseteq v\} \\ &= Inf\{m(\alpha) : m \in \mathcal{M}_Z, m \supseteq v\} \wedge Inf\{m(\beta) : m \in \mathcal{M}_Z, m \supseteq v\} \\ &= Lc(v)(\alpha) \wedge Lc(v)(\beta). \end{aligned}$$

Let α be a formula such that neither α or $\neg\alpha$ are classical tautologies. Then, since classical valuations m_1 and m_2 exist such that $m_1(\alpha) = 0$, $m_2(\neg\alpha) = 0$, we have that $Lc(s^0)(\alpha) = Lc(s^0)(\neg\alpha) = 0$ and, hence, that $Lc(s^0)(\neg\alpha) \neq 1-Lc(s^0)(\alpha)$. Moreover, from the equalities $Lc(s^0)(\alpha \vee \neg\alpha) = 1/2$ and $Lc(s^0)(\alpha) \vee Lc(s^0)(\neg\alpha) = 0$, we can conclude that $Lc(s^0)(\alpha \vee \neg\alpha) \neq Lc(s^0)(\alpha) \vee Lc(s^0)(\neg\alpha)$. □

Observe that (6.1) enables us to simplify the computation of $Lc(v)$ and $Lc(v)^{\perp}$. Indeed, let γ be a formula in the support of the initial valuation v and assume that γ is reduced to a conjunctive normal form $\delta_1 \wedge \ldots \wedge \delta_h$, where $\delta_1,\ldots,\delta_h$ are clauses. Then, since

$$\underline{\gamma}(\lambda_1,\ldots,\lambda_k) = \underline{\delta_1}(\lambda_1,\ldots,\lambda_k) \wedge \ldots \wedge \underline{\delta_h}(\lambda_1,\ldots,\lambda_k),$$

condition $\underline{\gamma}(\lambda_1,\ldots,\lambda_k) \geq v(\gamma)$ is equivalent to the set of conditions

$$\underline{\delta_j}(\lambda_1,\ldots,\lambda_k) \geq v(\gamma), \quad j = 1,\ldots,k.$$

This enables us to give a simple interpretation of the set $M(v)$ of models as a finite intersection of a finite amount of rectangles. Analogous arguments hold for $Lc(v)^{\perp}(\alpha)$.

We conclude this section by summarizing the main properties of Zadeh semantics.

Theorem 6.5. *Zadeh semantics $\mathcal{M}_Z$ is:*

- *balanced,*
- *closed under ultraproducts,*
- *logically compact,*
- *continuous.*

Moreover:

- *its associated logical consequence operator is continuous,*
- *every model m is maximal,*
- *$Inc(\mathcal{M}_Z) = 0.5$.*

7. ŁUKASIEWICZ (CONTINUOUS) LOGIC

In this section we consider the basic truth-functional logic defined by Łukasiewicz (see Łukasiewicz [1970] and, for a general treatment, Cignoli, D'Ottaviano, Mundici [2000]). The logical connectives are interpreted by the following valuation structure:

Definition 7.1. We denote by $\mathcal{U}$ the valuation structure $(U, \odot, \oplus, \sim)$, where U is the interval $[0,1]$ and, for any x, y in U,

$$x \odot y = 0 \vee (x + y - 1) \ ; \ x \oplus y = 1 \wedge (x+y) \ ; \ \sim(x) = 1 - x.$$

The operations $\odot$ and $\oplus$ are called *Łukasiewicz conjunction* and *Łukasiewicz sum*, respectively. We call *Łukasiewicz semantics* the related truth-functional semantics $\mathcal{M}_L$.

The so-defined logic is an extension of Zadeh logic. Indeed, it is easy to verify that

$$x \wedge y = x \odot (\sim(x) \oplus y) \text{ and } x \vee y = (x \odot \sim(y)) \oplus y.$$

In this case the notion of homomorphism is useless. In fact, the unique homomorphism in $\mathcal{U}$ is the identity (see, e.g. Cignoli, D'Ottaviano, Mundici [1995] Theorem 2.4.14).

Proposition 7.2. *The valuation structure $\mathcal{U}$ is rigid. Namely, the only homomorphism of $\mathcal{U}$ in $\mathcal{U}$ is the identity map.*

Proof. Let f be a homomorphism of $\mathcal{U}$. Then, since f is also a homomorphism of $\mathcal{U}_Z$, f is an order-preserving function such that $f(1/2) = 1/2$ and $f(\lambda) = 1-f(1-\lambda)$. Consequently, $f(0) \leq f(1/2) = 1/2$, and

$$f(0) = f(0 \oplus 0) = f(0) \oplus f(0) = f(0) + f(0).$$

This proves that $f(0) = 0$ and, consequently, that $f(1) = 1-f(0) = 1$. Let n be any natural number. Then, since

$$1 = f(1) = f(n/n) = f(1/n \oplus ... \oplus 1/n) = f(1/n) \oplus ... \oplus f(1/n),$$

we have that $f(1/n) \geq 1/n$. Furthermore, for every $m \leq n$,

$$f(m/n\}) = f(1/n \oplus ... \oplus 1/n) = f(1/n) \oplus ... \oplus f(1/n) \geq m/n.$$

On the other hand, since in particular $f((n-m)/n) \geq (n-m)/n$, we have

$$f(m/n) = 1-f(1- m/n) = 1-f((n-m)/n) \leq 1- (n-m)/n = m/n.$$

Thus $f(m/n) = m/n$ for any rational number m/n in U. Because f is order-preserving, this entails that f is the identity map. □

The following theorem follows from Proposition 4.4, Theorem 4.5 and the identities $x \odot \sim x = 0$ and $x \oplus \sim x = 1$.

Theorem 7.3. *Let $\mathcal{M}_L$ be the Łukasiewicz semantics. Then:*

(a) *$\mathcal{M}_L$ is continuous and closed under ultraproducts.*

(b) *All the elements of $\mathcal{M}_L$ are maximal in $\mathcal{M}_L$.*

(c) *Contradictions and tautologies exist.*

(d) $Inc(\mathcal{M}_L) = 0$.

(e) *The logical consequence operator $Lc : \mathcal{F}(\mathbf{F}) \rightarrow \mathcal{F}(\mathbf{F})$* associated with *$\mathcal{M}_L$ is continuous and logically compact.*

(f) *$\mathcal{M}_L$ is axiomatizable.*

Now we will examine the function $\underline{\alpha}$ obtained by interpreting a formula α in the algebra $\mathcal{U}$. Such functions are called *McNaughton functions* and are deeply investigated in a series of interesting papers related to the completeness theorem for Łukasiewicz logic. As an example, see Mundici [1992], Mundici [1994], Panti [1995]), Aguzzoli and Ciabattoni [2000].

Definition 7.4. We define a *rational partition* of the n-cube U^n as a finite set $\pi = \{S_1,...,S_t\}$ of simplexes such that:
(i) the coordinates of the vertices of each S_i are rational numbers,
(ii) any two simplexes of π intersect in a common face,
(iii) the n-cube U^n is the union of all simplexes of π.

Proposition 7.5. *Let α be a formula and $\underline{\alpha}$ the related interpretation. Then $\underline{\alpha}$ is continuous and a rational partition $\pi = \{S_1,...,S_t\}$ exists such that the restriction of $\underline{\alpha}$ to S_i is a linear polynomial function q_i with integral coefficients.*

Proof. The proof is by induction on the number of connectives occurring in the formula. □

The converse of Proposition 7.5 (McNaughton's theorem) gives a complete characterization of McNaughton functions and it was proved in McNaughton [1951]. A constructive version of McNaughton's theorem can be found in Mundici [1994].

Proposition 7.6. *Let v be an initial valuation whose support is the finite set $\{\gamma_1,...,\gamma_s\}$ of formulas. Then we can find (in an effective way) a rational partition $\pi = \{S_1,...,S_t\}$ and linear polynomials $q^1{}_1,...,q^1{}_s,...,q^t{}_1,...,q^t{}_s$* in such a way that,

$$M(v) = E_1 \cup ... \cup E_t$$

where E_i is the simplex defined by the linear system

$$\begin{array}{c} q_1{}^i(\lambda_1,...,\lambda_k) \geq v(\gamma_1) \\ \dots \\ q_s{}^i(\lambda_1,...,\lambda_k) \geq v(\gamma_s) \end{array} \tag{7.1}$$

with the constraint $(\lambda_1,...,\lambda_k) \in S_i$.

Proof. We know that $M(v)$ is the set of solutions of the system

$$\begin{array}{c} \underline{\gamma}_1(\lambda_1,...,\lambda_k) \geq v(\gamma_1) \\ \dots \\ \underline{\gamma}_s(\lambda_1,...,\lambda_k) \geq v(\gamma_s). \end{array} \tag{7.2}$$

For every $i \in \{1,...,s\}$, let $\pi_i = \{S^i{}_1,...,S^i{}_{t(i)}\}$ be a rational partition associated with γ_i via Proposition 7.4. Let $\pi = \{S_1,...,S_t\}$ be the rational partition generated by $\pi_1,...,\pi_s$. This means that an element S_i of π is a nonempty intersection $S^1{}_{i(1)} \cap ... \cap S^s{}_{i(s)}$. Then it is obvious that, in any S_i, (7.2) is equivalent to a linear system like (7.1). □

We conclude this section by giving a simple way to compute $Lc(v)$.

Theorem 7.7. *Let v be an initial valuation with rational values and assume that the support of v is finite. Then, for every formula α, we can find (in an effective* *way) a class* $\pi = \{E_1,...,E_t\}$ *of simplexes and linear polynomial* $q_1,...,q_t$ *in such a way that*

$$Lc(v)(\alpha) = Min(q_1,E_1) \wedge ... \wedge Min(q_t,E_t). \qquad (7.3)$$

Proof. See Mundici [1994]. □

Formula (7.3) gives an effective way to compute the fuzzy subset of logical consequences of an initial valuation v.

8. COMPARING TRUTH-FUNCTIONAL LOGIC WITH FUZZY LOGIC

The notion of deduction used in fuzzy logic is completely different from the traditional notion of deduction in truth-functional logics found in the literature. In fact, this latter notion is devised as a tool to produce a crisp set of formulas from a crisp set of formulas. In other words, the deduction apparatus is not different in nature from the deduction apparatus of a crisp logic. Namely, usually one proceeds as follows. Given a truth-functional semantics $\mathcal{M}$, a nonempty set D of truth values is fixed. The truth values in D are considered *"sufficiently near to truth"*, in a sense. We assume that D is an upper set containing 1, i.e.,

(a) $1 \in D$,

(b) $\lambda \in D,\ \mu \geq \lambda \Rightarrow \mu \in D$

and we say that the elements in D are *designed truth values*. The pair $<\mathcal{M},D>$ is called a *logical matrix*. Given a model m and a formula α, we say that *m is a model of α* and we write $m \models \alpha$, if $m(\alpha) \in D$, i.e., the truth value of α in m is *"sufficiently true"*. The (crisp) *set of valid formulas* is defined by setting

$$Valid(\mathcal{M}, D) = \{\alpha \in \mathbb{F} : m \models \alpha \text{ for every } m \in \mathcal{M}\}.$$

With the notations proposed in Chapter 5 and assuming that D belongs to the constraint frame $\mathcal{C}$, this means that

$$Valid(\mathcal{M}, D) = \{\alpha \in \mathbb{F} : Taut(\alpha) \subseteq D\} = \mathrm{C}(Taut, D). \qquad (8.1)$$

For example, if $U = [0,1]$ and $D = [0.8,1]$, we have $Valid(\mathcal{M}, D) = \mathrm{C}(Taut,0.8)$. So, in a sense, truth-functional logic is only interested to a fixed *"cut"* of the $\mathcal{C}$-subset of tautologies *Taut* and not to all the information expressed by *Taut*. In accordance, the aim of a deduction apparatus is only to produce exactly such a cut.

Furthermore, there is a basic difference for the notion of logical consequence. In fact, if X is a set of formulas and $\alpha \in \mathbb{F}$, we write

- $m \models X$ if $m \models \gamma$ for every $\gamma \in X$,
- $X \models \alpha$ if $m \models \alpha$ for every m such that $m \models X$.

The logical consequence operator $Lc_D : \mathcal{P}(\mathbb{F}) \to \mathcal{P}(\mathbb{F})$ is defined by setting

$$Lc_D(X) = \{\alpha \in \mathbb{F} : X \models \alpha\} \qquad (8.2)$$

for any crisp set of formulas X (see for example Tarski [1956], Wójcicki [1970] and Wojcicki [1973]). One proves that Lc_D is a closure operator. Obviously, differently from Lc, the operator Lc_D operates on crisp information to give crisp information. Nevertheless, we can try to approximate Lc by varying the set D of designed truth values. Indeed, for any $\lambda \in U$, we can refer to the set $[\lambda,1]$ of designed values, i.e., we select λ as a threshold of acceptability. In accordance, given $\alpha \in \mathcal{F}$ and a set X of formulas, we write

- $m \models_\lambda \alpha$ if $m(\alpha) \geq \lambda$,
- $m \models_\lambda X$ if $m \models_\lambda \beta$ for every $\beta \in X$,
- $X \models_\lambda \alpha$ if $m \models_\lambda \alpha$ for every m such that $m \models_\lambda X$.

Also, we set $Lc_\lambda = Lc_{[\lambda,1]}$, that is

$$Lc_\lambda(X) = \{\alpha \in \mathcal{F} : X \models_\lambda \alpha\}. \tag{8.3}$$

This enables us to associate a stratified logic to any truth-functional semantics.

Definition 8.1. Let $\mathcal{M}$ be a truth-functional semantics. Then we define *the stratified closure operator associated with* $\mathcal{M}$ as the closure operator $\underline{Lc}$ associated with the family $(Lc_\lambda)_{\lambda \in U}$ of closure operators.

The following theorem gives a negative answer to the question whether fuzzy logic can be reduced to *"traditional"* many-valued logic. Indeed, it shows that the stratified operator $\underline{Lc}$ associated with $\mathcal{M}$ (we can define in a traditional many-valued logic) is different from the logical consequence operator Lc defined by $\mathcal{M}$ (as defined in fuzzy logic).

Theorem 8.2. *Let $\mathcal{M}$ be a truth-functional semantics. Then, $\underline{Lc}$ coincides with the stratified operator Lc^* associated with Lc and therefore $\underline{Lc} \neq Lc$.*

Proof. Let $\mathcal{T}$ be the closure system generated by $\mathcal{M}$, and consider the family $\mathcal{H}(\mathcal{T},\lambda) = \{C(s,\lambda) : s \in \mathcal{T}\}$. Then $Co(\mathcal{H}(\mathcal{T},\lambda)) = Lc_\lambda$. Indeed,

$$\begin{aligned} Co(\mathcal{H}(\mathcal{T},\lambda))(X) &= \bigcap\{C(s,\lambda) : X \subseteq C(s,\lambda),\ s \in \mathcal{T}\} \\ &= \bigcap\{C(m,\lambda) : X \subseteq C(m,\lambda),\ m \in \mathcal{M}\} \\ &= \{\alpha \in \mathcal{F} : \forall\, m \in \mathcal{M}, X \subseteq C(m,\lambda) \Rightarrow \alpha \in C(m,\lambda)\} \\ &= \{\alpha \in \mathcal{F} : \text{every } [\lambda,1]\text{-model of } X \text{ is a } [\lambda,1]\text{-model of } \alpha\} \\ &= Lc_\lambda(X). \end{aligned}$$

Since by Proposition 5.1 in Chapter 7 it is $Co(\mathcal{H}(\mathcal{T},\lambda)) = \mathcal{K}(Co(\mathcal{T}),\lambda) = \mathcal{K}(Lc,\lambda)$, we can conclude that $\underline{Lc} = Lc^*$. Finally, observe that, as observed in Theorem 2.6, Lc is not stratified and therefore $Lc \neq Lc^*$. □

Such a theorem shows that the deduction apparatus usually proposed in truth-functional logics is not adequate. In fact, in our opinion, Lc is the correct semantics operator we have to consider in searching for a completeness theorem. Consequently, we need the more general notion of *"approximate reasoning"* and this notion is different from the notion of reasoning by cuts (that is, by thresholds).

CHAPTER 9

PROBABILISTIC FUZZY LOGICS

1. VAGUENESS AND UNCERTAINTY

The notion of *"vagueness"* must be sharply distinguished from the notion of *"uncertainty"*. Accordingly, fuzzy logic must be sharply distinguished from probabilistic logic. Indeed, a graded truth value for a formula α mustn't be confused with a measure of our degree of belief in α. To emphasize this difference, consider the following example:

Let α be the claim *"the rose on the table is red"* and imagine two different situations. In the first one we can freely examine the rose but, as a matter of fact, the color looks not exactly red. Then α is neither fully true nor fully false and we can express that by assigning to α a truth value, as an example 0.8, different from 0 and 1. This number is the truth degree of the claim that the color of the rose is red (fuzziness). In the second situation, imagine a world in which all the roses are either clearly red or clearly yellow. In such a world the claim α is either true or false but, inasmuch as we cannot examine the rose on the table, we are not able to know what is the case. Nevertheless, we have an opinion about the possible color of that rose and we could assign to α a number, as an example 0.8, as a subjective measure of our degree of belief that α is true (probability). This opinion could result from some general experience, from the percentage of the red roses in the world, from the taste of the possessor of the rose and so on, but not necessarily from a complete knowledge of that particular rose.

Now, in spite of such a basic distinction, we can try to consider belief theory as a particular chapter of fuzzy logic. Indeed, we can observe that in this case the vagueness is related to a metalogic notion as *"belief"*. In other words, while in *"the rose is red"* we have that *"red"* is a graded property for things (flowers), in *"α is believed"* we have that *"believed"* is a graded property for sentences.

In any case, as a matter of fact, the formalisms proposed in this book for vague notions are also very useful to manage information probabilistic in nature. In accordance, in this chapter we will consider three *"fuzzy logics"* related to probability theory. These logics are in accordance with classical logic in the sense that the valuation of a tautology is always 1 and if two formulas are logically equivalent, then they assume the same truth value. Consequently, we assume directly that the set of formulas coincides with the Lindenbaum algebra of the classical propositional calculus or, more generally, with a Boolean algebra $\boldsymbol{B}$ with the least element $\mathbf{0}$ and the greatest element $\mathbf{1}$. The fuzzy semantics we consider coincide with:

- the class $\mathcal{M}_{sa}$ of the constant-sum super-additive measures,
- the class $\mathcal{M}_{ul}$ of the upper-lower probabilities,

- the class $\mathcal{M}_p$ of the finitely additive probabilities.

The related classes of consistent theories coincide with

- the class $\mathcal{T}_{sa}$ of all super-additive measures,
- the class $\mathcal{T}_{ul}$ of all upper-lower probabilities,
- the class $\mathcal{T}_p$ of all lower envelopes.

These logics are also related to the logic of the necessities defined in Chapter 6. In fact, denote by $\mathcal{T}_n$ the class of consistent theories of this logic, i.e., the class of necessities. Then, because one proves that every necessity is a lower envelope (see for example Höhle [1983]), we have the following chain of strict inclusions:

$$\mathcal{T}_n \subseteq \mathcal{T}_p \subseteq \mathcal{T}_{ul} \subseteq \mathcal{T}_{sa}.$$

Also, we sketch a logic, Boolean in nature, which is related with the belief measures .

In the semantics we consider, **1** is a tautology, i.e., $m(\mathbf{1}) = 1$ for every model m and

$$m(\alpha) + m(\neg\alpha) \leq 1,$$

where we denote the unary operation in $\boldsymbol{B}$ by $\neg$. This means that the logics under consideration are with a negation. In particular,

$$m(\mathbf{0}) + 1 = m(\mathbf{0}) + m(\mathbf{1}) \leq 1,$$

and this proves that $m(\mathbf{0}) = 0$ and therefore that **0** is a contradiction. Consequently, the degree of inconsistency is zero and a theory is either totally consistent or totally inconsistent. As usual, given α and β in $\boldsymbol{B}$, we denote the formula $(\neg\alpha) \vee \beta$ by $\alpha \rightarrow \beta$ and we write $\alpha \mapsto \beta$ to denote that $\alpha \rightarrow \beta = \mathbf{1}$, that is $\alpha \leq \beta$.

2. LOGIC OF SUPER-ADDITIVE MEASURES

The first logic, we call *sa-logic*, admits as a semantics the class $\mathcal{M}_{sa}$ of the constant-sum super-additive measures (see Gerla [1997]). A *super-additive measure* in a Boolean algebra $\boldsymbol{B}$ is a map $p : \boldsymbol{B} \rightarrow [0,1]$ such that $p(\mathbf{1}) = 1$ and, for every $\alpha, \beta \in \boldsymbol{B}$ for which $\alpha \wedge \beta = \mathbf{0}$,

$$p(\alpha \vee \beta) \geq p(\alpha) + p(\beta). \tag{2.1}$$

A super-additive measure p is said to be a *constant-sum super-additive measure* if, for each $\alpha \in \boldsymbol{B}$,

$$p(\alpha) + p(\neg\alpha) = 1 \tag{2.2}$$

(see, e.g., Owen [1982]). Probability measures are examples of constant-sum super-additive measures but a constant-sum super-additive measure is not necessarily a probability as we shall see in the following remark:

Remark. A super-additive measure is also called *characteristic function* of an n-person game and this name is justified by the following interpretation. Consider a game with a set P of players and let $\boldsymbol{B} = \mathcal{P}(P)$. Every set $\alpha \in \boldsymbol{B}$ of players is interpreted as a *"coalition"* in the game and the number $p(\alpha)$ represents the (sure) gain of the coalition provided we set the gain arising from the coalition of all the players equal to 1. The meaning of (2.1) is that the gain of $\alpha \vee \beta$ is greater than or equal to the sum of the gain of α and the gain of β. Indeed, among the possible strategies of the coalition $\alpha \vee \beta$ there are the strategies in which the two groups α and β play separately. The inequality $p(\alpha \vee \beta) > p(\alpha) + p(\beta)$ means that the coalition $\alpha \vee \beta$ is convenient, while, if $p(\alpha \vee \beta) = p(\alpha) + p(\beta)$, the coalition is useless. In terms of game theory, the constant-sum condition means that the game is strictly competitive. An example of a constant-sum super-additive measure which is not a probability is furnished by a game in which the majority always wins. More specifically, let P be a finite set (the set of players) whose number of elements is odd and different from 1 and set $\boldsymbol{B} = \mathcal{P}(P)$. Then the function p defined by setting, for every subset α of P,

$$p(\alpha) = \begin{cases} 1 & \text{if } \alpha \text{ has more elements than } \neg\alpha, \\ 0 & \text{otherwise,} \end{cases}$$

is a constant-sum super-additive measure and, since $p(\{x\}) = 0$ for every x, p is not a probability.

The following proposition gives some properties of the fuzzy semantics $\mathcal{M}_{sa}$:

Proposition 2.1. *The class $\mathcal{M}_{sa}$ of the constant-sum super-additive functions is a balanced fuzzy semantics closed under ultraproducts.*

Proof. $\mathcal{M}_{sa}$ is balanced by definition. In order to prove that $\mathcal{M}_{sa}$ is closed under ultraproducts, we can apply Proposition 8.2 in Chapter 2. In fact,

- we can express condition $p(\mathbf{1}) = 1$ by setting
  $\mathcal{R} = \{1\}$, $p_0(x) = x$, $D = \{\mathbf{1}\}$,
- we can express the superadditivity by setting
  $\mathcal{R} = \{(\lambda_1, \lambda_2, \lambda_3) : \lambda_1 \geq \lambda_2 + \lambda_3\}$, $p_0(x,y) = x \vee y$,
  $p_1(x,y) = x$, $p_2(x,y) = y$, $D = \{(x,y) \in \boldsymbol{B} \times \boldsymbol{B} : x \wedge y = \mathbf{0}\}$,
- we can express the constant-sum condition by setting
  $\mathcal{R} = \{(\lambda_1,\lambda_2) : \lambda_1 + \lambda_2 = 1\}$, $p_0(x) = x$, $p_1(x) = \neg x$. □

Let's define a fuzzy H-system $\mathcal{S}_{sa}$, which we call the *sa-system*, whose only logical axiom is **1** and with two inference rules. The first is the *fuzzy disjunction rule* defined by setting for $\alpha_1, \alpha_2 \in \boldsymbol{B}$ such that $\alpha_1 \wedge \alpha_2 = \mathbf{0}$,

$$\frac{\alpha_1, \alpha_2}{\alpha_1 \vee \alpha_2} \quad ; \quad \frac{\lambda_1, \lambda_2}{\lambda_1 \oplus \lambda_2}$$

where $\oplus$ is Łukasiewicz disjunction. More formally, the disjunction rule is the fuzzy rule $d = (d',d'')$ such that

$Dom(d) = \{(\alpha_1,\alpha_2) \in \boldsymbol{B} \times \boldsymbol{B} : \alpha_1 \wedge \alpha_2 = \mathbf{0}\}$,
$d'(\alpha_1,\alpha_2) = \alpha_1 \vee \alpha_2$,
$d''(\lambda_1,\lambda_2) = \lambda_1 \oplus \lambda_2$.

The second inference rule is the *collapsing rule* which assert that if we have proven two disjoint formulas α_1 and α_2 to the degrees λ_1 and λ_2 and $\lambda_1 + \lambda_2 > 1$, then we may prove the contradiction (to degree 1). In this case it is evident that the whole theory collapses. In a sense, the collapsing rule is a *"consistency control"* device rather than an inference rule. More formally, we consider the rule $c = (c',c'')$ such that $Dom(c) = \{(\alpha_1,\alpha_2) : \alpha_1 \wedge \alpha_2 = \mathbf{0}\}$, $c'(\alpha_1,\alpha_2) = \mathbf{0}$ for any $\alpha_1, \alpha_2 \in Dom(c)$ and

$$c''(\lambda_1,\lambda_2) = \begin{cases} 1 \text{ if } \lambda_1 + \lambda_2 > 1 \\ \\ 0 \text{ in the case } \lambda_1 + \lambda_2 \leq 1. \end{cases}$$

We can picture this rule by setting for α_1 and α_2 formulas such that $\alpha_1 \wedge \alpha_2 = \mathbf{0}$

$$\frac{\alpha_1, \alpha_2}{\mathbf{0}} \quad ; \quad \frac{\lambda_1, \lambda_2}{c''(\lambda_1,\lambda_2)} \, .$$

The proofs in this system are very simple. For instance, let α be the formula $(\alpha_1 \vee \alpha_2) \vee \alpha_3$ where α_1, α_2 and α_3 are pairwise disjoint. Then, the following picture represents a proof π of α such that $Val(\pi,v) = v(\alpha_1) \oplus v(\alpha_2) \oplus v(\alpha_3)$:

$$\frac{\dfrac{\alpha_1, \alpha_2}{\alpha_1 \vee \alpha_2} \;,\; \alpha_3}{\alpha} \quad ; \quad \frac{\dfrac{v(\alpha_1), v(\alpha_2)}{v(\alpha_1) \oplus v(\alpha_2)} \;,\; v(\alpha_3)}{v(\alpha_1) \oplus v(\alpha_2) \oplus v(\alpha_3)} \, .$$

Theorem 2.2. *The class $\mathcal{T}_{sa}$ of consistent theories of S_{sa} coincides with the class of super-additive measures. The class of the consistent complete theories of S_{sa} coincides with $\mathcal{M}_{sa}$.*

Proof. Let τ be a consistent theory of the *sa*-system. Then, since τ contains the set of logical axioms, $\tau(\mathbf{1}) = 1$. In order to prove that τ satisfies (2.1), let α and β be two incompatible formulas. Then, the closure with respect to the disjunction rule entails that $\tau(\alpha \vee \beta) \geq \tau(\alpha) \oplus \tau(\beta)$. This inequality implies that τ is order-preserving. Furthermore, in the case $\tau(\alpha \vee \beta) < 1$, it is $\tau(\alpha) \oplus \tau(\beta) = \tau(\alpha) + \tau(\beta)$, and therefore we get (2.1). If $\tau(\alpha \vee \beta) = 1$ and, by way of contradiction, $\tau(\alpha) + \tau(\beta) > 1$, then, by the collapsing rule,

$$\tau(\mathbf{0}) = \tau(c'(\alpha,\beta)) \geq c''(\tau(\alpha), \tau(\beta)) = 1.$$

This contradicts the consistency of τ, and therefore we may conclude that $\tau(\alpha) + \tau(\beta) \leq 1 = \tau(\alpha \vee \beta)$. Thus, (2.1) is satisfied. Conversely, it is clear that every super-additive measure is a consistent theory. □

As a consequence, $\mathcal{D}(v)$ is the super-additive measure generated by v, i.e., the least super-additive measure greater or equal to v. In the next theorem we give a simple way to get $\mathcal{D}(v)$:

Theorem 2.3. *A fuzzy set of formulas v is consistent iff for every sequence $\alpha_1,\ldots,\alpha_n$ of pairwise disjoint formulas*

$$v(\alpha_1) + \ldots + v(\alpha_n) \leq 1. \tag{2.3}$$

If this is the case

$$\mathcal{D}(v)(\alpha) = \begin{cases} Sup\{v(\alpha_1)+...+v(\alpha_n) : (\alpha_1 \vee ... \vee \alpha_n) \mapsto \alpha,\ \alpha_i \wedge \alpha_j = \mathbf{0} \text{ for } i \neq j\} & if\ \alpha \neq \mathbf{1} \\ 1 & if\ \alpha = \mathbf{1}. \end{cases} \quad (2.4)$$

Proof. Assume (2.3) and define v' by setting $v'(\mathbf{1}) = 1$ and, if $\alpha \neq \mathbf{1}$,

$$v'(\alpha) = Sup\{v(\alpha_1) + ... + v(\alpha_n) : (\alpha_1 \vee ... \vee \alpha_n) \mapsto \alpha,\ \alpha_i \wedge \alpha_j = \mathbf{0} \text{ for } i \neq j\}.$$

We claim that $\mathcal{D}(v) = v'$. Indeed, (2.3) entails that the values of v' are in [0,1]. Moreover, for every pair α and β of disjoint formulas

$$\begin{aligned} v'(\alpha \vee \beta) &= Sup\{v(\alpha_1) + ... + v(\alpha_n) : (\alpha_1 \vee ... \vee \alpha_n) \mapsto \alpha \vee \beta,\ \alpha_i \wedge \alpha_j = \mathbf{0} \text{ for } i \neq j\} \\ &\geq Sup\{v(\alpha_1) + ... + v(\alpha_n) : \text{either } (\alpha_1 \vee ... \vee \alpha_n) \mapsto \alpha \text{ or} \\ &\qquad (\alpha_1 \vee ... \vee \alpha_n) \mapsto \beta,\ \alpha_i \wedge \alpha_j = \mathbf{0} \text{ for } i \neq j\} \\ &\geq Sup\{v(\alpha_1) + ... + v(\alpha_n) : (\alpha_1 \vee ... \vee \alpha_n) \mapsto \alpha,\ \alpha_i \wedge \alpha_j = \mathbf{0} \text{ for } i \neq j\} \\ &\quad + Sup\{v(\alpha_1) + ... + v(\alpha_n) : (\alpha_1 \vee ... \vee \alpha_n) \mapsto \beta,\ \alpha_i \wedge \alpha_j = \mathbf{0} \text{ for } i \neq j\} \\ &= v'(\alpha) + v'(\beta). \end{aligned}$$

Then v' is a super-additive function. It is obvious that $v' \supseteq v$. Let p be a super-additive measure such that $p \supseteq v$. Then, for every α and $\alpha_1,...,\alpha_n$ pairwise disjoint formulas such that $(\alpha_1 \vee ... \vee \alpha_n) \mapsto \alpha$,

$$v(\alpha_1) + ... + v(\alpha_n) \leq p(\alpha_1) + ... + p(\alpha_n)$$

and therefore $v'(\alpha) \leq p(\alpha)$. This demonstrates that v' is the intersection of the super-additive functions containing v and, hence, that $v' = \mathcal{D}(v)$. Trivially, this implies also that v is consistent. Conversely, if v is consistent, then the theory $\mathcal{D}(v)$ is a super-additive measure containing v and, consequently,

$$v(\alpha_1) + ... + v(\alpha_n) \leq \mathcal{D}(v)(\alpha_1) + ... + \mathcal{D}(v)(\alpha_n) \leq \mathcal{D}(v)(\alpha_1 \vee ... \vee \alpha_n) \leq 1. \quad \square$$

Corollary 2.4. *Let v be a consistent initial valuation v and $\lambda \neq 0$. Then the initial valuation $\lambda \cdot v$ is consistent and*

$$\mathcal{D}(\lambda \cdot v)(x) = \begin{cases} 1 & if\ \ x = \mathbf{1}, \\ \lambda \cdot \mathcal{D}(v) & otherwise. \end{cases}$$

Remark. In particular, we can assume that v is the characteristic function of a crisp set X and hence that $\lambda \cdot v$ is equal to $\lambda \wedge X$. This means that we assign a degree of belief at least equal to λ to any formula in X. In this case it is useful to consider the extended number $m(X) \in \boldsymbol{N} \cup \{\infty\}$ defined by setting

$$m(X) = \begin{cases} \infty & \text{if } \ \mathbf{0} \in X, \\ Max\{n \in \boldsymbol{N} : \ n \text{ pairwise disjoint elements in } X \text{ exist}\} & \text{otherwise.} \end{cases}$$

Indeed, in accordance with Theorem 2.3,

$$\lambda \wedge X \text{ consistent} \Leftrightarrow \lambda \leq 1/m(X).$$

Further, if $\lambda \wedge X$ is consistent, then

$$\mathcal{D}(\lambda \wedge X)(x) = \begin{cases} 1 & \text{if } x = \mathbf{1}, \\ \lambda \cdot m(\{z \in X : z \le x\}) & \text{otherwise.} \end{cases} \tag{2.3}$$

Suppose $m(X)$ is finite. Then, if we want to obtain the maximum possible information from the initial valuation X, it is very natural to consider the fuzzy subset $\lambda \wedge X$ with $\lambda = 1/m(X)$. The corresponding theory $\mathcal{D}(\lambda \wedge X)$ is defined by

$$\mathcal{D}(\lambda \wedge X)(x) = m(\{z \in X : z \le x\}) / m(X).$$

For instance, let $b_1, b_2, \ldots, b_n$ be pairwise disjoint elements of $\boldsymbol{B}$ and X the Boolean subalgebra of $\boldsymbol{B}$ generated by $P = \{b_1, b_2, \ldots, b_n\}$. Then $m(X) = n$ and, for $\lambda = 1/n$,

$$\mathcal{D}(\lambda \wedge X)(x) = |\{z \in X : z \le x\}| / n.$$

3. COMPLETENESS THEOREM

In order to prove the completeness theorem, we first observe that, as a direct consequence of (2.3) of Theorem 2.3, the following theorem holds :

Theorem 3.1. *The system S_{sa} is logically compact.*

The next two propositions enable us to extend any theory to a complete theory. Recall that in Section 5 of Chapter 5 we defined the extension of a fuzzy subset τ of formulas by a formula α as the fuzzy subset τ_α such that $\tau_\alpha(x) = \tau(x)$ if $x \neq \alpha$ and $\tau_\alpha(x) = \sim\tau(\neg x)$ if $x = \alpha$.

Proposition 3.2. *Let τ be a consistent theory of S_{sa}. Then, for every formula α*

- *the extension τ_α of τ by α is consistent,*
- *$\mathcal{D}(\tau_\alpha)(\neg\alpha) = \tau(\neg\alpha)$ and $\mathcal{D}(\tau_\alpha)(\alpha) = \sim\tau(\neg\alpha)$,*
- *α is decidable in τ_α.*

Proof. By the consistency of τ, $\tau(\alpha) + \tau(\neg\alpha) \le 1$ and, hence, $\tau_\alpha(\alpha) = 1-\tau(\neg\alpha)$. In order to prove that τ_α satisfies (2.3) we may confine ourselves to the case in which $\alpha_1, \ldots, \alpha_n$ are pairwise disjoint formulas such that $\alpha_1 = \alpha$. Now, since τ is order-preserving and $\neg\alpha \ge \alpha_2 \vee \ldots \vee \alpha_n$, we get $\tau(\neg\alpha) \ge \tau(\alpha_2 \vee \ldots \vee \alpha_n)$ and therefore $1 - \tau(\neg\alpha) + \tau(\alpha_2 \vee \ldots \vee \alpha_n) \le 1$. Hence,

$$\begin{aligned} \tau_\alpha(\alpha) + \tau_\alpha(\alpha_2) + \ldots + \tau_\alpha(\alpha_n) &= 1 - \tau(\neg\alpha) + \tau(\alpha_2) + \ldots + \tau(\alpha_n) \\ &\le 1 - \tau(\neg\alpha) + \tau(\alpha_2 \vee \ldots \vee \alpha_n) \le 1. \end{aligned}$$

This demonstrates that τ_α is consistent. Furthermore, since $\mathcal{D}(\tau_\alpha) \supseteq \tau_\alpha \supseteq \tau$,

$$1 \ge \mathcal{D}(\tau_\alpha)(\neg\alpha) + \mathcal{D}(\tau_\alpha)(\alpha) \ge \tau_\alpha(\neg\alpha) + (\tau_\alpha)(\alpha) = \tau(\neg\alpha) + 1-\tau(\neg\alpha) = 1,$$

and this demonstrates the remaining part of the proposition. □

Proposition 3.3. *Let τ be a consistent theory, α a formula and $(\alpha(n))_{n\in N}$ an enumeration of all the formulas in $\boldsymbol{B}$ such that $\alpha(1) = \neg\alpha$. Then by setting*

$$\tau_0 = \tau \ ; \ \tau_{n+1} = \mathcal{D}((\tau_n)_{\alpha(n+1)}) \ ; \ m = \bigcup_{n\in N} \tau_n$$

we get a complete theory extending τ *such that*

$$m(\alpha) = \tau(\alpha) \quad and \quad m(\neg\alpha) = 1-\tau(\alpha).$$

Proof. By Proposition 3.2 every τ_n is consistent, so, $\mathcal{D}$ being logically compact, m is consistent. The rest is evident. □

There is no difficulty to extend such a result to the case in which $\boldsymbol{B}$ is not enumerable. Indeed, the proof also works well by assuming that $\boldsymbol{B}$ is well ordered.

Theorem 3.4. *Every consistent theory is an intersection of a family of complete theories.*

Proof. Let τ be a consistent theory and α a formula. Then in order to prove the proposition it is sufficient to observe that Proposition 3.3 entails that for every formula α there is a complete theory m such that $m \supseteq \tau$ and $m(\alpha) = \tau(\alpha)$. □

This theorem enables us to prove the following completeness theorem:

Theorem 3.5. *The sa-system* $\mathcal{S}_{sa}$ *together with the semantics* $\mathcal{M}_{sa}$ *of the constant-sum super-additive measures define a fuzzy logic.*

Proof. We can apply Proposition 3.2 in Chapter 4. Indeed, by Theorem 3.1, $\mathcal{S}_{sa}$ is sound with respect to $\mathcal{M}_{sa}$. By Theorem 3.4 every theory τ of $\mathcal{S}_{sa}$ is an intersection of elements of $\mathcal{M}_{sa}$. This proves that $\mathcal{S}_{sa}$ and $\mathcal{M}_{sa}$ define a fuzzy logic. □

We call *logic of the super-additive measures* (in brief *sa-logic*) the logic defined by the *sa*-system $\mathcal{S}_{sa}$ and the semantics $\mathcal{M}_{sa}$.

4. LOGIC OF UPPER-LOWER PROBABILITIES

The second logic we are considering is an extension of the logic of the super-additive measures (see Gerla [1997]). The fuzzy semantics of such a logic is the class $\mathcal{M}_{ul}$ of the upper-lower probabilities. Recall that, given a fuzzy subset of formulas v, we denote by $v^{\perp}$ the fuzzy subset such that

$$v^{\perp}(x) = \sim v(\neg x) = 1 - v(\neg x)$$

for every formula x. A super-additive function p from $\boldsymbol{B}$ to U is called an *upper-lower probability* if $p^{\perp}$ is sub-additive, i.e., if, for every $\alpha, \beta \in \boldsymbol{B}$,

$$p^{\perp}(\alpha \vee \beta) \leq p^{\perp}(\alpha) + p^{\perp}(\beta) \tag{4.1}$$

or, equivalently

$$p(\alpha \wedge \beta) \geq p(\alpha) + p(\beta) - 1. \tag{4.2}$$

(As a matter of fact, in the literature the name upper-lower probability is attributed to the pair $(p,p^{\perp})$, see, e.g., Dempster [1967]). In the remark in Section 2 an example of an n-person game was given whose characteristic function is a super additive measure which is not an upper-lower probability.

Examples. Let S be a finite set, $p : \mathcal{P}(S) \to [0,1]$ a probability measure and $f : S \to \boldsymbol{R}$ be a map (a random variable). For instance, we can assume that S is a set of past experiments, that p is the frequency measure and $f(x)$ a physical quantity measured in x. Then we can define a map μ by setting, for every subset X of $\boldsymbol{R}$,

$$\mu(X) = p(\{x \in S : f(x) \in X\}). \tag{4.3}$$

In other words $\mu(X)$ is the frequency (equivalently, the probability) of the experiments in which the physical quantity $f(x)$ satisfied condition X. It is well known that the so defined function is a probability in $\mathcal{P}(\boldsymbol{R})$. Now, suppose we are not able to indicate the precise value of f. Then we may substitute f with a multivalued function $g : S \to \mathcal{P}([0,1])$ whose meaning is that, given $x \in S$, we are sure the actual value $f(x)$ belongs to $g(x)$. In this case, it is very natural to set

$$\mu(X) = p(\{x \in S : g(x) \subseteq X\}).$$

The number $\mu(X)$ is the frequency of the experiments x in which we are sure that $f(x)$ satisfied X (necessity). Trivially,

$$\mu^{\perp}(X) = p(\{x \in S : g(x) \cap X \neq \emptyset\}).$$

Then, $\mu^{\perp}(X)$ is the frequency of the experiments in which it is not excluded that X is satisfied (possibility). We have that μ is an upper and lower probability. Indeed, if X and Y are two disjoint subsets of $\boldsymbol{R}$, then

$$\begin{aligned}\mu(X \cup Y)) &= p(\{x \in S : g(x) \subseteq X \cup Y\})\\ &\geq p(\{x \in S : g(x) \subseteq X\} \cup \{x \in S : g(x) \subseteq Y\})\\ &= p(\{x \in S : g(x) \subseteq X\}) + p(\{x \in S : g(x) \subseteq Y\})\\ &= \mu(X) + \mu(Y).\end{aligned}$$

One similarly demonstrates that $\mu^{\perp}(X \cup Y) \leq \mu^{\perp}(X) + \mu^{\perp}(Y)$. It is obvious that μ is not a probability.

Proposition 4.1. *The class $\mathcal{M}_{ul}$ of the upper-lower probabilities is a fuzzy semantics closed under ultraproducts. Moreover, the balanced elements of $\mathcal{M}_{ul}$ coincide with the probabilities.*

Proof. To prove that $\mathcal{M}_{ul}$ is closed under ultraproducts we proceed as in Proposition 2.1. Let τ be a balanced element of $\mathcal{M}_{ul}$. Then, from $\tau = \tau^{\perp}$, (2.1) and (4.1) it follows that τ is a probability. Conversely, it is self-evident that every probability is a balanced element of $\mathcal{M}_{ul}$. □

Observe also that $\mathcal{M}_{ul} \cup \{s^1\}$ is a closure system. To axiomatize the fuzzy semantics $\mathcal{M}_{ul}$, we define the *upper-lower deduction system*, in brief *ul*-system, as the extension $\mathcal{S}_{ul}$ of $\mathcal{S}_{sa}$ obtained by adding the *conjunction rule* defined by

$$\frac{\alpha_1 , \alpha_2}{\alpha_1 \wedge \alpha_2} \quad ; \quad \frac{\lambda_1 , \lambda_2}{\lambda_1 \odot \lambda_2}$$

where $\odot$ is Łukasiewicz conjunction. More formally, the conjunction rule $c = (c',c'')$ is defined by setting

$$c'(\alpha,\beta) = \alpha \wedge \beta \quad ; \quad c''(x,y) = x \odot y.$$

To give an example of a proof in S_{ul}, we consider a formula α and we assume that $\alpha_1, \alpha_2, \alpha_3$ exist such that $\alpha = (\alpha_1 \vee \alpha_2) \wedge \alpha_3$ and $\alpha_1 \wedge \alpha_2 = \mathbf{0}$. Then the picture,

$$\frac{\dfrac{\alpha_1\ ,\ \alpha_2}{\alpha_1 \vee \alpha_2} \quad , \quad \alpha_3}{(\alpha_1 \vee \alpha_2) \wedge \alpha_3} \qquad ; \qquad \frac{\dfrac{v(\alpha_1)\ ,\ v(\alpha_2)}{v(\alpha_1) \oplus v(\alpha_2)} \quad , \quad v(\alpha_2)}{(v(\alpha_1) \oplus v(\alpha_2)) \odot v(\alpha_3)}$$

represents a proof of α whose valuation is the number $(v(\alpha_1) \oplus v(\alpha_2)) \odot v(\alpha_3)$.

Theorem 4.2. *The ul-system S_{ul} together with the fuzzy semantics $\mathcal{M}_{ul}$ defines a logically compact fuzzy logic extending the sa-logic. In such a logic, the class $\mathcal{T}_{ul}$ of the consistent theories of S_{ul} coincides with the class of the ul-probabilities. Moreover, p is a consistent complete theory of S_{ul} iff p is a probability.*

Proof. The proof is trivial. We observe only that the closure with respect to the conjunction rule is equivalent to (4.2). □

Note that, while every probability is a maximal theory, we do not know if the converse is true.

Theorem 4.3. *No balanced semantics for the system S_{ul} exists.*

Proof. Assume that $\mathcal{M}$ is a balanced semantics for S_{ul}. Then $\mathcal{M}$ is equivalent to $\mathcal{M}_{ul}$, i.e.,

(i) every element of $\mathcal{M}$ is an intersection of elements of $\mathcal{M}_{ul}$,

(ii) every element of $\mathcal{M}_{ul}$ is an intersection of elements of $\mathcal{M}$.

Since the models in $\mathcal{M}$ are balanced, (i) entails that every element of $\mathcal{M}$ is a probability. Then, by (ii) every upper-lower probability is an intersection of a family of probabilities and hence is a lower envelope (see next section). This is a contradiction. In fact, an upper-lower probability which is not an envelope exists (see, e.g., Papamarcou and Fine [1986], pag. 712). □

Given a fuzzy set v of formulas, by Proposition 4.2 the *ul*-probability generated by v coincides with the theory $\mathcal{D}(v)$ in the system S_{ul}, and it may be obtained as usual by the formula

$$\mathcal{D}(v)(\alpha) = Sup\{Val(\pi,v) : \pi \text{ is a proof of } \alpha\}.$$

Unlike the logic of the super-additive measures, we do not know a simpler formula to get $\mathcal{D}(v)$. In any case, we can obtain $\mathcal{D}$ by a step-by-step process. To this aim, we first observe that a fuzzy subset $p : \boldsymbol{B} \to U$ such that $p(\mathbf{1}) = 1$ is a *ul*-probability if and only if

$$p((x \vee y) \wedge z) \geq p(x) + p(y) + p(z) - 1, \tag{4.4}$$

for every $x, y \in \boldsymbol{B}$ such that $x \wedge y = \mathbf{0}$ and $z \in \boldsymbol{B}$.

Proposition 4.4. *Let J be the operator defined by setting, for every initial valuation v and every formula α,*

$$J(v)(\alpha) = Sup\{v(x) + v(y) + v(z) - 1 : x \wedge y = \mathbf{0} \text{ and } ((x \vee y) \wedge z) \mapsto \alpha\}. \tag{4.5}$$

Then $(\mathcal{F}(\boldsymbol{B}), J, \mathbf{1})$ *is a step-by-step deduction system whose associated system is* $\mathcal{S}_{ul}$.

5. PROBABILITY LOGIC: SEMANTICS

The last fuzzy logic considered in this chapter is an extension of the logic of the upper-lower probabilities which we are calling *probability logic* or *logic of envelopes* (see Gerla [1994], Gerla [1994] and Gerla [1997]). The class $\mathcal{M}_p$ of the finitely additive probabilities in a given Boolean algebra $\boldsymbol{B}$ is its semantics. This choice is rather natural since a probability represents a complete information about a random phenomena, i.e., a complete theory. Moreover, in accordance with a result of A. Horn and A. Tarski, we will consider finitely additive rather than σ-additive probabilities. Indeed, Theorem 3.2 in Horn and Tarski [1948] demonstrates that

an atomless countable Boolean algebra has no σ-additive probability.

So, there is no σ-additive probability in the Lindenbaum algebra of a logic and we cannot identify the complete probabilistic theories with the σ-additive probabilities.

In probability logic the information furnished by an initial valuation v is that, for every formula α, "the probability of α is at least $v(\alpha)$". In accordance, the models of v are the probabilities greater than or equal to v. Furthermore, the probabilistic theory $Lc(v)$ generated by v is the greatest lower bound (i.e., the intersection) of the probabilities greater than or equal to v. Now, in the literature a map that may be obtained as the least upper bound of a family of probabilities is named a lower envelope (see, e.g., Papamarcou and Fine [1986]), in brief envelope. Then we can say also that $Lc(v)$ is the *envelope generated* by v. The semantics $\mathcal{M}_p$ is balanced, so:

- to give a fuzzy set of axioms v is equivalent to giving an interval constraint for the unknown probability distribution,
- the operator Lc enables us to improve the initial interval constraint, more specifically, to give the best interval constraint given v.

This means that the research of a fuzzy H-system fitting this semantics well is equivalent to the research of an algorithm able to improve initial valuations probabilistic in nature. Such a problem is strictly related to the construction of diagnostic systems and, for example, is examined in Weichselberger and Pöhlmann [1990] under the assumption that $\boldsymbol{B}$ is finite and that only the atoms of $\boldsymbol{B}$ are valued.

Remark. There is a growing literature in probability logic where logical systems for probability logic are usually achieved by extending the first order language in a two-sorted language (see, e.g., Halpern [1991] and Bacchus [1990]). A sort is for the elements of the domain we want to consider, the other sort is devoted to represent numbers in [0,1], that is to express semantic concepts. Consequently, the deductive systems must contain axioms for the structure of [0,1]. The price paid for this is either the necessity of allowing as truth values nonstandard elements in arbitrary ordered fields (Bacchus) or the impossibility to have a complete axiomatization (Halpern). Instead, fuzzy logic enables us to manage probabilistic information in such a way that no direct reference to semantic objects appears within the formalized language.

To give a suitable consistency condition we introduce a new family of connectives in $\boldsymbol{B}$. Let $h \in \boldsymbol{N}$ and $k \in \boldsymbol{N}_0$. We define the *h-k-connective* as the h-ary operation C^k on $\boldsymbol{B}$ defined by assuming that $C^0 : \boldsymbol{B}^h \rightarrow \boldsymbol{B}$ is the map constantly equal to $\mathbf{1}$ and, in the case $k \neq 0$, by setting, for any $\alpha_1,...,\alpha_h$ in $\boldsymbol{B}$,

$$C^k(\alpha_1,...,\alpha_h) = Sup\{\alpha_{i(1)} \wedge ... \wedge \alpha_{i(k)} : i(1),...,i(k) \text{ are pairwise distinct elements in } \{1,...,n\}\}.$$

If $\boldsymbol{B}$ coincides with a power set $\mathcal{P}(S)$, then $C^k(\alpha_1,...,\alpha_h)$ is the set of elements belonging at least to k elements of the sequence $\alpha_1,...,\alpha_h$. Moreover, an interesting interpretation in terms of a game is the following: Consider a betting game such that:

- a player can bet on sequences $\alpha_1,...,\alpha_h$ of events in $\boldsymbol{B}$,
- the bank pays a fixed stake of one unit for any realized event α_i.

In this case $C^k(\alpha_1,...,\alpha_h)$ is the set of elementary events for which the multiple bet $\alpha_1,...,\alpha_h$ enables us to win at least k. A direct verification shows that:

$$C^0(\alpha_1,...,\alpha_h) \geq C^1(\alpha_1,...,\alpha_h) \geq ... \geq C^h(\alpha_1,...,\alpha_h),$$
$$C^1(\alpha_1,...,\alpha_h) = \alpha_1 \vee ... \vee \alpha_h ,$$
$$C^h(\alpha_1,...,\alpha_h) = \alpha_1 \wedge ... \wedge \alpha_h ,$$
$$C^{h+1}(\alpha_1,...,\alpha_h) = C^{h+2}(\alpha_1,...,\alpha_h) = ... = \mathbf{0} ,$$
$$C^k(\alpha_1,...,\alpha_h) \geq C^k(\alpha_1,...,\alpha_{h-1}) ,$$
$$C^k(\alpha_1,...,\alpha_h) = C^k(\alpha_1,...,\alpha_h,\mathbf{0}) = C^{k+1}(\alpha_1,...,\alpha_h,\mathbf{1}).$$

The link between the h-k-connectives and the probabilities is expressed by the following proposition (see Horn and Tarski [1948]):

Proposition 5.1. *A map p such that* $p(\mathbf{1}) = 1$ *and* $p(\mathbf{0}) = 0$ *is a probability iff, for every* $\alpha_1,...,\alpha_h$ *in* $\boldsymbol{B}$,

$$p(\alpha_1) +... + p(\alpha_h) = p(C^1(\alpha_1,...,\alpha_h)) + ... + p(C^h(\alpha_1,...,\alpha_h)). \tag{5.1}$$

Proof. Assume that (5.1) is satisfied. Then, for $h = 2$,

$$p(\alpha_1) + p(\alpha_2) = p(\alpha_1 \vee \alpha_2) + p(\alpha_1 \wedge \alpha_2) \tag{5.2}$$

and p is a probability. Let p be a probability. Then, in the case $h = 1$ condition (5.1) is trivial. To proceed by induction on h, observe that

$$C^i(\alpha_1,...,\alpha_h) = [\alpha_h \cap C^{i-1}(\alpha_1,...,\alpha_{h-1})] \cup C^i(\alpha_1,...,\alpha_{h-1}).$$

Therefore, since any probability satisfies (5.2),

$$p(C^i(\alpha_1,...,\alpha_h)) = p(\alpha_h \cap C^{i-1}(\alpha_1,...,\alpha_{h-1})) + p(C^i(\alpha_1,...,\alpha_{h-1})) - p(\alpha_h \cap C^i(\alpha_1,...,\alpha_{h-1})).$$

Then,

$$\begin{aligned}
&p(C^1(\alpha_1,...,\alpha_h)) + ... + p(C^h(\alpha_1,...,\alpha_h)) \\
&\quad = p(\alpha_h \cap C^0(\alpha_1,...,\alpha_{h-1})) + ... + p(\alpha_h \cap C^{h-1}(\alpha_1,...,\alpha_{h-1})) \\
&\qquad + p(C^1(\alpha_1,...,\alpha_{h-1})) + ... + p(C^h(\alpha_1,...,\alpha_{h-1})) \\
&\qquad - (p(\alpha_h \cap C^1(\alpha_1,...,\alpha_{h-1})) + ... + p(\alpha_h \cap C^h(\alpha_1,...,\alpha_{h-1})) \\
&\quad = p(\alpha_h) + p(C^1(\alpha_1,...,\alpha_{h-1})) + ... + p(C^h(\alpha_1,...,\alpha_{h-1})) - p(\alpha_h \cap C^h(\alpha_1,...,\alpha_{h-1})) \\
&\quad = p(\alpha_h) + p(C^1(\alpha_1,...,\alpha_{h-1})) + ... + p(C^h(\alpha_1,...,\alpha_{h-1})).
\end{aligned}$$

Since, by induction hypothesis,

$$p(C^1(\alpha_1,...,\alpha_{h-1})) + ... + p(C^h(\alpha_1,...,\alpha_{h-1})) = p(\alpha_1) + ... + p(\alpha_{h-1}),$$

(5.1) is proven. □

Given the finite sequence $\alpha_1,...,\alpha_h$ of formulas, we set

$$M(\alpha_1...,\alpha_h) = Max\{k \in N : C^k(\alpha_1,...,\alpha_h) \neq \mathbf{0}\}, \quad (5.3)$$

or, equivalently,

$$M(\alpha_1...,\alpha_h) = Min\{k \in N : C^k(\alpha_1...,\alpha_h) = \mathbf{0}\} - 1. \quad (5.4)$$

In a bet interpretation $C^k(\alpha_1,...,\alpha_h) \neq \mathbf{0}$ means that it is possible to win k units. Therefore the number $M(\alpha_1...,\alpha_h)$ represents the maximum amount of money a player betting on $\alpha_1...,\alpha_h$ can win. The function M enables us to characterize the satisfiable fuzzy sets of formulas (see Gerla [1992]a).

Theorem 5.2. *The semantics $\mathcal{M}_p$ of the probabilities is closed under ultraproducts and, hence, logically compact. Moreover, an initial valuation v is satisfiable iff, for $\alpha_1,...,\alpha_h$ in $\mathbf{B}$,*

$$v(\alpha_1) + ... + v(\alpha_h) \leq M(\alpha_1...,\alpha_h). \quad (5.5)$$

Proof. In order to apply Proposition 8.2 in Chapter 2, observe that we can express condition $p(\mathbf{1}) = 1$ by setting $\mathcal{R} = \{1\}$, p_0 equal to the identity map and $D = \{\mathbf{1}\}$. Moreover, we can express additivity by setting additively

$\mathcal{R} = \{(\lambda_1,\lambda_2,\lambda_3) : \lambda_1 = \lambda_2 + \lambda_3\}$,

$p_0(x,y) = x \vee y$, $p_1(x,y) = x$, $p_2(x,y) = y$,

$D = \{(x,y) \in \mathcal{F} \times \mathcal{F} : x \wedge y = \mathbf{0}\}$.

To prove the second part of the proposition, suppose v is satisfiable and let p be a probability containing v. Then, if we set $m = M(\alpha_1...,\alpha_h)$, from

$$C^{m+1}(\alpha_1...,\alpha_h) = ... = C^h(\alpha_1...,\alpha_h) = \mathbf{0},$$

we have

$$v(\alpha_1) + ... + v(\alpha_h) \leq p(\alpha_1) + ... + p(\alpha_h)$$
$$= p(C^1(\alpha_1...,\alpha_h)) + ... + p(C^m(\alpha_1...,\alpha_h)) \leq m.$$

Conversely, assume (5.5). Then, since $\mathcal{M}_p$ is logically compact, we have to prove that, for every $\beta_1,...,\beta_t$, a probability $p : \mathbf{B} \to [0,1]$ exists such that

$$p(\beta_1) \geq v(\beta_1), \ldots, p(\beta_t) \geq v(\beta_t).$$

Let $\mathbf{B}'$ be the finite Boolean algebra generated by $\beta_1...,\beta_t$ and $e_1,...,e_n$ the atoms of $\mathbf{B}'$. Then $e_1,...,e_n$ are pairwise disjoint, $e_1 \vee ... \vee e_n = \mathbf{1}$ and $p(e_1) + ... + p(e_n) = 1$. Set $a(i,j) = 1$ if $e_i \leq \beta_j$ and otherwise $a(i,j) = \mathbf{0}$. Then the above inequalities become

$$\begin{cases} a(1,1)\cdot p(e_1) + ... + a(n,1)\cdot p(e_n) \geq v(\beta_1), \\ \quad \cdots \\ a(1,t)\cdot p(e_1) + ... + a(n,t)\cdot p(e_n) \geq v(\beta_t). \end{cases}$$

By setting $b(i,j) = a(i,j) - v(\beta_j)$, since $v(\beta_j) = p(e_1)\cdot v(\beta_j) + ... + p(e_n)\cdot v(\beta_j)$, we have

$$\begin{cases} b(1,1)\cdot p(e_1) + ... + b(n,1)\cdot p(e_n) \geq 0, \\ \quad \cdots \\ b(1,t)\cdot p(e_1) + ... + b(n,t)\cdot p(e_n) \geq 0. \end{cases}$$

This induces a search for a non-zero solution of the following system:

$$\begin{cases} b(1,1)\cdot x_1 + \ldots + b(n,1)\cdot x_n \geq 0, \\ \ldots \\ b(1,t)\cdot x_1 + \ldots + b(n,t)\cdot x_n \geq 0. \end{cases} \tag{5.6}$$

with constraints $x_1 \geq 0, \ldots, x_n \geq 0$ and $x_1 + \ldots + x_n = 1$. In turn, this is equivalent to finding a numerical vector $c = (c_1,\ldots,c_n)$ such that

$$\begin{cases} c \times b^1 \geq 0, \\ \ldots \\ c \times b^t \geq 0, \\ c \times u^1 \geq 0, \\ \ldots \\ c \times u^n \geq 0, \\ c \times b < 0, \end{cases}$$

where we denote

- by b the numerical vector $(-1,\ldots,-1)$
- by u^i the vector $(\delta_{1i},\ldots,\delta_{ni})$, where, as usual, $\delta_{ij} = 1$ if $i = j$ and, otherwise, $\delta_{ij} = 0$,
- by b^j the vector $(b(1,j),\ldots,b(n,j))$.

Indeed, if this is the case, by setting $m_i = c_i/(c_1+\ldots+c_n)$ we get a positive solution of (5.6) such that $m_1+\ldots+m_n = 1$. The existence of the vector c is established by the fundamental theorem of linear inequalities (see, e. g., Theorem 7.1 in Schrijver [1986]). Such a theorem claims that:

If a numerical vector b is not a nonnegative linear combination of linearly independent vectors from $a_1,\ldots,a_m$, then a vector c exists such that

$$c \times b < 0, c \times a_1 \geq 0, \ldots, c \times a_m \geq 0.$$

Then we only have to prove that $b = (-1,\ldots,-1)$ is not a nonnegative linear combination of linearly independent vectors from $b^1,\ldots,b^t$, $u^1,\ldots,u^n$. Indeed, otherwise the open set $\{(y_1,\ldots,y_t) : y_i \geq 0, y_1\cdot b^1+\ldots+y_t\cdot b^t < 0\}$ should be nonempty. Then such a set contains a vector with rational components and hence a vector $(\lambda_1,\ldots,\lambda_t)$ with integer components. Consider the sequence $\alpha_1,\ldots,\alpha_h$ obtained by considering λ_1 formulas equal to β_1, λ_2 formulas equal to β_2 and so on. Set $m = M(\alpha_1,\ldots,\alpha_h)$ and let e_i be an atom belonging to m formulas of the sequence $\alpha_1,\ldots,\alpha_h$. Then

$$a(i,1)\cdot\lambda_1 + \ldots + a(i,t)\cdot\lambda_t = m$$

and therefore, taking (5.5) into account, the number

$$\begin{aligned} & a(i,1)\cdot\lambda_1 + \ldots + a(i,t)\cdot\lambda_t - (v(\alpha_1) + \ldots + v(\alpha_h)) \\ & = a(i,1)\cdot\lambda_1 + \ldots + a(i,t)\cdot\lambda_t - (v(\beta_1)\cdot\lambda_1 + \ldots + v(\beta_t)\cdot\lambda_t) \\ & = (a(1,1) - v(\beta_1))\cdot\lambda_1 + \ldots + (a(i,t) - v(\beta_t))\cdot\lambda_t \end{aligned}$$

is nonnegative. Since such a number coincides with the i-component of $\lambda_1 b^1 + \ldots + \lambda_t b^t$ this is absurd.

The values $m_1, \ldots, m_n$ define a probability in $\boldsymbol{B}'$. To get a probability in $\boldsymbol{B}$ we have to "concentrate" these masses in suitable "points" of $e_1,\ldots,e_n$. More specifically, for every $i = 1,...,n$, let $\mathcal{F}_i$ be a prime filter of $\boldsymbol{B}$ containing e_i and define p by setting $p(\mathbf{0}) = 0$ and $p(\alpha) = \Sigma\ \{m_i : \alpha \in \mathcal{F}_i\}$ for every $\alpha \neq \mathbf{0}$. Since $m_1 + ... + m_n = 1$, we get $p(\mathbf{1}) = 1$. Let $\alpha, \beta \in \boldsymbol{B}$ with $\alpha \wedge \beta = \mathbf{0}$. Then

$$p(\alpha \vee \beta) = \Sigma\{m_i : \alpha \vee \beta \in \mathcal{F}_i\} = \Sigma\ \{m_i : \alpha \in \mathcal{F}_i\} + \Sigma\{m_i : \beta \in \mathcal{F}_i\} = p(\alpha) + p(\beta)$$

and this demonstrates that p is a probability. Moreover,

$$p(\beta_j) = \Sigma\ \{m_i : \beta_j \in \mathcal{F}_i\} = \Sigma\ \{m_i : e_i \leq \beta_j\} = a(1,j) \cdot m_1 + ... + a(n,j) \cdot m_n \geq v(\beta_j)$$

and this concludes the proof. □

Observe that in (5.5) we may confine ourselves to formulas in *Supp*(v). Indeed, suppose the sequence $\alpha_1...,\alpha_h$ satisfies (5.5) and that $v(\alpha_{h+1}) = 0$. Then, because

$$v(\alpha_1) + ... + v(\alpha_h) + v(\alpha_{h+1}) = v(\alpha_1) + ... + v(\alpha_h) \leq M(\alpha_1,...,\alpha_h) \leq M(\alpha_1,...,\alpha_{h+1}),$$

(5.5) is also satisfied by $\alpha_1,...,\alpha_{h+1}$. This means that in the case *Supp*(v) finite, it is very simple to see whether such a condition is satisfied or not. For instance if *Supp*(v) = $\{\alpha_1,...,\alpha_m\}$ and $\alpha_1,...,\alpha_m$ are pairwise disjoint, then v is satisfiable iff $v(\alpha_1) + ... + v(\alpha_m) \leq 1$.

From Theorem 5.2 it is possible to derive some results about the extendibility of a set function in a probability in a simple way. For instance, we may restate the following extension theorem (Horn and Tarski [1948]):

Theorem 5.3. *Let* $\boldsymbol{B}$ *be a Boolean algebra,* $\boldsymbol{B}'$ *a subalgebra of* $\boldsymbol{B}$ *and* $p' : \boldsymbol{B}' \to [0,1]$ *a finitely additive probability defined on* $\boldsymbol{B}'$. *Then* p' *may be extended to a finitely additive probability defined in the whole algebra* $\boldsymbol{B}$.

Proof. Let v be the extension of p' to $\boldsymbol{B}$ obtained by setting $v(x) = 0$ for every $x \in \boldsymbol{B} - \boldsymbol{B}'$. Then, since v satisfies (5.5) in its support, there exists a probability p defined on $\boldsymbol{B}$ extending v. Thus, for every $x \in \boldsymbol{B}'$,

$$p(x) \geq v(x) = p'(x) = 1 - p'(\neg x) = 1 - v(\neg x) \geq 1 - p(\neg x) = p(x)$$

and therefore $p(x) = p'(x)$. Thus, p is an extension of the set function p'. □

The following consistency condition given in Weichselberger and Pöhlmann [1990] is a direct consequence of Theorem 5.2:

Corollary 5.4. *Assume that* $\boldsymbol{B}$ *is a finite Boolean algebra whose atoms are* $e_1,...,e_n$ *and that* $v : \boldsymbol{B} \to [0,1]$ *is an initial valuation such that Supp*(v) *is contained in* $\{e_1,...,e_n,\neg e_1,...,\neg e_n\}$. *Then*

- *m is a model of v iff* for $i = 1,...,n$,

$$m(e_i) \in [v(e_i), v^{\perp}(e_i)]\ ,$$

- *v is consistent iff*

$$v(e_1) + ... + v(e_n) \leq 1 \leq v^{\perp}(e_1) + ... + v^{\perp}(e_n).$$

6. PROBABILITY LOGIC: HILBERT SYSTEM

To find suitable inference rules for probability logic, we must first give some characterizations of the envelopes (see also Biacino [1992], Gerla [1992]a).

Proposition 6.1. *Let p be a fuzzy set of formulas such that* $p(\mathbf{1}) = 1$. *Then the following are equivalent:*

(a) *p is an envelope.*

(b) $p(\alpha_1) + ... + p(\alpha_h) \le (M(\alpha_1,...,\alpha_h) - k + 1) \cdot p(C^k(\alpha_1,...,\alpha_h)) + k - 1$ *for every* $h \in \boldsymbol{N}$, $k \in \boldsymbol{N}_0$ *and* $\alpha_1,...,\alpha_h$ *sequence of formulas.*

(c) *Every extension* p_α *of p via a formula* α *is satisfiable.*

(d) *For every* $\beta \in \boldsymbol{B}$, *there is a probability q containing p such that* $q(\beta) = p(\beta)$.

Proof. (a) ⇒ (b). Every probability p satisfies (b). Indeed, if we set $m = M(\alpha_1,...,\alpha_h)$, then

$$\begin{aligned} p(\alpha_1) + ... + p(\alpha_h) & \\ &= [p(C^1(\alpha_1,...,\alpha_h)) + ... + p(C^{k-1}(\alpha_1,...,\alpha_h))] + \\ &\quad + [p(C^k(\alpha_1,...,\alpha_h)) + ... + p(C^m(\alpha_1,...,\alpha_h))] \\ &\le k - 1 + p(C^k(\alpha_1,...,\alpha_h)))(m-k+1). \end{aligned}$$

Let $(p_i)_{i\in I}$ be a family of probabilities and $p = \text{Inf}_{i\in I}\, p_i$. Then, for any $i \in I$,

$$\begin{aligned} p(\alpha_1) + ... + p(\alpha_h) &\le p_i(\alpha_1) + ... + p_i(\alpha_h) \\ &\le (M(\alpha_1,...,\alpha_h) - k + 1) \cdot p_i(C^k(\alpha_1,...,\alpha_h)) + k - 1, \end{aligned}$$

and therefore

$$\begin{aligned} p(\alpha_1) + ... + p(\alpha_h) &\le Inf_{i\in I}\,((M(\alpha_1,...,\alpha_h) - k + 1) \cdot p_i(C^k(\alpha_1,...,\alpha_h)) + k - 1) \\ &= (M(\alpha_1,...,\alpha_h) - k + 1) \cdot (Inf_{i\in I}\, p_i(C^k(\alpha_1,...,\alpha_h))) + k - 1 \\ &= (M(\alpha_1,...,\alpha_h) - k + 1) \cdot p(C^k(\alpha_1,...,\alpha_h)) + k - 1. \end{aligned}$$

(b) ⇒ (c). First, we demonstrate some consequence of (b). By assuming that α_1 and α_2 are disjoint and that $k = 1$ we obtain $p(\alpha_1) + p(\alpha_2) \le p(\alpha_1 \vee \alpha_2)$. In particular, for $\alpha_1 = \mathbf{1}$, and $\alpha_2 = \mathbf{0}$ we obtain $p(\mathbf{0}) = 0$. Moreover, from $\alpha \le \beta$ it follows that $p(\alpha) \le p(\alpha) + p(\beta \wedge \neg\alpha) \le p(\alpha \vee (\beta \wedge \neg\alpha)) = p(\beta)$, and this proves that p is increasing. Finally, by setting $k = M(\alpha_1,...,\alpha_h)$ in (b), we have the satisfiability condition

$$p(\alpha_1) + ... + p(\alpha_h) \le M(\alpha_1,...,\alpha_h). \tag{6.1}$$

Now, since $p(\alpha) + p(\neg\alpha) \le 1$, the extension p_α is defined by setting $p_\alpha(x) = p(x)$ if $x \ne \alpha$ and $p_\alpha(x) = 1-p(\neg\alpha)$ if $x = \alpha$. Then, in view of Theorem 5.2, to prove that p_α is satisfiable we have to prove that, for every $\alpha_1,...,\alpha_h,\alpha_{h+1},...,\alpha_{h+t}$ such that $\alpha_1,...,\alpha_h$ are different from α and $\alpha_{h+1} = ... = \alpha_{h+t} = \alpha$,

$$p(\alpha_1) + ... + p(\alpha_h) + t\cdot(1-p(\neg\alpha)) \le M(\alpha_1,...,\alpha_{h+t}). \tag{6.2}$$

If $\alpha = \mathbf{0}$, since $p(\mathbf{1}) = 1$, (6.2) coincides with (6.1). If $\alpha \ne \mathbf{0}$ we proceed by induction on t. In the case $t = 0$ (6.2) again coincides with (6.1). Suppose (6.2) for $t = n$. Then

$$p(\alpha_1) + ... + p(\alpha_h) + (n+1)\cdot(1-p(\neg\alpha)) \le M(\alpha_1,...,\alpha_{h+n}) + 1-p(\neg\alpha).$$

Suppose $M(\alpha_1,...,\alpha_{h+n+1}) = M(\alpha_1,...,\alpha_{h+n}) + 1$. Then, since $1-p(\neg\alpha) \le 1$, we have

$$p(\alpha_1) + ... + p(\alpha_h) + (n+1)\cdot(1-p(\neg\alpha)) \le M(\alpha_1,...,\alpha_{h+n+1})$$

and (6.2) holds for $t = n + 1$. Suppose $M(\alpha_1,...,\alpha_{h+n+1}) = M(\alpha_1,...,\alpha_{h+n})$. Then,

$$M(\alpha_1,...,\alpha_h) \ne 0,\ n < m,\ M(\alpha_1,...,\alpha_{h+n+1}) = m \ \text{ and } \ \neg\alpha \ge C^{m-n}(\alpha_1,...,\alpha_h).$$

Therefore

$$1-p(\neg\alpha) \le 1-p(C^{m-n}(\alpha_1,...,\alpha_h))$$

and

$$(n+1)(1-p(\neg\alpha)) \le (n+1)(1-p(C^{m-n}(\alpha_1,...,\alpha_h))).$$

Consequently, by setting $k = m - n$ in (b),

$$p(\alpha_1) + ... + p(\alpha_h) + (n+1)\cdot(1-p(\neg\alpha))$$
$$\leq (n+1)\cdot p(C^{m-n}(\alpha_1,...,\alpha_h))+m-n-1+ (n+1)\cdot(1- p(C^{m-n}(\alpha_1,...,\alpha_h))$$
$$= m = M(\alpha_1,...,\alpha_{h+n+1}).$$

Thus, (6.2) holds for $t = n+1$.

(c) $\Rightarrow$ (d). Given a formula β, let q be a probability such that $q \geq p_{\neg\beta}$. Then

$$q(\neg\beta) \geq p_{\neg\beta}(\neg\beta) \geq 1-p(\beta)$$

and

$$q(\beta) = 1-q(\neg\beta) \leq p(\beta).$$

Because $p(\beta) \leq q(\beta)$, we conclude that $q(\beta) = p(\beta)$.

(d) $\Rightarrow$ (a). This is obvious. □

We can write (b) of Proposition 6.1 as follows:

$$p(C^k(\alpha_1,...,\alpha_h)) \geq \frac{p(\alpha_1) + ... + p(\alpha_h) - k + 1}{M(\alpha_1,...,\alpha_h) - k + 1}$$

and this suggests the following inference rules. Recall that the normalization function is defined by setting $n(x) = 0$ if $x < 0$, $n(x) = 1$ if $x > 1$, and $n(x) = x$ in the case $x \in [0,1]$. Then, given h, m, k integer numbers such that $h \geq m \geq k$, we define the *h-m-k*-rule of inference as the rule $r = (r',r'')$ defined on

$$D(h,m) = \{(\alpha_1,...,\alpha_h) : M(\alpha_1,...,\alpha_h) = m\}$$

by

$$r'(\alpha_1,...,\alpha_h) = C^k(\alpha_1,...,\alpha_h) \quad ; \quad r''(x_1,...,x_h) = n\left(\frac{x_1 + ... + x_h - k + 1}{m\text{-}k+1}\right).$$

The meaning of the *h-m-k-rules* is clear only in some particular case. For example the *h*-1-1-rule is the (extended) disjunction rule defined by

$$\frac{\alpha_1,...,\alpha_h}{\alpha_1 \vee ... \vee \alpha_h} \quad ; \quad \frac{\lambda_1,...,\lambda_h}{\lambda_1 \oplus ... \oplus \lambda_h}$$

where $\alpha_1,...,\alpha_h$ are pairwise disjoint. By setting $h = m = k$ we get the (extended) conjunction rule

$$\frac{\alpha_1,...,\alpha_h}{\alpha_1 \wedge ... \wedge \alpha_h} \quad ; \quad \frac{\lambda_1,...,\lambda_h}{\lambda_1 \odot ... \odot \lambda_h}$$

where $\alpha_1 \wedge ... \wedge \alpha_h \neq \mathbf{0}$. Note that the *h-m-k*-rules are not independent. For instance, the closure with respect to the 2-1-1-rule entails the closure with respect to any of the above *h*-1-1-rules. It is an open question to find a smaller and more intuitive system of rules.

Moreover, Theorem 5.2 suggests considering, for every h and m, $h \geq m$, the *h-m-collapsing rule* $c = (c',c'')$ defined on $D(h,m)$ by setting $c'(\alpha_1,...,\alpha_h) = \mathbf{0}$ for every $\alpha_1,...,\alpha_h$ and

$$c''(\lambda_1,...,\lambda_h) = \begin{cases} 1 & \text{if } \lambda_1+...+\lambda > m, \\ 0 & \text{otherwise.} \end{cases}$$

The h-m-collapsing rule says that if we have proven the formulas $\alpha_1,\ldots,\alpha_h$ with degrees $\lambda_1,\ldots,\lambda_h$ that violate the consistency condition (5.5), then we can prove the contradiction **0**. As an example, the 2-1-collapsing rule entails that if α_1 and α_2 are disjoint formulas proven with degree $\lambda_1 = 1/2$ and $\lambda_2 = 2/3$, then we can prove that **0** is true.

Definition 6.2. We call a *probabilistic deduction system* on the Boolean algebra ***B***, the fuzzy *H*-system $\mathcal{S}_p$ whose inference rules are the h-m-k-rules and the h-m-collapsing rules and whose set of axioms is $\{\mathbf{1}\}$. We call any theory of this deductive system a *probabilistic theory*.

In probabilistic deduction systems a proof proceeds step-by-step as follows.
Suppose we have previously proven $\alpha_1,\ldots,\alpha_h$ with degree $\lambda_1,\ldots,\lambda_h$, then:

IF $\lambda_1+\ldots+\lambda_h > M(\alpha_1,\ldots,\alpha_h)$, **THEN** the available information is inconsistent,
OTHERWISE, conclude that $C^k(\alpha_1,\ldots,\alpha_h)$ holds at least to degree

$$\frac{\lambda_1 + \ldots + \lambda_h - k + 1}{M(\alpha_1,\ldots,\alpha_h) - k + 1}.$$

7. COMPLETENESS THEOREM

Now, we are ready to demonstrate a completeness theorem, i.e., that the semantics $\mathcal{M}_p$ of the finitely additive probabilities and the system $\mathcal{S}_p$ define a fuzzy logic we call *probability logic* or *the logic of the lower envelopes*.

Proposition 7.1. *Given a probabilistic theory τ, the following are equivalent*
(a) τ is consistent.
(b) $\tau(\mathbf{0}) = 0$.
(c) *τ is satisfiable in the semantics $\mathcal{M}_p$ of the probabilities.*

Proof. (a) $\Rightarrow$ (*b*). Let α be a formula such that $\tau(\alpha) \neq 1$. Then, since by the 2-1-1-rule, $\tau(\alpha) = \tau(C^1(\alpha,\mathbf{0})) \geq \tau(\alpha) \oplus \tau(\mathbf{0})$, we can conclude that $\tau(\mathbf{0}) = 0$.
(b) $\Rightarrow$ (c). Let $\alpha_1,\ldots,\alpha_h$ be formulas and set $m = M(\alpha_1,\ldots,\alpha_h)$. Then, the closure with respect to the h-m-collapsing rule entails that

$$\tau(c'(\alpha_1,\ldots,\alpha_h)) = \tau(\mathbf{0}) = 0 \geq c''(\tau(\alpha_1),\ldots,\tau(\alpha_h)).$$

Thus, $\tau(\alpha_1) + \ldots + \tau(\alpha_h) \leq M(\alpha_1,\ldots,\alpha_h)$ and, by Proposition 5.2, this proves that τ is satisfiable. Implication (c) $\Rightarrow$ (a) is obvious. □

Theorem 7.2. *The class $\mathcal{T}_p$ of the consistent probabilistic theories coincides with the class of envelopes. Consequently, the probabilistic deductive system $\mathcal{S}_p$ together with the semantics of the probabilities $\mathcal{M}_p$ define a fuzzy logic extending the ul-logic.*

Proof. If τ is an envelope, then, since τ is satisfiable, τ is closed under the collapsing rules. Let h, m, k be integers such that $h \geq m \geq k$ and let $\alpha_1,...,\alpha_h$ be formulas such that $M(\alpha_1,...,\alpha_h) = m$. Then, by (b) of Proposition 6.1,

$$\tau(C^k(\alpha_1,...,\alpha_h)) \geq (\tau(\alpha_1) + ... + \tau(\alpha_h) - k + 1)/(m - k + 1) \quad (7.1)$$

and this proves the closure with respect to the h-m-k-rule. Conversely, let τ be a consistent theory. Then, since by Proposition 7.1 τ is satisfiable,

$$\tau(\alpha_1) + ... + \tau(\alpha_h) \leq M(\alpha_1,...,\alpha_h)$$

and

$$\frac{(\tau(\alpha_1) + ... + \tau(\alpha_h) - k + 1)}{(m - k + 1)} \leq 1.$$

Thus, the closure with respect to the h-m-k-rule is expressed by (7.1) and therefore, by (b) of Proposition 6.1, τ is an envelope. □

In accordance with such a theorem, $\mathcal{D}(v)$ is the lower envelope generated by v. The following theorem shows that complete theories coincide with probabilities, i.e., with the elements of $\mathcal{M}_p$.

Theorem 7.3. *Given a fuzzy set τ of formulas, the following are equivalent:*

(a) *τ is a complete theory.*
(b) *τ is probability (i.e., a model).*
(c) *τ is a maximal theory.*
(d) *τ is a maximal consistent fuzzy set of formulas.*
(e) *τ is a consistent fuzzy set of formulas such that $\tau(\alpha) + \tau(\neg\alpha) = 1$ for every formula α.*

Proof. (a) $\Rightarrow$ *(b)*. If τ is a theory such that $\tau(\alpha) + \tau(\neg\alpha) = 1$, then τ is consistent, trivially. Hence, by Theorem 7.2, τ is an envelope. Let q be a probability containing τ. Then, for every formula α, $\tau(\alpha) \leq q(\alpha)$ and therefore $q(\alpha) = 1 - q(\neg\alpha) \leq 1 - \tau(\neg\alpha) = \tau(\alpha)$. So $q = \tau$ and this shows that τ is a probability.
(b) $\Rightarrow$ (c) Straightforward.
(c) $\Rightarrow$ (d) Straightforward.
(d) $\Rightarrow$ (e). The consistency of τ entails that a probability q exists containing τ. Since τ is maximal in the class of consistent fuzzy subsets of formulas, $\tau = q$. Consequently, τ is a probability and (e) is proven.
(e) $\Rightarrow$ (a). We have that $\tau(\alpha) \leq \mathcal{D}(\tau)(\alpha)$ and, since $\mathcal{D}(\tau)$ is an envelope,

$$\mathcal{D}(\tau)(\alpha) \leq 1 - \mathcal{D}(\tau)(\neg\alpha) \leq 1 - \tau(\neg\alpha) = \tau(\alpha).$$

So, $\mathcal{D}(\tau) = \tau$ and this proves that τ is a theory. □

Theorem 7.3 emphasizes the fact that the concept of probability corresponds to complete information about a random world while Theorem 7.2 says that incomplete information is represented by an envelope. Therefore, being a characteristic of any logic to allow deductions under incomplete information, the concept of probability (= complete information) is not sufficient, in our opinion, to give a foundation of probabilistic logic. So, the notion of envelope seems to be necessary. This also suggests that finite additivity is a requirement which is too strong for the subjective approach to probabilistic valuations. Indeed, the incompleteness of the information is a feature of the individual estimates that usually cannot be avoided (see also the bet interpretation in Section 9).

Proposition 7.4. *The logic of envelopes is not with optimal proofs.*

Proof. Let $S = [0,1]^2$ be the unitary square and m the usual (Jordan) measure in S. Let $\boldsymbol{B}$ be the Boolean algebra of the measurable subsets of S and let v be the function defined by setting $v(x) = l_1 \times l_2$ if x is a rational rectangle whose sides measure l_1 and l_2 and, otherwise, $v(x) = 0$. Then, since $m \geq v$ and any measure $m' \geq v$ contains m, $m = \mathcal{D}(v)$. Let α be a circle contained in S. Then, $\mathcal{D}(v)(\alpha) = m(\alpha)$ is not a rational number. Since the valuation $Val(\pi,v)$ of any proof π of α is a rational number, we cannot get $\mathcal{D}(v)$ via a unique optimal proof. □

8. REFUTATIONS IN PROBABILITY LOGIC

In Theorem 5.2 we give a simple condition to verify whether an initial valuation is satisfiable or not. In Theorem 6.3 in Chapter 5, we propose a refutation approach to fuzzy deduction by showing that

$$Lc(v) = Inc(v,\neg\alpha) = Sup\{\lambda \in U : v \cup \{\neg\alpha\}^{1-\lambda} \text{ is not satisfiable}\}.$$

In this section we will combine these two results to give a simple formula to calculate $\mathcal{D}(v)$. To this aim, set

$$M(\alpha_1,...,\alpha_h/\alpha) = M(\alpha_1 \wedge \alpha,...,\alpha_h \wedge \alpha),$$

that is

$$M(\alpha_1,...,\alpha_h/\alpha) = Max\{k \in N : C^k(\alpha_1,...,\alpha_h) \text{ is compatible with } \alpha\}. \tag{8.1}$$

We can interpret the number $M(\alpha_1,...,\alpha_h/\alpha)$ as the maximum deal of money a player betting on $\alpha_1,...,\alpha_h$ can win "given α". Moreover, consider the function $d(\alpha_1,...,\alpha_h/\alpha)$ defined by

$$d(\alpha_1,...,\alpha_h/\alpha) = M(\alpha_1,...,\alpha_h) - M(\alpha_1,...,\alpha_h/\alpha) \tag{8.2}$$

whose meaning is obvious. Moreover, given a fuzzy set of formulas v and the formulas $\alpha_1,...,\alpha_h$, we define the quantity $Cons(\alpha,\alpha_1,...,\alpha_h/v)$ by setting $Cons(\alpha,\alpha_1,...,\alpha_h/v) = 1$ if $d(\alpha_1,...,\alpha_h/\alpha) = 0$ and, otherwise,

$$Cons(\alpha,\alpha_1,...,\alpha_h/v) = \frac{M(\alpha_1,...,\alpha_h) - (v(\alpha_1)+...+v(\alpha_h))}{d(\alpha_1,...,\alpha_h/\alpha)} \wedge 1\,. \tag{8.3}$$

We define $Inc(\alpha,\alpha_1,...,\alpha_h / v)$ by

$$Inc(\alpha,\alpha_1,...,\alpha_h / v) = \sim Cons(\alpha,\alpha_1,...,\alpha_h / v).$$

Trivially, $Inc(\alpha,\alpha_1,...,\alpha_h/ v) = 0$ if $d(\alpha_1,...,\alpha_h/\alpha) = 0$ and, otherwise,

$$Inc(\alpha,\alpha_1,...,\alpha_h/ v) = \frac{v(\alpha_1)+...+v(\alpha_h) - M(\alpha_1,...,\alpha_h/\alpha)}{d(\alpha_1,...,\alpha_h/\alpha)} \vee 0\,. \tag{8.4}$$

We could interpret $Cons(\alpha, \alpha_1,...,\alpha_h / v)$ as the degree of consistency of α with the finite subset $v \cap \{\alpha_1,...,\alpha_h\}$ of v and $Inc(\alpha,\alpha_1,...,\alpha_h / v)$ as its degree of inconsistency. Such interpretation is not completely justified. In any case,

Definitions (8.3) and (8.4) enable us to write the formulas in the subsequent Theorem 8.2 in a simple way.

Lemma 8.1. *For any* $\alpha_1,...,\alpha_h,\ \alpha$ *in* $\boldsymbol{B}$ *and* $t \in N$

$$M(\alpha_1,...,\alpha_h,\alpha^t) = Max\{\ M(\alpha_1,...,\alpha_h),\ t + M(\alpha_1,...,\alpha_h/\alpha)\}, \quad (8.5)$$

and

$$d(\alpha_1,...,\alpha_h/\alpha) = Max\{t \in N : M(\alpha_1,...,\alpha_h,\alpha^t) = M(\alpha_1,...,\alpha_h)\} \quad (8.6)$$

where $\alpha_1,...,\alpha_h,\alpha^t$ *is the sequence* $\alpha_1,...,\alpha_h,\alpha_{h+1},...\alpha_{h+t}$ *such that* $\alpha_{h+1} = ... = \alpha_{h+t} = \alpha$.

Proof. To demonstrate (8.5) observe that

$$M(\alpha_1,...,\alpha_h,\alpha^t) \geq M(\alpha_1,...,\alpha_h) \text{ and } M(\alpha_1,...,\alpha_h,\alpha^t) \geq t + M(\alpha_1,...,\alpha_h/\alpha)$$

and therefore that

$$M(\alpha_1,...,\alpha_h,\alpha^t) \geq Max\{M(\alpha_1,...,\alpha_h),\ t + M(\alpha_1,...,\alpha_h/\alpha)\}.$$

Moreover, in calculating $Max\{i \in N : C^i(\alpha,...\alpha_h,\alpha^t) \neq 0\}$ we have to divide the set of indexes such that $C^i(\alpha,...\alpha_h,\alpha^t) \neq 0$ in two subsets A and B. A is the set of indexes i obtained by the presence of α. In such a case $i \leq t + M(\alpha_1,...,\alpha_h/\alpha)$. B is the set B of remaining indexes. So, if $i \in B$, then $i \leq M(\alpha_1,...,\alpha_h)$. In accordance,

$$M(\alpha_1,...,\alpha_h,\alpha^t) = Sup(B) \vee Sup(A) \leq Max\{M(\alpha_1,...,\alpha_h),\ t + M(\alpha_1,...,\alpha_h/\alpha)\}.$$

In order to demonstrate 8.6, assume that $t \leq d(\alpha_1,...,\alpha_h/\alpha)$. Then

$$M(\alpha_1,...,\alpha_h) \leq M(\alpha_1,...,\alpha_h,\alpha^t)$$
$$\leq Max\{\ M(\alpha_1,...,\alpha_h),\ d(\alpha_1,...,\alpha_h/\alpha) + M(\alpha_1,...,\alpha_h/\alpha)\} = M(\alpha_1,...,\alpha_h),$$

and hence $M(\alpha_1,...,\alpha_h) = M(\alpha_1,...,\alpha_h,\alpha^t)$. Assume that $t > d(\alpha_1,...,\alpha_h/\alpha)$. Then

$$M(\alpha_1,...,\alpha_h,\alpha^t) \geq t + M(\alpha_1,...,\alpha_h/\alpha)$$
$$> d(\alpha_1,...,\alpha_h/\alpha) + M(\alpha_1,...,\alpha_h/\alpha) = M(\alpha_1,...,\alpha_h).$$

This proves (8.6). □

Now we are able to prove the main results in this section.

Theorem 8.2. *Let* α *be a formula and* v *a satisfiable initial valuation. Then*

$$\mathcal{D}(v)(\alpha) = Sup\{Inc(\neg\alpha,\alpha_1,...,\alpha_h\,/\,v) : \alpha_1,...,\alpha_h \in Supp(v)\}. \quad (8.7)$$

$$\mathcal{D}(v)^{\perp}(\alpha) = Inf\{Cons(\alpha,\alpha_1,...,\alpha_h\,/\,v) : \alpha_1,...,\alpha_h \in Supp(v)\}. \quad (8.8)$$

Proof. At first recall that, by Theorem 6.3 in Chapter 5,

$$\mathcal{D}(v)(\alpha) = Inc(v, \neg\alpha)$$

where, for any formula x,

$$Inc(v, x) = Sup\{\lambda \in U : v \cup \{x\}^{1-\lambda} \text{ is not satisfiable}\}.$$

Assume that $v \cup \{x\}^{1-\lambda}$ is not satisfiable. Then, by Theorem 5.2, $t \neq 0$ and $\alpha_1,...,\alpha_h \in Supp(v)$ exist such that

$$v(\alpha_1)+...+v(\alpha_h) + t\cdot(1-\lambda) > M(\alpha_1,...,\alpha_h,\ x^t). \quad (8.9)$$

Also, it is not restrictive to assume that

$$t \geq d(\alpha_1,...,\alpha_h/x). \quad (8.10)$$

Indeed, in the case $t < d(\alpha_1,...,\alpha_h/x)$ set $l = d(\alpha_1,...,\alpha_h/x)$. Then, since by (8.6),

$$M(\alpha_1,...,\alpha_h,x^t) = M(\alpha_1,...,\alpha_h,x^l) = M(\alpha_1,...,\alpha_h),$$

we get

$$v(\alpha_1) + ... + v(\alpha_h) + l\cdot(1-\lambda) > v(\alpha_1) + ... + v(\alpha_h) + t\cdot(1-\lambda)$$

$$> M(\alpha_1,...,\alpha_h, x^t) = M(\alpha_1,...,\alpha_h, x^l)$$

which means that l satisfies both (8.9) and (8.10). Now, by (8.5), (8.9),

$$v(\alpha_1)+...+v(\alpha_h) + t\cdot(1-\lambda) > t + M(\alpha_1,...,\alpha_h/x)$$

and therefore

$$v(\alpha_1)+...+v(\alpha_h) - M(\alpha_1,...,\alpha_h/x) > \lambda\cdot t \geq \lambda\cdot d(\alpha_1,...,\alpha_h/x). \tag{8.11}$$

From this inequality it follows that $v(\alpha_1)+...+v(\alpha_h) > M(\alpha_1,...,\alpha_h/x)$ and therefore, by the consistency of v, that $M(\alpha_1,...,\alpha_h) > M(\alpha_1,...,\alpha_h/x)$. Then $d(\alpha_1,...,\alpha_h/x) \neq 0$ and (8.11) becomes

$$\lambda < (v(\alpha_1)+...+v(\alpha_h) - M(\alpha_1,...,\alpha_h/x)) / d(\alpha_1,...,\alpha_h/x) = Inc(x,\alpha_1,...,\alpha_h/ v).$$

This proves that

$$Inc(v,x) \leq Sup\{Inc(x,\alpha_1,...,\alpha_h / v) : \alpha_1,...,\alpha_h \in Supp(v)\}.$$

Conversely, let

$$\lambda < Sup\{Inc(x, \alpha_1,...,\alpha_h / v) : \alpha_1,...,\alpha_h \in Supp(v)\}.$$

Then $\alpha_1,...,\alpha_h$ exist such that $\lambda < Inc(x,\alpha_1,...,\alpha_h / v)$. Consequently, $d(\alpha_1,...,\alpha_h/x) \neq 0$ and

$$v(\alpha_1) + ... + v(\alpha_h) - M(\alpha_1,...,\alpha_h/x) > \lambda\cdot d(\alpha_1,...,\alpha_h/x).$$

By adding $M(\alpha_1,...,\alpha_h)$ to both the sides of such an inequality, we obtain

$$v(\alpha_1) + ... + v(\alpha_h) + (1-\lambda)\cdot d(\alpha_1,...,\alpha_h/x) > M(\alpha_1,...,\alpha_h).$$

Set $t = d(\alpha_1,...,\alpha_h/x)$. Then, because $M(\alpha_1,...,\alpha_h) = M(\alpha_1,...,\alpha_h,x^t)$, we may rewrite the inequality as

$$v(\alpha_1) + ... + v(\alpha_h) + t\cdot(1-\lambda) > M(\alpha_1,...,\alpha_h,x^t),$$

and this entails that $v \cup \{x\}^{1-\lambda}$ is not satisfiable. Then $\lambda \leq Inc(v,x)$ and, hence,

$$Inc(v,x) \geq Sup\{Inc(x, \alpha_1,...,\alpha_h / v) : \alpha_1,...,\alpha_h \in Supp(v)\}.$$

Thus $Inc(v,x) = Sup\{Inc(x, \alpha_1,...,\alpha_h / v) : \alpha_1,...,\alpha_h \in Supp(v)\}$ concluding the proof of (8.7).

Equation (8.8) is a direct consequence of (8.7). □

Note that each interval

$$[Inc(\neg\alpha, \alpha_1,...,\alpha_h / v), Cons(\alpha, \alpha_1,...,\alpha_h / v)]$$

gives an approximation of the unknown probability $p(\alpha)$ and this suggests that (8.7) and (8.8) are useful tools to the questions examined in Weichselberger and Pohlmann [1990].

9. TWO REMARKS: PROBABILITY OF FORMULAS, SUBJECTIVE PROBABILITY

Probability logic in propositional language. In all the logics defined in this chapter we could start from the set $\mathbb{F}$ of formulas of the classical propositional calculus instead of a Boolean algebra. If we want to do so, we must consider only *transparent* models, i.e., valuations $m : \mathbb{F} \to U$ of the formulas such that

$$\alpha \equiv \beta \Rightarrow m(\alpha) = m(\beta),$$

where $\alpha \equiv \beta$ means that α is logically equivalent to β. For instance, consider the case of the *ul*-logic. Then we can define the fuzzy semantics $\mathcal{M}_{ul}(\mathbb{F})$ as the class of the fuzzy subsets $m : \mathbb{F} \to U$ of $\mathbb{F}$ such that

(i) $m(\alpha) = 1$ for every tautology α,

(ii) $\alpha \to \beta$ tautology $\Rightarrow$ $m(\alpha) \le m(\beta)$,
(iii) α inconsistent with β $\Rightarrow m(\alpha \vee \beta) \ge m(\alpha) + m(\beta)$,
(iv) $m(\alpha \wedge \beta) \ge m(\alpha) + m(\beta) - 1$.
Such a semantics is strictly related to the class of upper-lower probabilities.

Theorem 9.1. *The fuzzy subset of formulas* $m : \mathcal{F} \to U$ *satisfies* (i) - (iv) *iff there exist a Boolean algebra* $\boldsymbol{B}$*, a truth-functional Boolean valuation* $B_v : \mathcal{F} \to \boldsymbol{B}$ *and a upper-lower probability* $p : \boldsymbol{B} \to U$ *such that* $m = p \circ B_v$.

Proof. Suppose that $m : \mathcal{F} \to U$ satisfies (i) - (iv) and let $\boldsymbol{B}$ be the Lindenbaum algebra associated with $\mathcal{F}$. Then, the function $B_v : \mathcal{F} \to \boldsymbol{B}$ defined by setting $B_v(\alpha) = [\alpha]$ is a Boolean valuation. Define $p : \boldsymbol{B} \to U$ by setting, for every $[\alpha] \in \boldsymbol{B}$, $p([\alpha]) = m(\alpha)$. By (ii) such a definition is correct. Then p is an *ul*-probability such that $m(\alpha) = p(B_v(\alpha))$. The converse part is trivial. □

A fuzzy deduction apparatus $\mathcal{S}_{ul}(\mathcal{F})$ for $\mathcal{M}_{ul}(\mathcal{F})$ is achieved by assuming the set of tautologies as logical axioms and by considering a *fuzzy disjunction rule*

$$\frac{\alpha,\ \beta}{\alpha \vee \beta} \quad ; \quad \frac{\lambda,\ \mu}{\lambda \oplus \mu}$$

where α is inconsistent with β, a *fuzzy Modus Ponens rule*

$$\frac{\alpha,\ \alpha \to \beta}{\beta} \quad ; \quad \frac{\lambda,\ \mu}{\lambda \odot \mu}$$

and a *collapsing rule*

$$\frac{\alpha,\ \beta}{\alpha \wedge \beta} \quad ; \quad \frac{\lambda,\ \mu}{c(\lambda,\mu)}$$

where α is inconsistent with β and c is the function defined by setting $c(\lambda,\mu) = 1$ if $\lambda + \mu > 1$ and, otherwise, $c(\lambda,\mu) = 0$.

Theorem 9.2. *The class of consistent theories of* $\mathcal{S}_{ul}(\mathcal{F})$ *coincides with the semantics* $\mathcal{M}_{ul}(\mathcal{F})$*. Then,* $(\mathcal{M}_{ul}(\mathcal{F}), \mathcal{S}_{ul}(\mathcal{F}))$ *is a fuzzy logic (in Hilbert's style).*

Proof. If τ is a consistent theory of $\mathcal{S}_{ul}(\mathcal{F})$, then (i) is obvious. In order to prove (ii), assume that the formula $\alpha \to \beta$ is a tautology. Then,

$$\tau(\beta) \ge \tau(\alpha) \odot \tau(\alpha \to \beta) = \tau(\alpha) \odot 1 = \tau(\alpha).$$

In order to prove (iii) it is sufficient to observe that if α is inconsistent with β, then $\tau(\alpha) + \tau(\beta) \le 1$. Indeed, otherwise by the collapsing rule

$$\tau(\alpha \wedge \beta) \ge c(\tau(\alpha), \tau(\beta)) = 1$$

and, as for every formula γ the formula $\alpha \wedge \beta \to \gamma$ is a tautology, $\tau(\gamma) = 1$. Then

$$\tau(\alpha \vee \beta) \ge \tau(\alpha) \oplus \tau(\beta) = \tau(\alpha) + \tau(\beta).$$

To prove (iv) observe that $\beta \to (\alpha \to (\alpha \wedge \beta))$ is a tautology and therefore

$\tau(\alpha \to (\alpha \wedge \beta)) \geq \tau(\beta)$. Then

$$\tau(\alpha \wedge \beta) \geq \tau(\alpha) \odot \tau(\alpha \to (\alpha \wedge \beta)) \geq \tau(\alpha) \odot \tau(\beta).$$

Conversely, assume that (i) - (iv) are satisfied by τ. Then it is self-evident that τ contains the set of logical axioms and that τ is closed under the disjunction and collapsing rules. By observing that both the formulas $\alpha \wedge \beta \to \alpha$ and $(\beta \wedge (\beta \to \alpha)) \to \alpha \wedge \beta$ are tautologies, by (iv) we have

$$\tau(\alpha) \geq \tau(\alpha \wedge \beta) \geq \tau(\beta \wedge (\beta \to \alpha)) \geq \tau(\beta) + \tau(\beta \to \alpha) - 1.$$

As $\tau(\alpha)$ is positive, we also have that $\tau(\alpha) \geq \tau(\beta \to \alpha) \odot \tau(\beta)$. □

Subjective approach to probability. As we observed earlier, it is possible to give a bet interpretation to the notions and results of probability logic. In the game under consideration a player can bet on the occurrences of the events in $\boldsymbol{B}$, and the bank pays a fixed stake of one unit for every realized event. $M(\alpha_1,...,\alpha_h)$ represents the maximum amount of money a player betting on $\alpha_1,...,\alpha_h$ can win.

Moreover, we interpret any initial valuation $v : \boldsymbol{B} \to [0,1]$ as a *"betting function"* assuming that, for every event α, $v(\alpha)$ is the money a player have to pay to bet on α. In accordance, $v(\alpha_1) + ... + v(\alpha_h)$ is the amount of money a player betting on a multiple bet $\alpha_1,...,\alpha_h$, has to pay. We say that a bet $\alpha_1,...,\alpha_h$ is *"player-reasonable"* if it satisfies (5.5), i.e., the player has the possibility of receiving at least the amount of money that he has disbursed. We say that a betting function is player-reasonable if all the multiple bets are player-reasonable. As an example, due to zero rule, the betting function of the roulette game is not player-reasonable. Thus, Proposition 5.3 says that v is satisfiable iff v is player-reasonable.

Of course, the term "player-reasonable" refers to the player's point of view and not to the bank's point of view. A notion of "reasonable" from the point of view of the bank is obtained by considering the function

$$m(\alpha_1,...,\alpha_h) = Max\{t \in N_0 : C^t(\alpha_1,...,\alpha_h) = \mathbf{1}\}.$$

The number $m(\alpha_1,...,\alpha_h)$ represents the minimum amount of money a player betting on $\alpha_1,...,\alpha_h$ is sure to win. We say that v is *"bank-reasonable"* if

$$m(\alpha_1,...,\alpha_h) \leq p(\alpha_1) + ... + p(\alpha_h)$$

for every multiple bet $\alpha_1,...,\alpha_h$. Then v is bank-reasonable provided that there is no multiple bet for which a player surely wins.

Proposition 9.3. *A map p is a probability iff*

$$m(\alpha_1,...,\alpha_h) \leq p(\alpha_1) + ... + p(\alpha_h) \leq M(\alpha_1,...,\alpha_h). \quad (9.1)$$

Proof. Let p be a probability. Then (9.1) is a consequence of (5.1). Conversely, assume that p satisfies (9.1). Then, by Theorem 5.2, p is satisfiable and a probability q exists extending p. Furthermore, since for every α,

$$p(\alpha) + p(\neg\alpha) \geq m(\alpha,\neg\alpha) = 1,$$

we have

$$q(\alpha) = 1 - q(\neg\alpha) \leq 1 - p(\neg\alpha) \leq p(\alpha).$$

Thus, $p(\alpha) = q(\alpha)$ and p is a probability. □

A condition like (9.1) was proposed in.Vickers [1976] under the name *sum condition*. Thus, in accordance with the subjective point of view, we have the following theorem (see the Dutch Book Theorem in de Finetti [1964]):

Theorem 9.4. *The probabilities coincide with the betting functions which are both player-reasonable and bank-reasonable. In other words, a probability is a betting function for which no surely losing or surely winning bet is possible.*

10. BELIEF LOGIC AND BOOLEAN LOGIC

Another very interesting class of probabilistic valuations is the class of belief measures (see Shafer [1976]). These measures are defined in a Boolean algebra but we prefer to define them in the set $\mathcal{F}$ of formulas of classical propositional calculus in the following way. A *generalized belief measure* is a map $p' : \mathcal{F} \to [0,1]$ such that:

i) α logically equivalent to $\beta \Rightarrow p'(\alpha) = p'(\beta)$;

ii) α tautology $\Rightarrow p'(\alpha) = 1$;

iii) for any integer n and every n-tuple of formulas $\alpha_1, \dots, \alpha_n$

$p'(\alpha_1 \vee \dots \vee \alpha_n) \geq \sum\{(-1)^{|I|+1} p'(\alpha_{i(1)} \wedge \dots \wedge \alpha_{i(k)}), I = \{i(1), \dots, i(k)\} \subseteq \{1, \dots, n\}\}$.

If we assume also that

iv) α contradiction $\Rightarrow p'(\alpha) = 0$,

then we say that p' is a *belief measure*. The probabilities and the necessities are examples of belief measures. In turn, any belief measure is a lower envelop. The following proposition shows that it is not possible to consider the class of belief measures as the class of theories in a fuzzy logic.

Proposition 10.1. *No fuzzy logic exists whose class of theories coincides either with the class of belief measures or with the class of generalized belief measures.*

Proof. It is sufficient to observe that the set $\mathcal{T}_b$ of belief measures is not closed under intersections. Indeed, otherwise, since $\mathcal{T}_b$ contains the class of probabilities, $\mathcal{T}_b \supseteq \mathcal{T}_p$. This contradicts the fact that lower envelopes that are not belief measures exist. A similar argument holds for the generalized belief measures.

In spite of this result, we can relate the belief measures with a fuzzy deduction system Boolean in nature. Let $\boldsymbol{B}$ be the Boolean algebra $\mathcal{P}(S)$ where S is a set whose elements we interpret as *past cases* we have stored in a statistical setting. Also, we can interpret these elements as *worlds* in a Kripke semantics. We consider the deduction apparatus obtained as a canonical extension by $\cap$ of the classical propositional calculus (see Section 8 of Chapter 6). Then the deduction operator $\mathcal{D}^{\cap} : \boldsymbol{B}^{\mathcal{F}} \to \boldsymbol{B}^{\mathcal{F}}$ is defined by setting, for any $\boldsymbol{B}$-set of formulas v, $\mathcal{D}^{\cap}(v)(\alpha) = S$ if α is a tautology and, otherwise,

$$\mathcal{D}^{\cap}(v)(\alpha) = \bigcup\{v(\alpha_1) \cap \dots \cap v(\alpha_n) : \alpha_1, \dots, \alpha_n \vdash \alpha\}. \tag{10.1}$$

Notice that while we interpret $v(\alpha)$ as the set of the past cases in which the available information claims that α is true, $\mathcal{D}^{\cup}(v)(\alpha)$ represents the set of past cases in which the information is sufficient to prove α. We call *degree of inconsistency* of v the Boolean value

$$Inc(v) = \bigcap\{\mathcal{D}^{\cap}(v)(\alpha) : \alpha \in \mathcal{F}\}. \qquad (10.2)$$

Obviously, given any contradiction α, $Inc(v) = \mathcal{D}^{\cap}(v)(\alpha)$, i.e. $Inc(v)$ is the set of past cases in which the information is inconsistent. We say that v is *totally consistent* if $Inc(v) = \varnothing$ and that v is *complete* if, for every formula α,

$$\mathcal{D}^{\cap}(v)(\alpha)\cup\mathcal{D}^{\cap}(v)(\neg\alpha) = S. \qquad (10.3)$$

Proposition 10.2. *Let v be a **B**-valuation and α and β formulas. Then*

i) $\alpha \vdash \beta \Rightarrow \mathcal{D}^{\cap}(v)(\alpha) \subseteq \mathcal{D}^{\cap}(v)(\beta)$.

ii) $\alpha \equiv \beta \Rightarrow \mathcal{D}^{\cap}(v)(\alpha) = \mathcal{D}^{\cap}(v)(\beta)$.

iii) $\mathcal{D}^{\cap}(v)(\alpha\wedge\beta) = \mathcal{D}^{\cap}(v)(\alpha)\cap\mathcal{D}^{\cap}(v)(\beta)$.

iv) $\mathcal{D}^{\cap}(v)(\alpha\vee\beta) \supseteq \mathcal{D}^{\cap}(v)(\alpha)\cup\mathcal{D}^{\cap}(v)(\beta)$.

Proof. i), ii), and iv) are evident. To prove iii), observe that,

$$\begin{aligned}
&\mathcal{D}^{\cap}(v)(\alpha)\cap\mathcal{D}^{\cap}(v)(\beta)\\
&= (\bigcup\{v(\alpha_1)\cap\ldots\cap v(\alpha_n) : \alpha_1,\ldots,\alpha_n \vdash \alpha\})\cap(\bigcup\{v(\beta_1)\cap\ldots\cap v(\beta_m) : \beta_1,\ldots,\beta_m \vdash \beta\}\\
&= \bigcup\{v(\alpha_1)\cap\ldots\cap v(\alpha_n)\cap v(\beta_1)\cap\ldots\cap v(\beta_m) : \alpha_1,\ldots,\alpha_n \vdash \alpha \text{ and } \beta_1,\ldots,\beta_m \vdash \beta\}\\
&= \bigcup(\{v(x_1)\cap\ldots\cap v(x_n) : x_1,\ldots,x_h \vdash \alpha \text{ and } x_1,\ldots,x_h \vdash \beta\})\\
&= \bigcup\{v(x_1)\cap\ldots\cap v(x_n) : x_1,\ldots,x_h \vdash \alpha\wedge\beta\} = \mathcal{D}^{\cap}(v)(\alpha\wedge\beta).
\end{aligned}$$

Such a Boolean deduction apparatus is related with the Boolean logic defined in Section 3 of Chapter 5. Recall that a Boolean model is an element m of $\boldsymbol{B}^{\mathcal{F}}$ such that

$$m(\alpha \wedge \beta) = m(\alpha) \cap m(\beta) \;\; ; \;\; m(\alpha \vee \beta) = m(\alpha) \cup m(\beta) \;\; ; \;\; m(\neg\alpha) = S - m(\alpha).$$

Given a Boolean model m and $\alpha \in \mathcal{F}$, $m(\alpha)$ represents the set of past cases in which α is true. We denote by $\mathcal{M}$ the class of Boolean models and by Lc the related logical consequence operator. Such an operator is different from $\mathcal{D}^{\cap}$. Indeed, if v is satisfiable and α is a contradiction, then $Lc(v)(\alpha) = \varnothing$ while $\mathcal{D}^{\cap}(v) = Inc(v) \neq \varnothing$, in general.

Theorem 10.3. *Let $m \in \mathcal{M}$ and X a subset of S. Then $m\cup X$ is a complete theory of $\mathcal{D}^{\cap}$ such that $Inc(m\cup X) = X$. Moreover,*

$$\tau \textit{ is complete and totally consistent theory} \Leftrightarrow \tau \in \mathcal{M}.$$

Proof. To prove that $m\cup X$ is a fixed point for $\mathcal{D}^{\cap}$, observe that, if α is a tautology, then $\mathcal{D}^{\cap}(m\cup X)(\alpha) = S = (m\cup X)(\alpha)$. If α is not a tautology, then

$$\begin{aligned}
\mathcal{D}^{\cap}(m\cup X)(\alpha) &= \bigcup\{(m(\alpha_1)\cap \ldots \cap m(\alpha_n))\cup X : \alpha_1\wedge \ldots \wedge\alpha_n \vdash \alpha\}\\
&\subseteq \bigcup\{m(\alpha_1\wedge\ldots \wedge\alpha_n) \cup X : m(\alpha_1\wedge \ldots \wedge\alpha_n)\subseteq m(\alpha)\} = m(\alpha)\cup X.
\end{aligned}$$

It is trivial that $m\cup X$ is complete and that $Inc(m\cup X) = X$.

Assume that τ is a totally consistent and complete theory. Then, by iii) of Proposition 10.2, $\tau(\alpha\wedge\beta) = \tau(\alpha)\cap\tau(\beta)$. Moreover, $\tau(\alpha)\cap\tau(\neg\alpha) = \tau(\alpha\wedge\neg\alpha) = \varnothing$, and therefore, by (10.3), $\tau(\neg\alpha) = S-\tau(\alpha)$. Finally,

$$\tau(\alpha\vee\beta) = \tau(\neg(\neg\alpha\wedge\neg\beta)) = S-\tau(\neg\alpha\wedge\neg\beta) = S-\tau(\neg\alpha)\cap\tau(\neg\beta)$$

$= (S-\tau(\neg\alpha))\cup(S-\tau(\neg\beta)) = \tau(\alpha)\cup\tau(\beta)$.

It is evident that if $m \in \mathcal{M}$, then m is a complete and totally consistent theory. □

Assume that $p : \boldsymbol{B} \to [0,1]$ is a probability. Then we associate any initial valuation v with the map $p_v : \mathcal{F} \to [0,1]$ defined by setting:

$$p_v(\alpha) = p(\mathcal{D}^{\frown}(v)(\alpha)). \tag{10.4}$$

In other words, $p_v(\alpha)$ is the frequency of the past cases in which the available information enables us to prove α..

Theorem 10.4. (Biacino and Gerla [2000]b). *The map* p_v *defined by* (10.4) *is a generalized belief measure. If* v *is totally consistent, then* p_v *is a belief measure. If* v *is also complete, then* p_v *is a probability.*

Proof. Since $\mathcal{D}^{\frown}(v)(\alpha_1\vee\ldots\vee\alpha_n) \supseteq \mathcal{D}^{\frown}(v)(\alpha_1)\cup\ldots\cup\mathcal{D}^{\frown}(v)(\alpha_n)$ and any probability is a belief measure,

$$p_v(\alpha_1\vee\ldots\vee\alpha_n) = p(\mathcal{D}^{\frown}(v)(\alpha_1\vee\ldots\vee\alpha_n)) \geq p(\mathcal{D}^{\frown}(v)(\alpha_1)\cup\ldots\cup\mathcal{D}^{\frown}(v)(\alpha_n))$$
$$\geq \Sigma\{(-1)^{|I|+1}p(\mathcal{D}^{\frown}(v)(\alpha_{i(1)})\cap\ldots\cap\mathcal{D}^{\frown}(v)(\alpha_{i(k)})), I = \{i(1),\ldots, i(k)\} \subseteq \{1,\ldots, n\}\}$$
$$= \Sigma\{(-1)^{|I|+1}p(\mathcal{D}^{\frown}(v)(\alpha_{i(1)}\wedge\ldots\wedge\alpha_{i(k)})), I = \{i(1),\ldots, i(k)\}\subseteq \{1, \ldots, n\}\}$$
$$= \Sigma\{(-1)^{|I|+1}p_v(\alpha_{i(1)}\wedge\ldots\wedge\alpha_{i(k)}), I = \{i(1),\ldots, i(k)\}\subseteq \{1, \ldots, n\}\}.$$

Assume that v is totally consistent and that α is a contradiction. Then $p_v(\alpha) = p(\mathcal{D}^{\frown}(v)(\alpha)) = p(\emptyset) = 0$. Assume that v is complete. Then, since $\mathcal{D}^{\frown}(v)$ is a Boolean model,

$$p_v(\alpha\vee\beta) = p(\mathcal{D}^{\frown}(v)(\alpha\vee\beta)) = p(\mathcal{D}^{\frown}(v)(\alpha)\cup\mathcal{D}^{\frown}(v)(\beta))$$
$$= p(\mathcal{D}^{\frown}(v)(\alpha)) + p(\mathcal{D}^{\frown}(v)(\beta)) - p(\mathcal{D}^{\frown}(v)(\alpha)\cap\mathcal{D}^{\frown}(v)(\beta))$$
$$= p_v(\alpha)+p_v(\beta) - p_v(\alpha\wedge\beta).$$ □

It is an open question whether we can represent any (generalized) belief measure in $\mathcal{F}$ by an equation as (10.4).

11. QUALITATIVE PROBABILITY LOGICS

We can interpret the crisp part of the probabilistic logics considered in this chapter as a *"qualitative"* approach to probability logic.

Definition 11.1. We say that a set $\mathcal{F}$ of formulas is *consistent* if

$$\alpha \in \mathcal{F},\ \beta \in \mathcal{F} \Rightarrow \alpha\wedge\beta \neq \mathbf{0}$$

We say that $\mathcal{F}$ is a *class of probable formulas* if $\mathcal{F}$ is consistent and upper. If X also satisfies

$$\neg\alpha \notin \mathcal{F} \Rightarrow \alpha \in \mathcal{F},$$

then we say that $\mathcal{F}$ is a *complete class of probable formulas*.

Such a nomenclature (only slightly different from the nomenclature proposed in Walley and Fine [1979]) is justified by the fact that, given any probability p, the cuts $O(p,\lambda)$ and $C(p,\lambda)$ are complete classes of probable formulas for every $\lambda \geq 1/2$ and for every $\lambda > 1/2$, respectively. Trivially, a proper filter is a class of probable formulas and an ultrafilter is a complete class of probable formulas. Given a finite set S with an odd number of elements, the class

$$\mathcal{F} = \{X \in \mathcal{P}(S) : |X| > |-X|\}$$

is an example of complete class of probable formulas which is not a filter. The class of probable formulas together with the whole set $\boldsymbol{B}$ defines a closure system contained in the closure system of the upper subsets of $\boldsymbol{B}$. Recall that, given any subset X of $\boldsymbol{B}$, $Upp(X)$ denotes the upper set generated by X (see Section 5 of Chapter 6). The class of probable formulas defines the closure operator *Prob* defined by

$$Prob(X) = \begin{cases} \boldsymbol{B} & \text{if } X \text{ is inconsistent,} \\ Upp(X) & \text{if } X \text{ is consistent.} \end{cases}$$

The following propositions show that the logic of the super additive measures is not an extension of the classical logic but of the *qualitative logic of the probable events* as defined in Walley and Fine [1979].

Proposition 11.2. *A set X of formulas is consistent in the sa-logic iff it is consistent in accordance with Definition* 11.1. *Let $\mathcal{D}$ be the deduction operator of the sa-logic. Then, since*

$$\mathcal{D}(X) = Prob(X),$$

$\mathcal{D}$ is an extension of Prob. Nevertheless, $\mathcal{D}$ is not the canonical extension of Prob.

Proof. By Theorem 2.3 the characteristic function c_X of X is consistent iff there is no pair α, β of disjoint formulas in X (indeed, otherwise $c_X(\alpha) + c_X(\beta) = 1 + 1 > 1$). Moreover, $\mathcal{D}(X)(\mathbf{1}) = \mathbf{1}$ and, if $\alpha \neq \mathbf{1}$,

$$\mathcal{D}(X)(\alpha) = Sup\{c_X(\alpha') : \alpha' \leq \alpha\}.$$

Hence, $\mathcal{D}(X)$ is the characteristic function of $Prob(X)$. In order to prove that $\mathcal{D}$ is not the canonical extension of $Prob$, observe that

$$\begin{aligned} Prob^*(v)(\alpha) &= Sup\{\lambda \in U : \alpha \in Prob(\mathrm{C}(v,\lambda))\} \\ &= (Sup\{\lambda \subset U : \exists\beta \leq \alpha,\ v(\beta) \geq \lambda\}) \vee (Sup\{\lambda \in U : \mathrm{C}(v,\lambda) \text{ is inconsistent}\}) \\ &= (Sup\{v(\beta) : \beta \leq \alpha\}) \vee (Sup\{v(\beta) \wedge v(\neg\beta) : \beta \in \boldsymbol{B}\}). \end{aligned}$$

It is evident that, in general, this value is different from $\mathcal{D}(v)(\alpha)$. □

Theorem 11.3. *The consistent crisp theories of the sa-system coincide with the classes of probable formulas. The complete crisp theories coincide with the complete classes of probable formulas.*

Proof. Let τ be a consistent crisp theory and $\mathcal{F} = \{\alpha \in \boldsymbol{B} : \tau(\alpha) = 1\}$. Then, since τ is increasing, $\mathcal{F}$ is upper. If α is an element of $\mathcal{F}$, then $1 = \tau(\alpha \vee \neg\alpha) \geq \tau(\alpha) + \tau(\neg\alpha) = 1 + \tau(\neg\alpha)$, and this entails that $\tau(\neg\alpha) = 0$, that is, $\neg\alpha \notin \mathcal{F}$. This proves that τ is a characteristic function of a set of probable formulas. Assume that τ is complete and that $\neg\alpha \notin \mathcal{F}$, i.e., $\tau(\neg\alpha) = 0$. Then, because $\tau(\alpha) + \tau(\neg\alpha) = 1$, we have that $\tau(\alpha) = 1$ and $\alpha \in \mathcal{F}$. This proves that $\mathcal{F}$ is a complete class of probable formulas.

Conversely, let τ be the characteristic function of a set $\mathcal{F}$ of probable formulas. Then, since $\mathbf{1} \in \mathcal{F}$, $\tau(\mathbf{1}) = 1$. Let α and β be two disjoint formulas, and suppose $\tau(\alpha \vee \beta) < \tau(\alpha) + \tau(\beta)$. Then in the case $\alpha \in \mathcal{F}$, since $\neg\alpha \geq \beta$ and $\neg\alpha \notin \mathcal{F}$, we cannot have $\beta \in \mathcal{F}$. This entails that $\tau(\alpha \vee \beta) = 1$, $\tau(\alpha) = 1$ and $\tau(\beta) = 0$ and, hence, $\tau(\alpha \vee \beta) = \tau(\alpha) + \tau(\beta)$. The same equation holds if $\beta \in \mathcal{F}$. In the case $\alpha \notin \mathcal{F}$ and $\beta \notin \mathcal{F}$, it is $\tau(\alpha \vee \beta) \geq \tau(\alpha) + \tau(\beta) = 0$. Thus τ is a super additive measure. Trivially, if $\mathcal{F}$ is complete, then τ is complete. □

Proposition 11.4. *sa*-logic is not stratified. In fact, $\mathcal{D}^*$ is the canonical extension of *Prob* and therefore is different from $\mathcal{D}$.

Proof. If $\lambda \wedge X$ is consistent, taking (3.3) into account we have

$$\mathcal{K}(\mathcal{D},\lambda)(X) = \mathrm{C}(\mathcal{D}(\lambda \wedge X),\lambda) = \{x \in \boldsymbol{B} : \lambda \cdot m(\{z \in X : z \leq x\}) \geq \lambda\} \cup \{\mathbf{1}\}$$
$$= \{x \in \boldsymbol{B} : m(\{z \in X : z \leq x\}) \geq 1\} \cup \{\mathbf{1}\}$$
$$= \{x \in \boldsymbol{B} : \text{there exists } z \in X \text{ such that } z \leq x\} \cup \{\mathbf{1}\} = Prob(X).$$

If $\lambda \wedge X$ is not consistent, that is $m(X) < \lambda$, then $\mathcal{K}(\mathcal{D},\lambda)(X) = B$. In brief,

$$\mathcal{K}(\mathcal{D},\lambda)(X) = \begin{cases} Prob(X) & \text{if } m(X) \leq 1/\lambda, \\ \boldsymbol{B} & \text{otherwise.} \end{cases}$$

Obviously, the family $(\mathcal{K}(\mathcal{D},\lambda))_{\lambda \in U}$ is order-preserving. Assume that s is consistent and that $\lambda \neq 0$. Then, if $x_1,\ldots,x_n$ are pairwise disjoint elements of $\mathrm{C}(s,\lambda)$, then $1 \geq s(x_1) + \ldots + s(x_n) \geq n \cdot \lambda$ and $n \leq 1/\lambda$. Consequently $m(\mathrm{C}(s,\lambda)) \leq 1/\lambda$ and

$$\mathcal{K}(\mathcal{D},\lambda)(\mathrm{C}(s,\lambda)) = Prob(\mathrm{C}(s,\lambda)).$$

Thus, $\mathcal{D}^*(s) = Prob^*(s)$ and this demonstrates that $\mathcal{D}^* \neq \mathcal{D}$. □

To examine the crisp part of *ul*-logic, we denote by *Fil* the closure operator which associates any subset X of $\boldsymbol{B}$ with the filter $Fil(X)$ generated by X.

Theorem 11.5. *The crisp theories of ul-logic coincide with the filters and the complete crisp theories with the ultrafilters in* $\boldsymbol{B}$. *Namely, the deduction operator* $\mathcal{D}$ *of ul-logic is an extension but not a canonical extension of Fil.*

Proof. The first part of the theorem follows from the results in Section 4. In order to prove that $\mathcal{D}$ is not a canonical extension, note that the canonical extension of *Fil* gives, for every fuzzy subset v, the necessity generated by v. □

The next theorem, whose proof is matter of routine, shows that the crisp part of probability logic coincides with the crisp part of *ul*-logic.

Theorem 11.6. *The crisp theories of probability logic coincide with the filters, and the complete crisp theories with the ultrafilters in* $\boldsymbol{B}$.

CHAPTER 10

FUZZY CONTROL AND APPROXIMATE REASONING

1. INFORMATION BY WORDS VERSUS INFORMATION BY NUMBERS

Traditional control techniques are possible only in the case of complete understanding of the physical nature of the problem and only after a suitable mathematical treatment leading to a usable model. This enables us to obtain a numerical function f whose intended meaning is that f(x) is the correct control given x. Unfortunately, this is not the case for a majority of real systems. Difficulties can arise, for instance, from poor understanding of the underlying phenomena (and therefore from a lack of theory), or from the complexity of the resulting mathematical model. In such cases fuzzy control, as devised in Zadeh [1965], [1975]a, [1975]b and in Mamdani [1981], is a very useful tool. To explain the idea, we can distinguish two phases in the building of a fuzzy controller.

In the first phase we individuate some "experts" on the control under consideration. We can imagine an expert as a clever "old hand" with no theoretical knowledge or mathematical competence. Moreover, we ask the experts about their controller-experience and we obtain a set of answers as

"in the case that x is big, usually I set y slow".

We represent the resulting information by a collection of IF-THEN rules like

IF x is *Big* THEN y is *Slow*.

We can also attach a label as *"sure"*, *"questionable"*, to each rule. This system of rules, in a sense, represents a rough description giving a qualitative, incomplete and approximate information of the objective function f. The information is qualitative since no number occurs. Instead of numbers, labels as *"Big"*, *"Slow"* are used to represent a granulation of the quantities under consideration. The information is incomplete because in using words such as *"Big"*, *"Slow"* an expert cannot communicate exactly what he means, and therefore he cannot transfer his whole experience. Finally, the information is approximate since each expert possesses only an approximate idea of the objective function f. Nevertheless, some kind of useful information is conveyed by the system of IF-THEN rules. In fact:

- only a particular set of rules is proposed (eventually weighed) among the whole set of possible rules,
- the interpretation of the labels is *"flexible"* but not completely arbitrary.

Fuzzy control translates such an information in a fuzzy function f by a suitable procedure. Such a function depends on the initial interpretation of the labels as *"Big"* and *"Slow"* by fuzzy subsets and of the labels as *"sure"* and *"questionable"* by elements in [0,1]. A defuzzification process associates f with a classical function f' which is the output of the first phase.

The second phase is devoted to completing and to improving the information conveyed by the IF-THEN system by a *learning process*. Indeed, if the obtained function f' is a suitable approximation of f, then the process terminates. Otherwise we proceed by successive *"tunings"* of the interpretations (i.e. the fuzzy subsets and the weights) until a correct interpretation of the labels and therefore a satisfactory approximation f' is obtained.

Note that, in spite of the fact that such a problem looks to be only technical in nature and that the success of fuzzy control is mainly in engineering applications, its basic ideas are very interesting from a philosophical viewpoint and they are of central importance for fuzzy logic. In particular the key ideas of *"fuzzy granulation"*, *"computing with words"* and *"learning"* give important suggestions to understand why natural language is so efficient in storing, elaborating and transmitting information (see Zadeh [1997]). This especially if we are able to give a logical meaning to fuzzy control: but:

how is fuzzy control related to fuzzy logic in narrow sense?

This is an open question and, in my opinion, the answer is not simple. An interesting analysis and some proposals can be found in Hájek [1998].

In this chapter we will propose an answer to this question based on a simple fuzzy logic and which is related to the notion of a fuzzy Herbrand model. More precisely, in Section 2 we will expose fuzzy control as usually proposed in fuzzy literature (geometrical approach). Successively, we will relate fuzzy control with fuzzy logic as formalized in this book, in particular with a suitable fuzzification of logic programming (logical approach). Then, in Sections 3 we will define the basic notions of logic programming. In Section 4 we extend these notions by defining the fuzzy programs and the least fuzzy Herbrand model of a fuzzy program. In the successive sections, we relate fuzzy logic programming with fuzzy control. For more information on fuzzy control, see Babuška [1998] and Gottwald [1993].

2. CONTROL BY TRIANGULAR NORMS

As claimed in Zadeh [1997], the basic point of fuzzy control is that we can

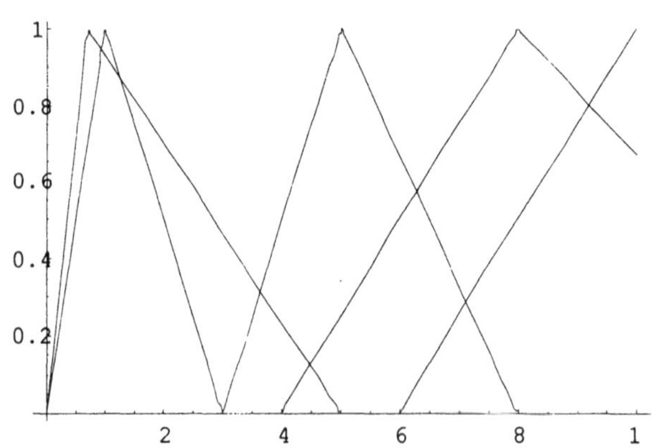

Picture 1: fuzzy granules of X

"granulate" the set X of possible inputs and the set Y of possible outputs by a finite number of fuzzy subsets we call *fuzzy quantities* or *fuzzy granules*. As an example, assume that $X = [0,10]$ is the set of possible temperatures and $Y = [0,5]$ the set of

possible speeds of a ventilator. Then a granulation of X can be furnished by the fuzzy quantities *"little"*, *"small"*, *"medium"*, *"big"*, *"verybig"* (see Picture 1). A granulation of Y can be given by the fuzzy quantities *"slow"*, *"moderate"*, *"fast"*, *"veryfast"* (see Picture 2). As in the classical case, any pair of fuzzy quantities defines a *fuzzy point*, i.e., a two-dimensional fuzzy granule, obtained as the Cartesian product of these quantities. As an example the pair $(small, fast)$ defines the fuzzy point $small \times fast : X \times Y \to U$ where the Cartesian product can be defined by any triangular norm. The set of two-dimensional granules obtained in such a way gives a *"granulation"* of $X \times Y$. As an example, assume that X and Y are intervals and that a crisp granulation is obtained by partitioning X and Y into intervals. Then the resulting granulation of $X \times Y$ is a set of rectangles.

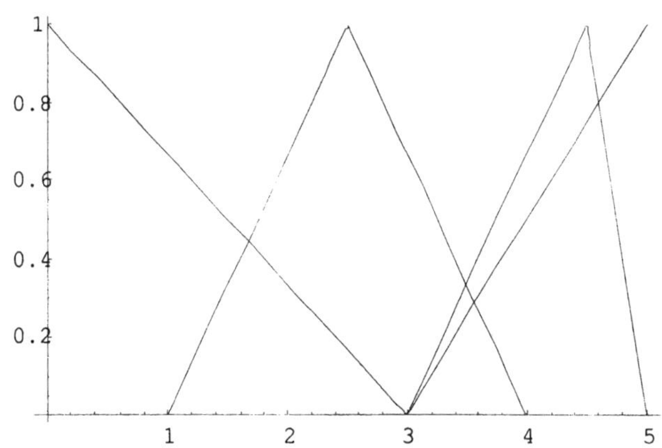

Picture 2: fuzzy granules of Y

The basic question is to approximate the objective function $f : X \to Y$ by a finite number of granules. Such an approximation is described by a system $\mathcal{S}$ of fuzzy IF-THEN rules like

$$\begin{array}{l} \text{IF } x \text{ is } \textit{Little} \text{ THEN } y \text{ is } \textit{Slow}, \\ \text{IF } x \text{ is } \textit{Small} \text{ THEN } y \text{ is } \textit{Fast}, \\ \text{IF } x \text{ is } \textit{Medium} \text{ THEN } y \text{ is } \textit{Moderate}, \\ \text{IF } x \text{ is } \textit{Big} \text{ THEN } y \text{ is } \textit{Veryfast}, \\ \text{IF } x \text{ is } \textit{Verybig} \text{ THEN } y \text{ is } \textit{Moderate}, \end{array} \tag{2.1}$$

where *"Little"*, *"Slow"*, *"Small"*, *"Fast"*, *"Medium"*, *"Moderate"*, *"Big"*, *"Veryfast"*, *"Verybig"* are labels for the fuzzy granules

$$\begin{array}{ll} little : X \to [0,1], & slow : Y \to [0,1], \\ small : X \to [0,1], & fast : Y \to [0,1], \\ medium : X \to [0,1], & moderate : Y \to [0,1], \\ big : X \to [0,1], & veryfast : Y \to [0,1], \\ verybig : X \to [0,1]. & \end{array}$$

This system is associated with the fuzzy function

$$f = little \times slow \cup small \times fast \cup medium \times moderate$$
$$\cup\, big \times veryfast \cup verybig \times moderate.$$

The following is a justification of such a procedure. Classically, a function f can be partially represented by a table like

x	y
x_1	y_1
...	...
x_n	y_n

where $x_1,...,x_n$ are elements in X and $y_1,...,y_n$ elements in Y, and therefore by the "union" of the finite set $(x_1,y_1),..., (x_n,y_n)$ of points. Similarly, a fuzzy approximation of f can be roughly represented by a table like

x	y
Little	*Slow*
Small	*Fast*
Medium	*Moderate*
Big	*Veryfast*
Verybig	*Moderate*

i.e., by the union of the finite set

little×slow, *small×fast*, *medium×moderate*, *big×veryfast*, *verybig×moderate*

of fuzzy points. This means that a rule like "IF x is *Small* THEN y is *Fast*" is not a logical implication but a reading of the ordered pair (*Small*, *Fast*) in the table. Moreover, we interpret an expression like "x is *Small*" as a shortcut for "x is equal to the fuzzy quantity *Small*". This is in accordance with the fact that these fuzzy subsets are considered quantities (fuzzy granules) and not vague predicates. In the subsequent sections we will try the opposite approach. The following is a general and precise definition where we assume that $U = [0,1]$:

Definition 2.1. A *system $\mathcal{S}$ of IF-THEN fuzzy rules* is a system of rules like

$$\begin{array}{c}\text{IF } x \text{ is } A_1 \text{ THEN } y \text{ is } B_1\\ \cdots \\ \text{IF } x \text{ is } A_n \text{ THEN } y \text{ is } B_n\end{array} \tag{2.2}$$

where $i = 1,...,n$ and where the labels A_i and B_i are interpreted by the fuzzy granules $a_i : X \to U$ and $b_i : Y \to U$. We associate any rule with the Cartesian product $a_i \times b_i : X \times Y \to U$. We associate the whole system of rules with the fuzzy function $f : X \sim> Y$ defined by

$$f = \bigcup_{i=1,...,n} a_i \times b_i. \tag{2.3}$$

Obviously, the definition of f depend on the norm $\odot$ we use to define the Cartesian product. We can also assign to each i-rule a "*weigh*" $\lambda_i \in U$ expressing our confidence degree on the rule. In such a case we substitute the union in (2.3) by a *weighted* union, i.e.,

$$f(x,y) = Max_{i=1,...,n}\, \lambda_i \odot a_i(x) \odot b_i(y). \tag{2.4}$$

A further step is the *defuzzification process* enabling us to associate a classical function f' with the fuzzy function f. Usually the defuzzification process is obtained by the *centroid method* by setting, for every $r \in X$,

$$f'(r) = \frac{\int_Y f(r,y)\cdot y\,dy}{\int_Y f(r,y)\,dy}. \tag{2.5}$$

Obviously, such a method is possible only in the case that for any $r \in X$ an element y in Y exists such that $f(r,y) \neq 0$, i.e., any element $r \in X$ belongs to the domain of f to a degree different from 0. In Picture 3 both the fuzzy function f and the result f' of the defuzzification process are represented (the triangular norm is the minimum).

The final phase is a *learning process* in which the weights and the interpretation of the labels are changed until we can accept f' as a good approximation of the objective function f.

Observe that the importance of the granulation cannot be underestimated. As an example, L. Zadeh claims :

> ... *the effectiveness and success of fuzzy logic in dealing with real-world problems rest in large measure on the use of the machinery of fuzzy information granulation. This machinery is unique to fuzzy logic and differentiates it from all other methodologies.*

This is so because

> ... *fuzzy information granulation is an integral part of human cognition* (Zadeh [1997]).

Obviously, granulation is related with the basic role played by words and natural language. Indeed, the words are *"labels"* for granules and the computation of the objective function f is intended, in a sense, as a *"computation by words"*.

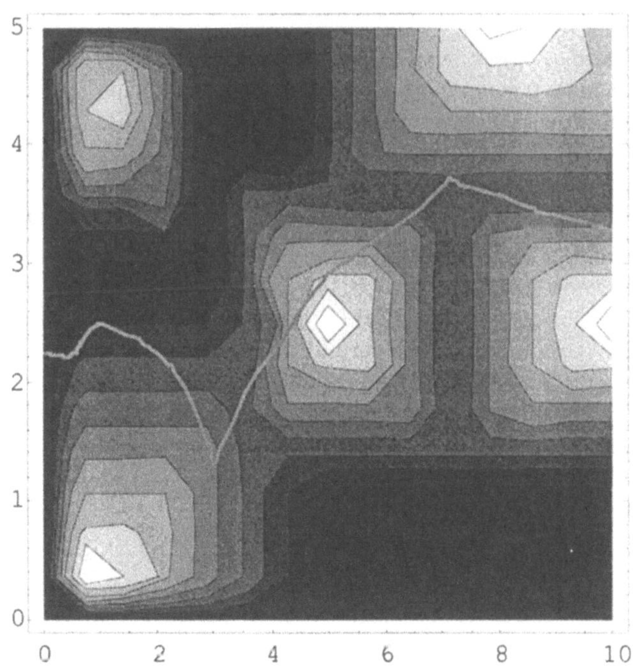

Picture 3: the fuzzy function and the result of the defuzzyfication process.

3. PROGRAMS AND HERBRAND MODELS

This section is devoted to recall the basic notions in logic programming (see Lloyd [1987]). Let $\mathcal{L}$ be a first order language with some constants and let $\mathbb{F}$ be the related set of formulas. A *ground term* of $\mathcal{L}$ is a term not containing variables, the set $U_{\mathcal{L}}$ of ground terms of $\mathcal{L}$ is called the *Herbrand universe* for $\mathcal{L}$. If $\mathcal{L}$ is function free, then $U_{\mathcal{L}}$ is the set of constants. A *ground atom* is an atomic formula not containing variables and the set $B_{\mathcal{L}}$ of ground atoms is called the *Herbrand base* for $\mathcal{L}$. We call an *Herbrand interpretation* any subset M of $B_{\mathcal{L}}$. The name is justified by the fact that M defines a model of $\mathcal{L}$ in which:

- the domain is the Herbrand universe $U_{\mathcal{L}}$,
- every constant in $\mathcal{L}$ is interpreted by itself,
- any n-ary function symbol f in $\mathcal{L}$ is interpreted as the map associating the terms $t_1,...,t_n$ in $U_{\mathcal{L}}$ with $f(t_1,...,t_n)$,
- any n-ary predicate symbol r is interpreted by the n-ary relation r' defined by setting

$$(t_1,...,t_n) \in r' \Leftrightarrow r(t_1,...,t_n) \in M.$$

A *definite program clause* is either an atomic formula or a formula of the form $\beta_1 \wedge ... \wedge \beta_n \to \beta$ where β, $\beta_1,...,\beta_n$ are atomic formulas. We denote the set of definite program clauses by *PC*. A *ground instance* of a program clause is a closed formula achieved from this clause by suitable substitutions of the free variables with closed terms. A *definite program* is a set $\mathbb{P}$ of definite program clauses. We denote the set of ground instances of clauses in $\mathbb{P}$ by $Ground(\mathbb{P})$.

Let $\mathbb{P}$ be a definite program. Then the operator $J_{\mathbb{P}}: \mathcal{P}(B_{\mathcal{L}}) \to \mathcal{P}(B_{\mathcal{L}})$, defined by setting for any subset X of $B_{\mathcal{L}}$,

$$J_{\mathbb{P}}(X) = \{\alpha \in B_{\mathcal{L}} : \alpha_1 \wedge ... \wedge \alpha_n \to \alpha \in Ground(\mathbb{P}),\ \alpha_1,...,\alpha_n \in X\} \cup X \quad (3.1)$$

is called the *immediate consequence operator* associated with the program $\mathbb{P}$. We denote the closure operator generated by $J_{\mathbb{P}}$ by $\mathcal{H}_{\mathbb{P}}$.

Definition 3.1. We call *Herbrand model* of $\mathbb{P}$ any fixed point of $J_{\mathbb{P}}$ (equivalently, of $\mathcal{H}_{\mathbb{P}}$). Given a set X of ground atoms, we say that $\mathcal{H}_{\mathbb{P}}(X)$ is the *least Herbrand model for $\mathbb{P}$ containing X*. We denote the model $\mathcal{H}_{\mathbb{P}}(\varnothing)$ by $M_{\mathbb{P}}$ and we call it the *least Herbrand model for $\mathbb{P}$.*

Trivially,

$$M_{\mathbb{P}} = \cup_{n\in N}(J_{\mathbb{P}})^n(\varnothing). \quad (3.2)$$

Let $\mathcal{D}$ be the deduction operator of first order calculus and $\vdash$ the associate consequence relation. Then we have the following:

Theorem 3.2. *For every $X \in \mathcal{P}(B_{\mathcal{L}})$*

$$\mathcal{H}_{\mathbb{P}}(X) = \{\alpha \in B_{\mathcal{L}} : \mathbb{P} \cup X \vdash \alpha\} = \mathcal{D}(\mathbb{P} \cup X) \cap B_{\mathcal{L}}. \quad (3.3)$$

In particular,

$$M_{\mathbb{P}} = \{\alpha \in B_{\mathcal{L}} : \mathbb{P} \vdash \alpha\}. \quad (3.4)$$

The above definitions can be extended in an obvious way to many-sorted languages (see, for example, Lloyd [1987]).

4. FUZZY PROGRAMS AND FUZZY HERBRAND MODELS

We are going to extend the notions in the previous section by starting from the canonical extension of a first order logic by a continuous triangular norm $\odot$ (see Section 8 in Chapter 6). In the resulting logic the fuzzy set a of logical axioms is the characteristic function of the set *Taut* of all logically-true formulas. Moreover, we have two fuzzy rules:

Generalization:

$$\frac{\alpha}{\forall x_i(\alpha)} \quad ; \quad \frac{\lambda}{\lambda} .$$

Fuzzy Modus Ponens:

$$\frac{\alpha\ ,\ \alpha \to \beta}{\beta} \quad ; \quad \frac{\lambda\ ,\ \mu}{\lambda \odot \mu} .$$

Furthermore, we can consider some *"derived"* rules. As an example, by denoting the universal closure of the formula α by $Q(\alpha)$, we can consider the

Extended Generalization:

$$\frac{\alpha}{Q(\alpha)} \quad ; \quad \frac{\lambda}{\lambda}$$

which we can obtain by an iterate application of Generalization Rule. We also have the

Extended fuzzy Modus Ponens:

$$\frac{\alpha_1, \dots ,\alpha_n\ ,\ \alpha_1 \wedge \dots \wedge \alpha_n \to \alpha}{\alpha} \quad ; \quad \frac{\lambda_1, \dots ,\lambda_n, \lambda}{\lambda_1 \odot \dots \odot \lambda_n \odot \lambda}$$

which we can obtain by observing that the formula

$$(\alpha_1 \wedge \dots \wedge \alpha_n \to \alpha) \to (\alpha_1 \to (\dots(\alpha_n \to \alpha)\dots)$$

is logically true. Finally, we have the

Particularization Rule

$$\frac{\alpha(x_1,\dots,x_n)}{\alpha(t_1,\dots,t_n)} \quad ; \quad \frac{\lambda}{\lambda}$$

where $t_1,\dots,t_n$ are ground terms. Such a rule can be obtained by observing that the formula $\alpha(x_1,\dots,x_n) \to \alpha(t_1,\dots,t_n)$ is logically true.

Recall that the so-defined logic is not truth-functional and that its semantics coincides with the class of all theories. Therefore an Herbrand interpretation cannot be defined by giving the list of true ground atoms as in the classical case. Then, we can call fuzzy Herbrand interpretation for $\mathcal{L}$ any fuzzy theory. Since in logic programming we are interested mainly in ground atoms, we call *fuzzy Herbrand interpretation for* $\mathcal{L}$ any restriction of a fuzzy theory to $B_{\mathcal{L}}$. Like the classical case, a fuzzy Herbrand interpretation defines a multi-valued interpretation of $\mathcal{L}$ in the Herbrand universe in which any n-ary predicate symbol r is interpreted by the fuzzy n-ary relation r' on $U_{\mathcal{L}}$ defined by setting

$$r'(t_1,\ldots,t_n) = m(r(t_1,\ldots,t_n)).$$

We call *fuzzy definite program* any fuzzy set $p : PC \to U$ of definite program clauses.

Definition 4.1. Let p be a fuzzy definite program. Then the *least fuzzy Herbrand model* for p is the fuzzy set $m_p : B_{\mathcal{L}} \to U$ defined by setting

$$m_p(\alpha) = \mathcal{D}(p)(\alpha) \tag{4.1}$$

for any $\alpha \in B_{\mathcal{L}}$.

Trivially, for $\alpha \in B_{\mathcal{L}}$

$$m_p(\alpha) = Sup\{Incl(\mathbb{P},p) : \mathbb{P} \in \mathcal{P}_f(Supp(p)) \text{ such that } \alpha \in M_{\mathbb{P}}\}. \tag{4.2}$$

Assume that the triangular norm under consideration is the minimum and denote the program $C(p,\lambda)$ by $\mathbb{P}(\lambda)$. Then

$$m_p(\alpha) = Sup\{\lambda \in U : \alpha \in M_{\mathbb{P}(\lambda)}\}.$$

In the case that $Supp(p)$ is finite, in the co-domain of p there is only a finite number of elements $\lambda_1 > \lambda_2 > \ldots > \lambda_n$. So, to calculate $m_p(\alpha)$ it is sufficient to calculate the least Herbrand models $M_{\mathbb{P}(\lambda 1)} \subseteq \ldots \subseteq M_{\mathbb{P}(\lambda n)}$ by a parallel process.

5. LOGIC APPROACH TO FUZZY CONTROL

The logical representation of fuzzy control is rather questionable (see for example Hájek [1999]) since, as a matter of fact an IF-THEN rule like

IF x is A THEN y is B

cannot be a logical implication like $A(x)\to B(y)$. For example, observe that, by generalization rule, $A(x)\to B(y)$ entails the formula $\forall x\forall y(A(x)\to B(y))$. In turn, by virtue of two rewriting rules for the reduction of a formula in prenex form (see, e.g., Mendelson [1964] Lemma 2.30) such a formula is equivalent to $\forall x(A(x) \to \forall yB(y))$ and therefore to $(\exists xA(x)) \to \forall yB(y)$. Then, we can admit $A(x) \to B(y)$ only in the cases that $\exists xA(x)$ is false or that $\forall yB(y)$ is true. This is unsatisfactory, obviously. We think that the users of an IF-THEN system of rules as a logical formula implicitly assumes a functional dependence of y from x while, as a matter of fact, such a dependence is not expressed by such a rule. Perhaps it would be better to give a name f to the objective function we refer to and to write a rule like $A(x)\to B(f(x))$ where such a dependence is expressed.

In this chapter, we will show that it is possible to give a logical meaning to a fuzzy system $\mathcal{S}$ of IF-THEN rules like

IF x is A_1 THEN y is B_1

. . . (5.1)

IF x is A_n THEN y is B_n

by translating it into a suitable system of clauses. To this aim, we consider A_i and B_i names for vague predicates and not labels for fuzzy granules. In accordance, we interpret "*x is* A_i" and "*y is* B_i" as "*x satisfies* A_i" and "*y satisfies* B_i", respectively. Moreover, we associate (5.1) with the fuzzy set p_1

$$\begin{array}{ll} A_1(x) \wedge B_1(y) \to Good(x,y) & [\lambda_1] \\ \ldots & \\ A_n(x) \wedge B_n(y) \to Good(x,y) & [\lambda_n] \end{array} \tag{5.2}$$

of signed clauses, where "*Good(x,y)*" is a new predicate whose intended meaning is

"*given x, y is a good value for the control variable*",

(see Gerla [2000]). The meaning of each λ_i is that the i-rule is accepted at level λ_i. Furthermore, by assuming that A_i and B_j are interpreted by the fuzzy subsets a_i and b_j, we consider, for $i,j = 1,\ldots,n$, $r \in X$ and $t \in Y$, the fuzzy set p_2 of signed ground atoms

$$\begin{array}{ll} A_i(r) & [a_i(r)] \\ B_j(t) & [b_j(t)]. \end{array}$$

In other words, we associate the IF-THEN system (5.1) with the fuzzy definite program $p = p_1 \cup p_2$, i.e., the fuzzy definite program

$$p(\alpha) = \begin{cases} \lambda_i & \text{if } \alpha \text{ is the clause } A_i(x) \wedge B_i(y) \to Good(x,y), \\ a_i(r) & \text{if } \alpha \text{ is the ground atom } A_i(r), \\ & \\ b_i(t) & \text{if } \alpha \text{ is the ground atom } B_i(t), \\ 0 & \text{otherwise.} \end{cases} \tag{5.3}$$

Observe that each element in X or in Y is considered as a constant. Therefore, the Herbrand universe of p is $X \cup Y$.

The following theorem shows that a basic part of fuzzy control process can be defined in logical terms. As we will see, it is more difficult to give a logical meaning to the centroid method.

Theorem 5.1. *The calculus of the fuzzy function in the fuzzy control system* (5.1) *is equivalent to the calculus of the least fuzzy Herbrand model of the associated fuzzy definite program p given by* (5.3). *More precisely, the interpretation of the predicate Good in this model coincides with the fuzzy function* (2.6).

Proof. Consider the fuzzy definite program associated with the system (5.1)

$$\begin{array}{ll} A_1(x) \wedge B_1(y) \to Good(x,y) & [\lambda_1] \\ \ldots & \\ A_n(x) \wedge B_n(y) \to Good(x,y) & [\lambda_n] \\ A_i(r) & [a_i(r)] \\ \ldots & \\ B_j(t) & [b_j(t)] \end{array}$$

where $\lambda_1, \ldots, \lambda_n$ are elements in U, r varies in X and t varies in Y. Then, given the constants r and t, we can try to prove the formula $Good(r,t)$. Consider the ground instance of the first rule,

$A_1(r) \wedge B_1(r) \rightarrow Good(r,t)$

and the ground atoms

$A_1(r)$,

$B_1(t)$.

Then, by the extended fuzzy Modus Ponens rule, we can prove $Good(r,t)$ to degree $\lambda_1 \odot a_1(r) \odot b_1(t)$. Likewise, from the second fuzzy clause we get a proof of $Good(r,t)$ capable of proving $Good(r,t)$ to degree $\lambda_2 \odot a_2(r) \odot b_2(t)$ and so on. It is evident that these are the only possible proofs of $Good(r,t)$. As a consequence, by setting

$$good(r,t) = m_p(Good(r,t)),$$

we get

$$good(r,t) = Max\{\lambda_1 \odot a_1(r) \odot b_1(t), \ldots, \lambda_n \odot a_n(r) \odot b_n(t)\}.$$

in accordance with (2.4). □

Observe that the calculus of the least fuzzy Herbrand model coincides with the calculus of the predicate *Good*. Indeed, since *Good* is the only predicate occurring in the head of a rule, we have complete information about all the predicates different from *Good*. In other words, while Pictures 1 and 2 are given, Picture 3 is calculated. These three pictures represent the least fuzzy Herbrand model of the fuzzy definite program given in (5.3).

As we will try to show in the next sections, such a logical approach gives the possibility of expressing the information of an expert in a more complete way.

6. THE LOGICAL INTERPRETATION SUGGESTS NEW TOOLS

The logical interpretation of an IF-THEN system of fuzzy rules gives a natural justification to several notions in fuzzy control. Also, it suggests some new tools. In the following, we will list some possibilities.

Degree of completeness. The completeness of a fuzzy system of rules $\mathcal{S}$ is represented by the fact that for any "situation" r, a good control t exists. In other words a system is complete provided that the formula

$$\forall x \exists y Good(x,y)$$

is satisfied. Now, since in a multi-valued logic the quantifiers $\forall$ and $\exists$ are interpreted by the operators *Inf* and *Sup*, respectively, we can propose the following definition:

Definition 6.1. The *degree of completeness* of a fuzzy system of rules $\mathcal{S}$ is the number

$$Compl(\mathcal{S}) = Inf_{x \in X} Sup_{y \in Y}\, good(x,y).$$

Equivalently, by denoting with *Dom(good)* the domain of *good*,

$$Compl(\mathcal{S}) = Inf_{x \in X}\, Dom(good)(x).$$

If each predicate b_i is normal, i.e., $b_i(\underline{y}) = 1$ for a suitable $\underline{y} \in Y$, then

$$Dom(good) = a_1 \cup \ldots \cup a_n.$$

In fact, for any $x \in X$,

$$a_1(x) \vee \ldots \vee a_n(x) \geq a_1(x) \odot b_1(y) \vee \ldots \vee a_n(x) \odot b_n(y)$$

and therefore $a_1(x) \vee \ldots \vee a_n(x) \geq good(x,y)$. Moreover, assume that $a_1(x) \vee \ldots \vee a_n(x) = a_i(x)$ and that $\underline{y}$ is an element in Y such that $b_i(\underline{y}) = 1$. Then

$$a_1(x) \vee \ldots \vee a_n(x) = a_i(x) \odot b_i(\underline{y}) \leq Sup\{a_1(x) \odot b_1(y) \vee \ldots \vee a_n(x) \odot b_n(y) : y \in Y\}.$$

Thus

$$Compl(\mathcal{S}) = Inf_{x \in X}(a_1(x) \vee \ldots \vee a_n(x))$$

and therefore,

$Compl(\mathcal{S}) = 1 \Leftrightarrow$ the set of fuzzy predicates in X is a covering of X.

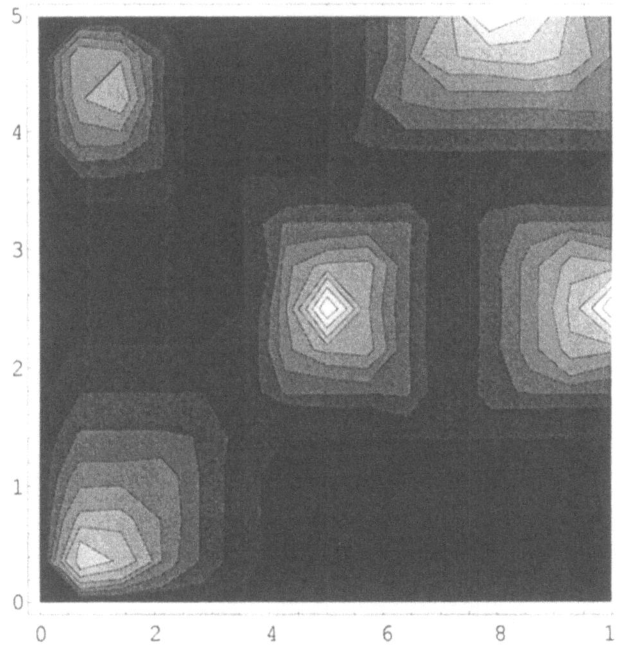

Picture 4 : the predicate *Clearly*(*Good*)

Linguistic modifiers. Another possibility is to define some *"linguistic modifiers"*. As an example, we can define the modifiers *"Clearly"* and *"Vaguely"* by associating to these predicates the functions $clearly : U \rightarrow U$ and $vaguely : U \rightarrow U$ defined by setting $clearly(x) = x^2$ and $vaguely(x) = x^{0.5}$ for any $x \in U$. In accordance, the predicate *"Vaguely(Good)"* is interpreted by the fuzzy subset

$$vaguelygood(x,y) = vaguely(good(x,y)) = good(x,y)^{0.5}$$

The predicate *Clearly*(*Good*) is interpreted by the fuzzy subset

$$clearlygood(x,y) = clearly(good(x,y)) = good(x,y)^2.$$

The predicates *"Clearly(Good)"* and *"Vaguely(Good)"* are represented in Picture 4 and 5, respectively. These linguistic modifiers can be applied also to predicates which are premises in a rule. As an example, we can consider rules as

$Vaguely(Little)(x) \wedge Fast(y) \rightarrow Good(x,y)$

$Little(x) \wedge Vaguely(Slow)(y) \rightarrow Good(x,y)$.

Rule chaining. The power of classical logic programming is mainly based on stratified deductions and recursion and the same holds for fuzzy logic programming. Then, in accordance with the proposed logical interpretation of fuzzy control, stratified deductions and recursion perhaps can give a more flexible and powerful approach to fuzzy control. Indeed, we can consider systems with several layers and chained rules in which the output of some rules may serve as an input of other rules. Some examples are in Section 7 but no systematic investigation was made in this promising direction.

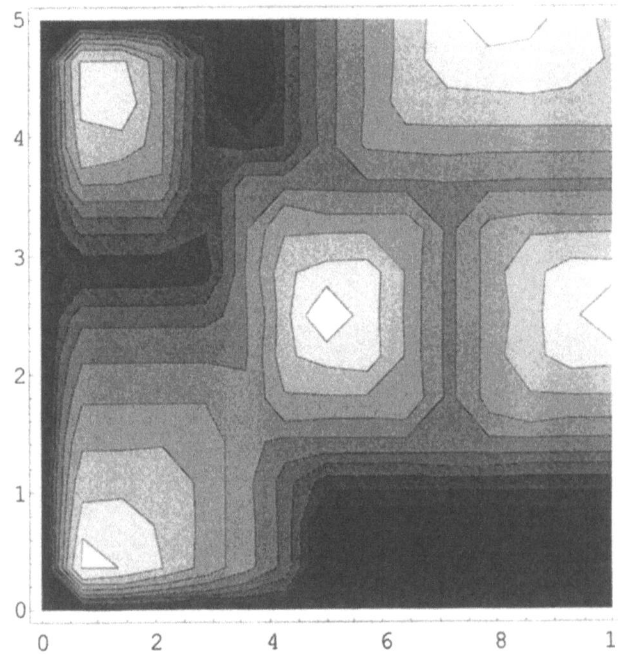

Picture 5: the predicate *Vaguely*(*Good*)

7. CONTROL BY IMPLICATION AND NEGATIVE INFORMATION

Some authors define fuzzy control by an implication operation and not by a triangular norm (see, for example, Klir and Yuan [1995]). Namely, let $\rightarrow$ be a binary operation in U able to interpret the logical connective "implies" and consider the system

$$\begin{array}{l} \text{IF } x \text{ is } C_1 \text{ THEN } y \text{ is } D_1 \\ \quad \ldots \\ \text{IF } x \text{ is } C_n \text{ THEN } y \text{ is } D_n \end{array} \tag{7.1}$$

of IF-THEN fuzzy rules. Also, assume that each C_i is interpreted by the fuzzy subset c_i and each D_i by the fuzzy subset d_i. Then in the implication-based fuzzy control any rule "IF x is C_i THEN y is D_i" is associated with the fuzzy relation

$c_i \to d_i$ defined by setting, for any $x \in X$ and $y \in Y$

$$(c_i \to d_i)(x,y) = c_i(x) \to d_i(y).$$

The whole system of rules is associated with the fuzzy function $g : X \sim> Y$ defined by

$$g = \bigcap_{i=1,\dots,n} c_i \to d_i. \tag{7.2}$$

In order to give a logical interpretation of such a procedure, in this section we assume that $\lambda \to \mu$ is equal to $\sim\lambda \oplus \mu$ where $\oplus$ is the Łukasiewicz disjunction. Moreover, we consider a new binary predicate name $Bad(x,y)$ and the following system of fuzzy rules:

$$\begin{array}{ll} C_1(x) \wedge (\neg D_1(y)) \to Bad(x,y) & [\lambda_1] \\ \dots & \\ C_n(x) \wedge (\neg D_n(y)) \to Bad(x,y) & [\lambda_n]. \end{array} \tag{7.3}$$

Theorem 7.1. *Consider the fuzzy control system* (7.1). *Then the calculus of the fuzzy function obtained by the implication procedure is equivalent to the calculus of the least fuzzy Herbrand model of the fuzzy definite program* (7.3) for $\lambda_1 = \dots = \lambda_n = 1$. *More precisely, the fuzzy function* (7.2) *coincides with the interpretation of the vague predicate* $\neg Bad$ *in such a model.*

Proof. Let *bad* be the interpretation of *Bad* in the least fuzzy Herbrand model of (7.3). Then, by proceeding as in Section 5, we have that

$$bad = \bigcup_{i=1,\dots,n} c_i \times (-d_i).$$

Equivalently,

$$\sim bad(x,y) = \bigcap_{i=1,\dots,n} \sim(c_i(x) \odot \sim d_i(y)) = \bigcap_{i=1,\dots,n} (\sim c_i(x) \oplus d_i(y)) = \bigcap_{i=1,\dots,n} c_i(x) \to d_i(y)$$

and therefore

$$\sim bad = \bigcap_{i=1,\dots,n} c_i \to d_i = g. \tag{7.4}$$

□

Observe that the logical approach to fuzzy control we propose in this chapter enables us to emphasize the different meanings of the norm-based and implication-based procedures (see also Dubois and Prade [1997] and Babuška [1998] pag. 21). Indeed, in a sense, fuzzy control by a triangular norm enables us to manage positive information (the fuzzy set of pairs (x,y) we consider good). Fuzzy control by an implication enables us to manage negative information (the fuzzy subset of pairs (x,y) we will avoid). Also, such a logical approach enables us to consider the two methods at the same time. In fact we can consider a fuzzy definite program containing both rules for the predicate *Good* and rules for the predicate *Bad*. In such a case, by adding to the language the predicate *Optimum* and the rule

$$Good(x,y) \wedge \neg Bad(x,y) \to Optimum(x,y),$$

we can compose the two different kinds of information into a two-layers IF-THEN system. It is immediate to verify that the fuzzy relation *optimum* interpreting *Optimum* is defined by setting

$$optimum(x,y) = \begin{cases} 0 & \text{if } good(x,y) \le bad(x,y), \\ good(x,y) - bad(x,y) & \text{otherwise.} \end{cases}$$

Clearly, we have to refer to the predicate *"Optimum"* and not *"Good"* in the successive defuzzification process.

This leads to the more general problem of using *"negative"* information in fuzzy logic programming. Unfortunately, this is an hard problem even in classical logic programming and it is faced, for example, by the *closed world rule* (see, e.g. Lloyd [1987]):

if a ground atom A is not a logical consequence of a program $\mathbb{P}$, then we are entitled to infer $\neg A$.

Obviously, this rule is rather questionable both from a semantic and a computational viewpoint. We can try to extend closed world rule to fuzzy logic programming by assuming that the negation $\neg A$ of a ground atom A is true at degree $1-\mathcal{D}(p)(A)$. As in the classical case, this *"rule"* creates several problems. As an example, if a proof π gives a lower bound $Val(\pi,p)$ for the truth value of A then $1-Val(\pi,p)$ gives an upper bound for the truth value of $\neg A$. On the other hand, the fuzzy logic deduction machinery as proposed in literature is not able to manage these upper bounds. Perhaps, the extended notion of fuzzy logic by constraints examined in Chapter 5 should be the appropriate tool to manage both positive and negative information.

In any case, if the negation occurs in the body of the rules of the simple fuzzy definite programs we associate with a fuzzy IF-THEN system, no difficulty arises.

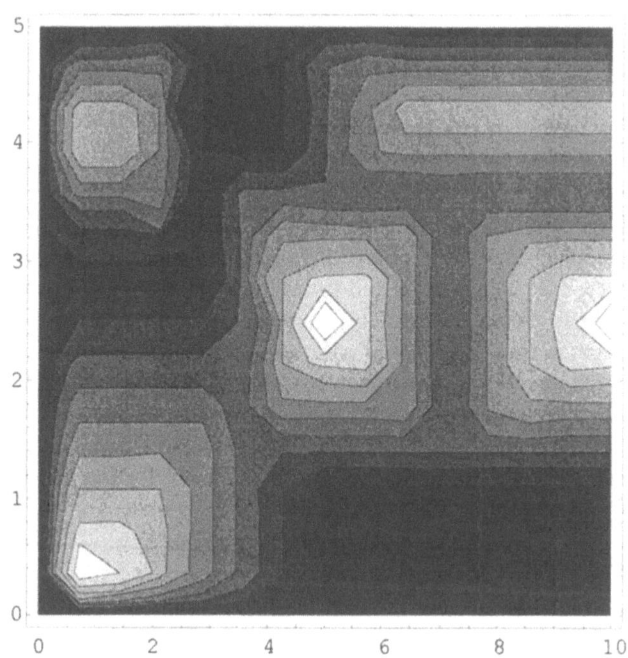

Picture 6 : the predicate *Safe*

Example 1 (Negative information for a safe control). Suppose we need to take into account that there are some control actions we have to avoid. For instance, assume that a too fast control y is considered dangerous. Then, we can express this by adding to the fuzzy definite program giving *Good* the following rule:

$$Clearly(Veryfast)(y) \rightarrow Dangerous(y).$$

Obviously, it should be possible to consider a more complex definition of the vague predicate *Dangerous*. Once the vague predicate *Dangerous* is defined, we obtain a two-layers system by introducing the new predicate *"Safe"* and the rule

$$Good(x,y) \wedge \neg(Dangerous(y)) \rightarrow Safe(x,y).$$

Denote the interpretations of *Dangerous* and *Safe* by *dangerous* : $Y \rightarrow U$ and *safe* : $X{\times}Y \rightarrow U$, respectively. Then, given $r \in X$ and $t \in Y$, the first clause enables us to calculate

$$dangerous(t) = \mathcal{D}(p)(Dangerous(t)) = clearly(veryfast(t)).$$

By the closed world rule

$$\mathcal{D}(p)(\neg Dangerous(t)) = 1 - \mathcal{D}(p)(Dangerous(t)) = 1 - clearly(veryfast(t)).$$

Then, by the second clause

$$safe(r,t) = \mathcal{D}(p)(Safe(r,t)) = good(r,t) \odot (1 - clearly(veryfast(t))).$$

Clearly, in such a case we have to refer to the predicate *"Safe"* and not *"Good"* in the successive defuzzification process. Picture 6 represents such a new predicate.

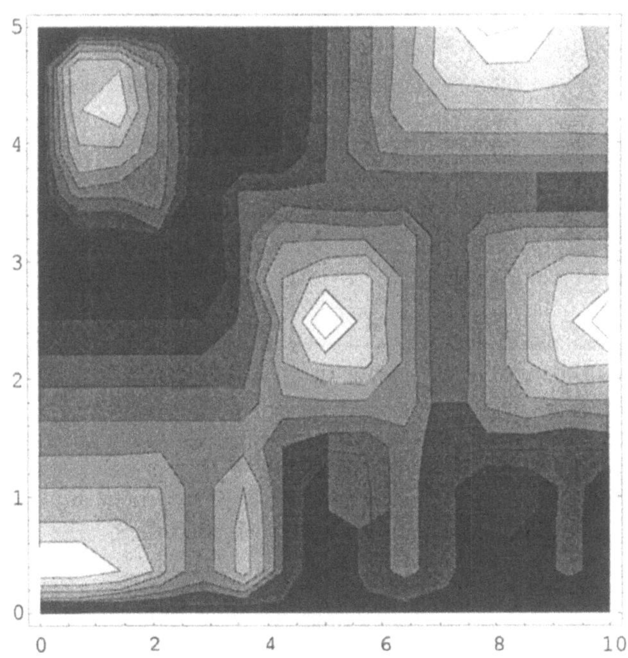

Picture 7 : adding the default rule

Example 2 (Negative information for a default rule). Another interesting use of the negation connective is the possibility of defining a *"default"* rule, i.e., a rule enabling us to choose a control in the case in which no condition $A_1,\dots,A_n$ in (5.2) is satisfied. As an example, assume that in this case an expert suggests a slow y. Then, by assuming that $Domain(x)$ is the formula $A_1(x) \vee ... \vee A_n$, we add the rule

$$\neg Domain(x) \wedge Slow(y) \rightarrow Good(x,y).$$

In such a case the degree of completeness of the system increases (see Picture 7).

Observe that in the just given examples, the fuzzy relation *safe* is contained in the fuzzy relation *good* while the default rule increases such a relation. This shows that, by adding new information, it is possible both:

- to increase the area of the fuzzy upper covering of the objective function f, in order to get completeness, i.e., to be sure that the whole set of points of f is covered,
- to decrease such an area, in order to get a more precise representation of the objective function f.

8. CONTROL BY SIMILARITY AND PROTOTYPES

In Section 10 of Chapter 2 a similarity in a set S was defined as a fuzzy relation $near : S \times S \rightarrow U$ such that, for any $x, y, z \in S$,

$$near(x,y) \odot near(y,z) \leq near(x,z),$$
$$near(x,x) = 1,$$
$$near(x,y) \leq near(y,x).$$

Consequently, we can define a similarity as a model of the clauses

$$Near(x,y) \wedge Near(y,z) \rightarrow Near(x,z), \quad \text{(transitive)}$$
$$Near(x,x), \quad \text{(reflexive)}$$
$$Near(x,y) \rightarrow Near(y,x) \quad \text{(symmetric).}$$

Let P be a set of elements of S called *prototypes*; also suppose S is equipped with a similarity called *near*. Then we define the fuzzy subset of elements which are similar to a prototype by setting

$$s(x) = Sup\{near(x,x') : x' \in P\}.$$

Conversely, given any normal fuzzy subset s of S, there exist a similarity and a set of prototypes such that the associated fuzzy subset coincides with s. Then, the notion of fuzzy subset can be reduced to the notion of similarity.

We can use such a notion to propose a system of rules that emphasizes the geometrical nature of fuzzy control. Consider a first order language with two predicate names "*Near*" and "*Sim*" to denote a similarity in X and Y respectively. Assume that the objective function f has been scheduled in the table

x	y
x_1	y_1
...	...
x_n	y_n

where $x_1,...,x_n$ are elements in X and in Y, respectively. Then we can consider the fuzzy definite program $p = p_1 \cup p_2$ where p_1 is the program

$$Near(x, x_1) \wedge Sim(y, y_1) \rightarrow Good(x, y),$$

. . .

$Near(x, x_n) \wedge Sim(y, y_n) \rightarrow Good(x, y)$,

and p_2 is a set of clauses to define *Near* and *Sim*. This fuzzy definite program defines the fuzzy relation *good* as a union of n fuzzy points centered in (x_1, y_1), ...,(x_n, y_n), respectively. Obviously, $good(x_i,y_i) = 1$ for any $i \in \{1,...,n\}$. Note that the centroid method gives a function f' such that $f'(x_i)$ is not necessarily equal to y_i. In fact, the value $f'(x_i)$ depends on all the rules in which an x_j occurs such that $near(x_i,x_j) \neq 0$. On the other side, the choice of the maximum in the defuzzification process always gives the property $f'(x_i) = x_i$.

As an example, consider the fuzzy definite program defined by the rules

$Near(x,0) \wedge Sim(y,0) \rightarrow Good(x,y)$,
$Near(x,1) \wedge Sim(y,3) \rightarrow Good(x,y)$,
$Near(x,2) \wedge Sim(y,3) \rightarrow Good(x,y)$,
$Near(x,4) \wedge Sim(y,4) \rightarrow Good(x,y)$,
$Near(x,6) \wedge Sim(y,4) \rightarrow Good(x,y)$,
$Near(x,8) \wedge Sim(y,5) \rightarrow Good(x,y)$,
$Near(x,10) \wedge Sim(y,1) \rightarrow Good(x,y)$.

In Picture 8 we represent the resulting fuzzy relation *good*. Such a relation is a union of 7 fuzzy points centered in (0,0), (1,3), (2,3), (4,4), (6,4), (8,5), (10,1).

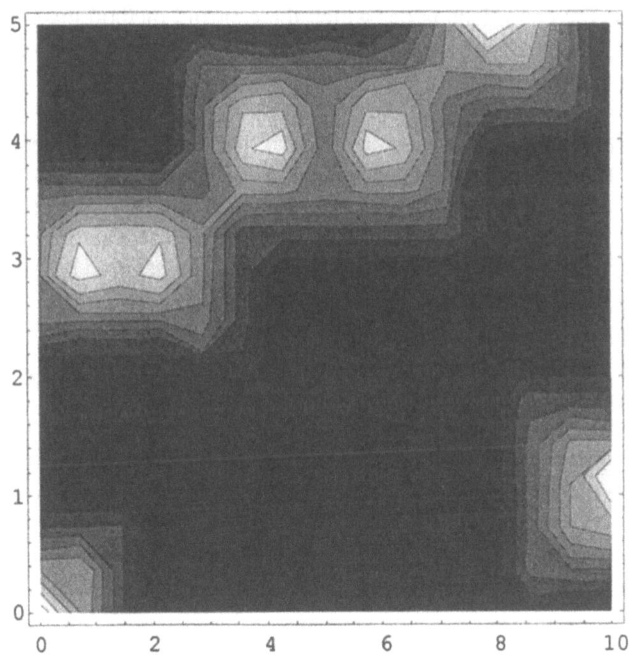

Picture 8 : control by similarity

More precisely, in this picture we assume that:

- X and Y are intervals [0,10] and [0, 5], respectively,
- the points 1, 2, 4, 5, 6, 8, 9, 10 are prototypes in X,
- the points 0, 1, 3, 4, 5 are prototypes in Y.
- the similarities are defined by setting

$$Near(x,x') = 1-(|x-x'|/2)\wedge 1 \quad ; \quad Sim(y,y') = 1-|y-y'|\wedge 1.$$

Observe that $f'(0) = 1.5833 \neq 0$. Indeed, 0 is near to 1 to a degree different from zero.

Also, we can use *"Near"* and *"Sim"* to improve the predicate *"Good"* defined in the previous sections. For example, we can add the recursive rule

$$Good(x,y') \wedge Sim(y,y') \rightarrow Good(x,y)$$

whose meaning is obvious. We can avoid the computational difficulties of recursion by introducing a new predicate *"Good*"* and the formula

$$Good(x,y') \wedge Near(y,y') \rightarrow Good^*(x,y).$$

9. LOGIC INTERPRETATION OF DEFUZZIFICATION: AN OPEN QUESTION

The centroid method gives rise to two questions. The first one is whether this method has a logical meaning. To this regard, observe that, in accordance with the interpretation of *good*(r,y) as a degree of preference (see Section 10), from a logical viewpoint would be better, given r, to take a value y that maximizes *good*(r,y). The second question is how to make explicit (by a suitable set of first order formulas) the conditions under which the centroid method gives satisfactory results. Indeed, in reality, the centroid method does not work well in several cases. The following is an example:

Example. Assume that a driver sees a yellow traffic light. Then, it depends on the distance from the traffic light and on the car speed the suitable way to adjust the speed. On the other hand some cases occur in which "rapidly increase the speed" and "roughly brake" are both good choices (the choice only depends on the driver's temperament). In accordance, both of the following rules seem to be valid:

$$High(x) \wedge Little(y) \wedge Big\_\text{positive}(z) \rightarrow Good(x,y,z)$$
$$High(x) \wedge Little(y) \wedge Big\_\text{negative}(z) \rightarrow Good(x,y,z)$$

where x is the speed, y the distance between the car and the traffic light and z the (positive or negative) acceleration. Let x and y be such that $high(x) = 1$ and $little(y) = 1$. Then, due to the symmetry of the fuzzy predicates *big_positive* and *big_negative*, the centroid method suggests an acceleration equal to zero. This means that the driver does not modify his speed and this leads to a probable disaster.

These two questions are related, perhaps, and we can formulate the following conjecture:

Conjecture. *The conditions under which the centroid method is correct can be expressed by a fuzzy subset p' of formulas. By adding the information p' to the fuzzy definite program p, a new fuzzy Herbrand interpretation* $m_{p\cup p'}$ *is defined. Let good' be the fuzzy predicate defined by setting*

$$good'(r,t) = m_{p\cup p'}(Good(r,t)).$$

Then,

i) *good'*(r,y) *has a unique maximum (with respect to y),*

ii) *such a maximum coincides with the centroid of good(r,y).*

We are not able to demonstrate this conjecture. We expose only some considerations and results as an hint for further investigations. A first observation is that the centroid method can be applied only in cases where if two controls y' and y'' are acceptable, then all the intermediate controls are acceptable. We can express such a property by adding to the program p the rule

$$Good(x,y') \wedge Good(x,y'') \wedge (y' \leq y \leq y'') \rightarrow Good(x,y).$$

In order to avoid recursion, we can also add a new predicate name *Good'* and the rule

$$Good(x,y') \wedge Good(x,y'') \wedge (y' \leq y \leq y'') \rightarrow Good'(x,y).$$

The corresponding fuzzy relation *good'* is defined by setting

$$good'(x,y) = Sup\{good(x,y') \odot good(x,y'') : y' \leq y \leq y''\}.$$

Now, in literature a fuzzy subset $s : \boldsymbol{R} \rightarrow U$ of the real line is called *convex* provided that, for every x, x_1, x_2,

$$x_1 \leq x \leq x_2 \Rightarrow s(x) \geq s(x_1) \odot s(x_2),$$

(see Zadeh [1965]). It is a matter of routine to prove that the class of convex fuzzy subsets is a closure system. The proof of the following proposition is trivial:

Proposition 9.1. *Assume that $\odot$ is the minimum and that $r \in X$. Then good'(r,y) is the convex closure of good(r,y).*

This is not sufficient to demonstrate the conjecture. For example, if *good'(r,y)* is constant with respect to y (and therefore convex), then *good'(r,y)* coincides with *good(r,y)* and such a function has not a unique maximum as required. A better result can be achieved by considering as T-norm the product and by substituting the operator *Sup* with the sum:

$$good'(r,y) = \Sigma_{y' \leq y \leq y''}\, good(x,y') \cdot good(x,y'').$$

More precisely, in order to get $good'(r,y) \leq 1$, the values assumed by *good* must be sufficiently small. This can be achieved by assuming that all the values $\lambda_1,\ldots,\lambda_5$ assigned to the clauses defining *good* are less or equal to $1/|Y|$. For such elements the sum coincides with the Łukasiewicz disjunction. A numerical simulation shows that in this case *good'* fits well, i.e., the maximum of *good'(r,y)* is approximately equal to the centroid of *good(r,y)*.

Some suggestions to give a logical meaning to the centroid method can be furnished by the following proposition:

Proposition 9.2. *Assume that $Y = [a,b]$, define $far : [a,b] \times [a,b] \rightarrow U$ by setting $far(x,y) = |x-y|/(a-b)^2$ and $good' : X \times Y \rightarrow U$ by setting, for any $x \in X$ and $y \in Y$,*

$$good'(x,y) = \left(\int_a^y far(y,y') \cdot good(x,y')dy'\right) \wedge \left(\int_y^b far(y,y') \cdot good(x,y')dy'\right).$$

Then, for every $r \in X$, the fuzzy subset good'(r,y) is a fuzzy interval of Y with only one maximum. Moreover, such a maximum is the centroid of good(r,y).

Proof. Set

$$h(y) = \int_a^y far(y,y')\cdot good(r,y')dy' \quad ;$$

$$k(y) = \int_y^b far(y,y')\cdot good(r,y')dy'.$$

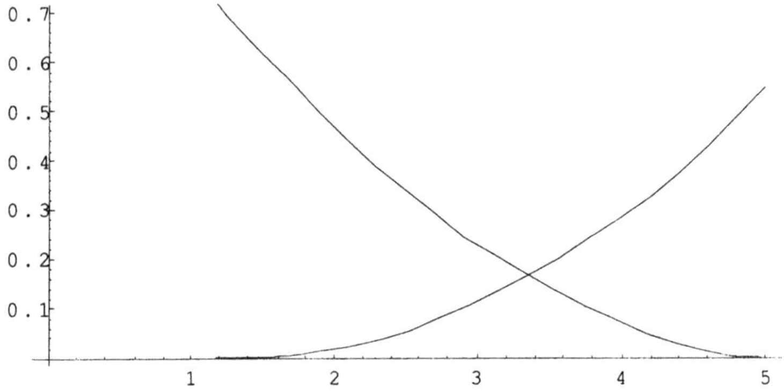

Picture 9

Then $h : [a,b] \to U$ is a strictly-increasing continuous function such that $h(a) = 0$ and that $k : [a,b] \to U$ is a strictly-decreasing continuous function such that $k(b) = 0$ (see Picture 9 where $r = 8$). Moreover, the maximum of $good^{*'}(r,y)$ is the unique point y_0 such that $h(y_0)=k(y_0)$, i.e., satisfying the equation

$$\int_a^y far(y,y')\cdot good(r,y')dy' = \int_y^b far(y,y')\cdot good(r,y')dy'.$$ Now, since

$$(h(y)-k(y))(b-a)^2 = \int_a^y |y-y'|\cdot good(r,y')dy' - \int_y^b |y-y'|\cdot good(r,y')dy'$$

$$= \int_a^y (y-y')\cdot good(r,y')dy' - \int_y^b (y'-y)\cdot good(r,y')dy'$$

$$= y\int_a^b good(r,y')dy' - \int_a^b good(r,y')\cdot y'dy',$$

the zero of such an equation is the centroid of $good(r,y)$. □

We can try to translate Proposition 9.2 in logical terms by noticing that, under the hypothesis that the values of *good* and *far* are not too big, the integral operator $\int$ coincides with the operator $\oplus : \mathcal{P}(U) \to U$ associated with the Łukasiewicz disjunction (see Remark 1 in Section 8 of Chapter 6). Indeed, in such a case let $Good_1$, $Good_2$ and $Good'$ be two-places predicates and add to the fuzzy definite program giving *Good*, the rules

$(y' \le y)$ & $Far(y',y)$ & $Good(x,y') \to Good_1(x,y)$,

$(y' \ge y)$ & $Far(y',y)$ & $Good(x,y') \to Good_2(x,y)$,

$Good_1(x,y) \wedge Good_2(x,y) \to Good'(x,y)$.

Moreover, interpret the resulting program in a multivalued logic in which $\exists$ is interpreted by the operator $\oplus$ associated with the Łukasiewicz disjunction, & is interpreted by the product and $\wedge$ by the minimum. Then it is possible to prove that the resulting interpretation of *Good'* coincides with the fuzzy relation *good'* defined in Proposition 9.2. Then this fuzzy relation has a unique maximum and such a maximum is the centroid of *good*(r,y).

This answer to the conjecture is rather unsatisfactory. In fact, the meaning of the proposed formulas is not clear and the logic we need is rather obscure. Then, the question whether a logical interpretation of the centroid method is possible, remains open.

10. THE PREDICATE *MAMD* AND SOME OBSERVATIONS

Now we will list some features of the logical approach to fuzzy control, by emphasizing the differences with the *"granular approach"*.

(a) *The fuzzy relation good is not a fuzzy function defined by cases.* Indeed, in our interpretation *"good"* is a fuzzy predicate enabling us to say, given r, if a control t is good or not. This is so because the aim of the fuzzy definite program p is not to calculate the objective function $\mathrm{f} : X \to Y$ representing the correct answer $t = \mathrm{f}(r)$ given the input r. The aim of p is only to define vague predicates such as *"Good"*, *"Stable"*, *"Dangerous"* which express our graded opinion (degree of preference, taste) on a possible control t, given an input r. Consequently, it is very natural to admit that two different elements t and t' exist in Y such that $good(r,t) = good(r,t')$. If we admit this, then we also must admit the following claim:

(b) *Partitions are not necessary.* By referring to our example, this means that it is not necessary for the class

$$C = \{little, small, medium, big, verybig\}$$

of fuzzy predicates defined in X to be a partition. Indeed, it is not necessary that these predicates are pairwise disjoint. For example, the predicate *"Small"* is a synonym of the predicate *"Little"*, and therefore, the related interpretations *small* and *little* almost completely overlap. Moreover, it is not necessary that C is a covering of X. In fact, it is possible that the available information is incomplete and therefore that there is an element r such that

$$little(r) = small(r) = medium(r) = big(r) = verybig(r) = 0$$

(see also the possibility of defining a default rule in Section 7).

(c) *The number good(r,t) represents a constraint.* Indeed, $good(r,t)$ is the degree to which $Good(r,t)$ can be proved and not the truth degree of $Good(r,t)$. This since, as it is usual in fuzzy logic, $good(r,t)$ represents the information

"given r, we can prove that the control t is good at least to degree good(r,t)".

Consequently, the number $good(r,t)$ represents only the information we can derive from the available information. By adding new information, as an example a clause $A(x) \wedge B(y) \to Good(x,y)$, it is possible that $good(r,t)$ assumes a new value. Namely, as the fuzzy logic we use is monotone, such a value can only increase. Only by assuming the point of view of logic programming and with reference to the least fuzzy Herbrand model associated with the given program we can claim that $good(r,t)$ is a truth value.

(d) ***The elements in U are degrees of preference rather than truth values.*** Indeed, the number *good*(r,t) represents the information

"given r, we are able to prove that t is preferred at least to degree good(r,t)".

If we admit such a point of view, no notion of model is necessary for the fuzzy logic we use. This leads to the question if fuzzy logic can be considered a basis for a logic of the judgement values, subjective in nature, while usually only logics of the truth values are proposed whose purpose is an objective description of the world (see also Section 10 in Chapter 3).

(e) ***A second-order logical approach to fuzzy control is reasonable.*** As an example, we can consider rules like:

"If x is small then the function f is lightly increasing",

"The function f has only a maximum",

"If x is small then f(x) is medium".

A system of rules as such defines a fuzzy predicate *Good*(f) in the class Y^X of possible functions from X to Y.

We conclude this section by emphasizing that

The reduction of fuzzy control to logic proposed in this chapter does not coincide with the one proposed in Hájek [1998].

In fact, in Hájek [1998] one defines the predicate $MAMD(x,y)$ by the axiom

$$MAMD(x,y) \leftrightarrow ((A_1(x) \wedge B_1(y)) \vee \ldots \vee (A_n(x) \wedge B_n(y))). \qquad (H)$$

It is evident that in an Herbrand model of such an axiom the predicate *MAMD* is interpreted in the same way as *Good*, i.e., by the fuzzy relation *good*. So, from an extensional point of view the two approaches look to be equivalent. Nevertheless, they are different in nature. As an example, as we observed earlier, in the approach we have proposed the number *good*(r,t) represents only the information (a lower constraint) we can derive from the available information. By adding new information (that is, new clauses) it is possible that *good*(r,t) assumes a new value. Since *good*(r,t) is just a lower bound for the exact value, this is not contradictory: it provides only more complete information. Instead, we cannot add this new information to Axiom (H) in which one establishes the exact truth value of $MAMD(x,y)$. Indeed, any new information on the predicate *MAMD* contradicts Axiom (H).

CHAPTER 11

EFFECTIVENESS IN FUZZY LOGICS

1. INTRODUCTION

The concepts of a decidable subset and a recursively enumerable subset are crucial for first order classical logic. In particular, they are basic tools for the proof of the famous limitative theorems about the undecidability and incompleteness of first order logic (see, for example, Shoenfield [1967]). Then, the question of a suitable extension of such concepts to fuzzy set theory arises. A first proposal in such a direction was made by E. S. Santos in an interesting series of papers. Indeed, Santos, starting from an idea of L. Zadeh (Zadeh [1968]), proposed the notions of fuzzy Turing machine, Markov normal fuzzy algorithm and fuzzy program. Santos proved that all these definitions determine the same notion of computability for fuzzy maps (see Santos [1970] and Santos [1976]). As in the classical case, a corresponding definition of recursively enumerable fuzzy subset is obtained by calling recursively enumerable any fuzzy subset which is the domain of a computable fuzzy map. Successively, a notion of recursive enumerability was proposed in Harkleroad [1984] where a fuzzy subset s is said to be recursively enumerable if the restriction of s to its support is a partial recursive function.

In a large series of papers L. Biacino and the author proposed a definition of recursive enumerability which is a proper extension of both definitions of Santos and Harkleroad. In this chapter, we will refer to the resulting theory. Background in recursion theory is required for understanding the arguments in this chapter (see, for example, Rogers [1976]).

2. RECURSIVELY ENUMERABLE FUZZY SETS

We say that a set S admits a *coding* if there exists a (intuitively) computable one-one map $c : S \rightarrow N$ from S onto the set N of positive natural numbers. For example, the set of rational numbers, the set of integers, and the set of formulas of a logic all admit a coding. We identify a set with a coding with N and this enables us to extend to this set all the notions of recursion theory usually defined in N (see Rogers [1976] page 27). As an example, assume that S_1 and S_2 are codified by the one-one computable functions $c_1 : S_1 \rightarrow N$ and $c_2 : S_2 \rightarrow N$. Then we say that a partial map $f : S_1 \rightarrow S_2$ is *partial recursive* provided that the map $f' : N \rightarrow N$ defined by setting,

$$f'(n) = c_2(f(c_1^{-1}(n)))$$

for any $n \in N$ such that $c_1^{-1}(n)$ belongs to the domain of f, is a partial recursive function.

Let S be a codified set, then a subset X of S is called *recursively enumerable* if a partial recursive function $f : S \to N$ exists whose domain is X. It is possible to prove that X is recursively enumerable iff either X is empty or a recursive sequence $f : N \to S$ exists such that X is the codomain of f, that is

$$X = \{f(1), f(2), \ldots\}.$$

The following proposition shows that we can define the recursive enumerability in terms of limit. Observe that the Cartesian product of two sets with a coding is a set with a coding. This means that, in particular, the notion of recursive map h from $S \times N$ to $\{0,1\}$ is defined.

Proposition 2.1. *A subset X of S is recursively enumerable if and only if there exists a recursive map $h : S \times N \to \{0,1\}$, increasing with respect to the second variable, such that, for any $x \in S$,*

$$c_X(x) = lim_{n \to \infty} h(x,n). \tag{2.1}$$

Proof. Assume that X is recursively enumerable. Then, in the case $X = \emptyset$ equation (2.1) is satisfied by setting h equal to the function constantly equal to zero. In the case $X \neq \emptyset$, let $f : N \to S$ be a recursive function whose codomain is X and define h by setting, for every $x \in S$,

$$h(x,n) = \begin{cases} 1 & \text{if } x \in \{f(1), \ldots, f(n)\}, \\ 0 & \text{otherwise.} \end{cases}$$

Then, h is recursive, increasing with respect to the second variable. Moreover, (2.1) is satisfied.

Conversely, suppose that (2.1) is satisfied where h is a recursive function increasing with respect to n. Then the map

$$f(x) = \begin{cases} 1 & \text{if } h(x,n) > 0 \text{ for a suitable } n, \\ \text{undefined} & \text{otherwise} \end{cases}$$

is a partial recursive function whose domain is X. Thus, X is recursively enumerable. □

Let $U = [0,1]$ and denote by $\ddot{U}$ the set of rational numbers in U, i.e.,

$$\ddot{U} = \{x \in Q : 0 \le x \le 1\}.$$

Then Proposition 2.1 suggests to extend the definition of recursive enumerability as follows (see Biacino and Gerla [1987] and [1988]):

Definition 2.2. A fuzzy subset $s : S \to U$ of S is *recursively enumerable* if a recursive map $h : S \times N \to \ddot{U}$ exists such that, for every $x \in S$, $h(x,n)$ is increasing with respect to n and

$$s(x) = lim_{n\to\infty} h(x,n). \tag{2.2}$$

Observe that, since h is increasing with respect to n, (2.2) is equivalent to

$$s(x) = Sup\{h(x,n) : n \in N\}.$$

As usual, we assume that a coding for all the partial recursive functions in S is given and we denote by $\phi_i : S \to S$ the partial recursive function in S whose code number is i. We denote the domain of ϕ_i by W_i obtaining in this way a coding for the class of recursively enumerable subsets of S. The following proposition gives some characterizations of the recursive enumerability:

Proposition 2.3. *For every fuzzy subset s of S the following are equivalent:*

(a) *s is recursively enumerable.*

(b) *The set* $K(s) = \{(x,\lambda) \in S \times \ddot{U} : s(x) > \lambda\}$ *is recursively enumerable.*

(c) *A recursive map* $k : \ddot{U} \to N$ *exists such that, for every* $\lambda \in \ddot{U}$,

$$O(s,\lambda) = W_{k(\lambda)}.$$

(d) *A recursive map* $k : S \times N \to \ddot{U}$ *exists such that, for every* $x \in S$,

$$s(x) = Sup\{k(x,n) : n \in N\}. \tag{2.3}$$

Proof. (a) $\Rightarrow$ (b). Let $h(x,n)$ be a recursive map increasing with respect to n and such that $s(x) = lim_{n\to\infty} h(x,n)$. Define the function $g : S \times \ddot{U} \to N$ by

$$g(x,\lambda) = \begin{cases} 1 & \text{if } h(x,n) > \lambda \text{ for a suitable } n \in N, \\ \text{divergent} & \text{otherwise.} \end{cases}$$

Moreover, denote the domain of g by $Dom(g)$. Then, since

$$(x,\lambda) \in K(s) \Leftrightarrow s(x) > \lambda \Leftrightarrow lim_{n\to\infty} h(x,n) > \lambda$$
$$\Leftrightarrow \exists\, n \in N \text{ such that } h(x,n) > \lambda \Leftrightarrow (x,\lambda) \in Dom(g),$$

we have that $K(s) = Dom(g)$. Since g is a partial recursive function, this proves that $K(s)$ is recursively enumerable.

(b) $\Rightarrow$ (c). Let g be a partial recursive map whose domain is $K(s)$. Then, by the s-m-n-theorem, a recursive map k exists such that $\phi_{k(\lambda)}(x) = g(x,\lambda)$. It is evident that

$$x \in O(s,\lambda) \Leftrightarrow (x,\lambda) \in K(s) \Leftrightarrow g \text{ converges in } (x,\lambda)$$
$$\Leftrightarrow \phi_{k(\lambda)} \text{ converges in } x \Leftrightarrow x \in W_{k(\lambda)}.$$

(c) $\Rightarrow$ (d). Note that

$$s(x) = Sup\{\lambda \in \ddot{U} : x \in O(s,\lambda)\} = Sup\{\lambda \in \ddot{U} : x \in W_{k(\lambda)}\}.$$

Set

$$W(\lambda, j) = \{x \in S : \phi_{k(\lambda)} \text{ converges in } x \text{ in fewer than } j \text{ steps}\},$$

let $\pi : N \to N \times \ddot{U}$ be a recursive one-one map and

$$g(x,n) = \begin{cases} \lambda & \text{if } \pi(n) = (j,\lambda) \text{ and } x \in W(\lambda,j), \\ 0 & \text{otherwise.} \end{cases}$$

Then, $g(x,n)$ is recursive and $s(x) = Sup\{g(x,n) : n \in N\}$.

(d) $\Rightarrow$ (a). Assume that k is a recursive map such that $s(x) = Sup\{k(x,n) : n \in N\}$ and set $h(x,n) = k(x,1) \vee ... \vee k(x,n)$. Then h is a recursive map increasing with respect to n and $s(x) = lim_{n\to\infty} h(x,n)$. □

The characterization given by (b) shows that the notion of recursive enumerability fits the definition of fuzzy point proposed in Wong [1974] well. The characterization given by (c) shows that we can identify $\mathcal{F}_e(S)$ with the lattice of "effective" continuous co-chains of recursively enumerable subsets.

Also, we can give a definition of recursive enumerability in terms of finite fuzzy subsets. Let $\mathcal{F}_f(S)$ be the class of finite fuzzy subsets of S whose values belong to $\ddot{U}$. It is evident that a coding exists for such a set and, therefore, that the notion of recursively enumerable class of elements in $\mathcal{F}_f(S)$ is defined.

Proposition 2.4. *Given a fuzzy subset s the following are equivalent* :
(i) *s is recursively enumerable.*
(ii) *s is a union of a computable order-preserving sequence of elements in* $\mathcal{F}_f(S)$.
(iii) *s is a limit of a computable directed sequence of elements in* $\mathcal{F}_f(S)$.
(iv) *s is a union of a recursively enumerable class of subsets in* $\mathcal{F}_f(S)$.

Proof. (i) $\Rightarrow$ (ii). Let $k(x,n)$ be a computable map order-preserving with respect to n such that $s(x) = Sup\{k(x,n) : n \in N\}$. Let $g : N \to \mathcal{F}_f(S)$ be the function defined by setting, for any integer n, $g(n)(x) = k(x,n)$ if $x \leq n$ and $g(n)(x) = 0$ in the case $x>n$. Then g is computable and $(g(n))_{n\in N}$ is an order preserving sequence of elements in $\mathcal{F}_f(S)$ such that $s = \bigcup_{n\in N} g(n)$.
(ii) $\Rightarrow$ (iii) and (iii) $\Rightarrow$ (iv). Straightforward.
(iv) $\Rightarrow$ (i). Assume that a computable map $g : N \to \mathcal{F}_f(S)$ exists such that $s = \bigcup_{n\in N} g(n)$ and set $k(x,n) = g(n)(x)$. Then k is a computable map such that $s(x) = Sup\{k(x,n) : n \in N\}$ and this proves that s is recursively enumerable. □

The following proposition shows that the intersection and the union of two recursively enumerable fuzzy subsets is a recursively enumerable fuzzy subset.

Proposition 2.5. *The class* $\mathcal{F}_e(S)$ *of recursively enumerable fuzzy subsets of S is closed under finite unions and intersections. More specifically,* $\mathcal{F}_e(S)$ *is a sublattice of the lattice* $\mathcal{F}(S)$ *of all fuzzy subsets of S extending the lattice of the (classical) recursively enumerable subsets.*

Proof. Let s_1 and s_2 be two recursively enumerable fuzzy subsets and $h_1 : S \times N \to \ddot{U}$, $h_2 : S \times N \to \ddot{U}$ two computable functions, increasing with respect to n, such that

$$s_1(x) = lim_{n\to\infty} h_1(x,n) \quad \text{and} \quad s_2(x) = lim_{n\to\infty} h_2(x,n)$$

for every $x \in S$. Then it is evident that

$$(s_1 \cup s_2)(x) = lim_{n\to\infty} (h_1(x,n) \vee h_2(x,n))$$

and

$$(s_1 \cap s_2)(x) = lim_{n\to\infty} (h_1(x,n) \wedge h_2(x,n)).$$

Furthermore, both the maps $h_1 \vee h_2$ and $h_1 \wedge h_2$ are computable and increasing with respect to n. Then, the first part of the proposition is proven.

Finally, observe that from Proposition 2.1 it follows that any recursively enumerable subset has a characteristic function which is a recursively enumerable fuzzy subset. □

The closed cuts of a recursively enumerable fuzzy subset are not necessarily recursively enumerable. In fact, the following theorem holds:

Theorem 2.6. *A subset of S is a closed cut of a recursively enumerable fuzzy subset iff it belongs to the Π_2-level of the arithmetical hierarchy.*

Proof. Let s be a recursively enumerable fuzzy subset and assume that $s(x) = lim_{n\to\infty}h(x,n)$, where $h(x,n)$ is a recursive map increasing with respect to n. Then, for any $\lambda \in \ddot{U}$,

$$x \in \mathrm{C}(s,\lambda) \Leftrightarrow s(x) \geq \lambda \Leftrightarrow lim_{n\to\infty} h(x,n) \geq \lambda$$
$$\Leftrightarrow \forall k \exists m \text{ such that } h(x,m) \geq \lambda - 1/k.$$

This proves that $\mathrm{C}(s,\lambda)$ belongs to the Π_2-level of the arithmetical hierarchy.

Let X be a subset of S belonging to the Π_2-level of the arithmetical hierarchy. By using a result of Hájek (see Hájek [1998], Theorem 6.3.4), it is possible to give an example of a recursively enumerable fuzzy subset s with a closed cut $\mathrm{C}(s,\mu)$ which is Π_2-complete. Then X is one-one reducible to $\mathrm{C}(s,\mu)$ by a recursive map f: $S \to S$, i.e., $x \in X$ iff $s(f(x)) \geq \mu$. This means that X is the cut of the recursively enumerable fuzzy subset $s \circ f$. □

The next theorem shows that $\mathcal{F}_e(S)$ is not related to the canonical extension of the class of recursively enumerable subsets.

Theorem 2.7. *A fuzzy subset exists which is not recursively enumerable in spite of the recursive enumerability of all its open and closed cuts. Moreover, a recursively enumerable fuzzy subset exists whose closed cuts are not recursively enumerable.*

Proof. Let μ be a real number which cannot be obtained as a limit of an increasing effectively computable sequence of rational numbers. Then the fuzzy subset s^μ constantly equal to μ is not recursively enumerable. Since a cut of s^μ coincides either with $\varnothing$ or with S, all the cuts of s^μ are recursively enumerable. The second part of the proposition was proved in Theorem 2.6. □

3. DECIDABILITY AND FUZZY COMPUTABILITY

In classical theory the notion of recursively enumerable subset enables us to define several basic concepts. For example, we can define *decidable* a subset X of $\boldsymbol{N}$ such that both X and its complement $-X$ are recursively enumerable. Moreover, we can define *computable* a function which is a recursively enumerable subset of $\boldsymbol{N} \times \boldsymbol{N}$.

Likewise, the proposed notion of recursively enumerable fuzzy subset enables us to obtain the definitions of decidable fuzzy subset and computable fuzzy functions.

Definition 3.1. A fuzzy subset *s of S* is *recursively co-enumerable* if its complement $-s$ is recursively enumerable. We say that *s* is *decidable* if it is both recursively enumerable and recursively co-enumerable (see Biacino and Gerla [1989]).

Proposition 3.2. *A fuzzy set s is recursively co-enumerable if and only if a recursive map* $k : S \times N \to \ddot{U}$ *exists such that, for every* $x \in S$, $k(x,n)$ *is decreasing with respect to n and*

$$s(x) = lim_{n\to\infty} k(x,n).$$

Proof. Let $d : S \times N \to \ddot{U}$ be a recursive function such that $d(x,n)$ is increasing with respect to n and $-s(x) = lim_{n\to\infty}\, d(x,n)$. Then, by setting $k(x,n) = 1 - d(x,n)$,

$$s(x) = 1 - lim_{n\to\infty}\, d(x,n) = lim_{n\to\infty}\, 1 - d(x,n) = lim_{n\to\infty}\, k(x,n),$$

where $k(x,n)$ is decreasing with respect to n. In the same way we can prove the converse implication. □

The following theorem, whose proof is trivial, gives a characterization of the decidable fuzzy subsets.

Theorem 3.3. *A fuzzy set s is decidable iff for every* $x \in S$, $s(x)$ *is the limit of an effectively computable nested sequence of intervals, i.e., iff two recursive maps* $h : S \times N \to \ddot{U}$ *and* $k : S \times N \to \ddot{U}$ *exist such that, for any* $x \in S$,

- $h(x,n)$ *is increasing and* $k(x,n)$ *is decreasing with respect to n,*
- *for every* $n \in N$,

$$h(x,n) \le s(x) \le k(x,n),$$

- *and*

$$lim_{n\to\infty}\, h(x,n) = s(x) = lim_{n\to\infty}\, k(x,n).$$

We say that a recursive function $f : S \times N \to \ddot{U}$ is *recursively convergent to s* if, for any $x \in S$, $s(x) = lim_{n\to\infty} f(x,n)$ and a recursive function $e : S \times N \to N$ exists such that, for every $x \in S$ and $p \in N$,

$$|\, f(x,n) - f(x,m)\,| < 1/p, \text{ for any } n, m \ge e(x, p).$$

This notion enables us to obtain an interesting characterization of decidability.

Theorem 3.4. *A fuzzy subset s is decidable iff there exists a recursive function* $f : S \times N \to \ddot{U}$ *recursively convergent to s.*

Proof. Assume that s is decidable and let h and k be as in Theorem 3.3. We claim that, for every $n, m \in N$ and $x \in S$,

$$|\, h(x,n) - h(x,m)\,| \le k(x, n \wedge m) - h(x, n \wedge m).$$

Indeed, since

$$k(x, n \wedge m) \ge k(x,n) \ge h(x,n) \text{ and } -h(x,m) \le -h(x,n \wedge m),$$

it is, for $n \geq m$

$$| h(x,n) - h(x,m) | = h(x,n) - h(x,m) \leq k(x, n \wedge m) - h(x,m)$$
$$\leq k(x, n \wedge m) - h(x,n \wedge m),$$

and the same holds for $m \geq n$. Set

$$e(x,p) = Min\{j \in \boldsymbol{N} : k(x,j) - h(x,j) < 1/p\}.$$

The map $e(x,p)$ is recursive and

$$| h(x,n) - h(x,m) | \leq k(x, n \wedge m) - h(x,n \wedge m) < 1/p \text{ for every } m, n \geq e(x,p),$$

and this proves that the recursive function h is recursively convergent to s.

Conversely, assume that f is a recursive function recursively convergent to s by the recursive function e and set, for every $n \in \boldsymbol{N}$, $m_n = e(x,n)$. We have

$$f(x,m_n) - 1/n \leq s(x) \leq f(x,m_n) + 1/n.$$

If we set

$u(x,n) = Sup\{f(x,m_i) - 1/i : i = 1,...,n\}$ and $v(x,n) = Inf\{f(x,m_i) + 1/i : i = 1,...,n\}$, then u is recursive and increasing with respect to n, and v is recursive and decreasing with respect to n. Moreover, since

$$lim_{n\to\infty}(f(x,m_n) - 1/n) = s(x) = lim_{n\to\infty}(f(x,m_n) + 1/n),$$

it is

$$lim_{n\to\infty}u(x,n) = s(x) = lim_{n\to\infty}v(x,n).$$ □

From such a proposition it follows that a decidable fuzzy subset assumes only recursive real numbers as values. In particular, by recalling that, for any $\lambda \in U$, s^λ is the fuzzy subset constantly equal to λ,

$$s^\lambda \text{ is decidable } \Leftrightarrow \lambda \text{ is a recursive real number.}$$

Proposition 3.5. *The class $\mathcal{F}_d(S)$ of decidable fuzzy subsets is closed under finite unions, finite intersections and complements. Then, $\mathcal{F}_d(S)$ is a sublattice of the lattice $\mathcal{F}_e(S)$ of all recursively enumerable fuzzy sets extending the lattice of decidable sets.*

Proof. Let s be a decidable fuzzy set and let h and k be two functions as in Theorem 3.3. Then, by setting $h' = 1-k$ and $k' = 1-h$, we obtain two recursive sequences the first increasing the latter decreasing such that

$$\text{-}s(x) = lim_{n\to\infty} h'(x,n) = lim_{n\to\infty} k'(x,n).$$

This proves that $-s$ is decidable. Assume that s_1 and s_2 are decidable. Then, as s_1 and s_2 are recursively enumerable, $s_1 \cap s_2$ is recursively enumerable. Moreover, as $-s_1$ and $-s_2$ are recursively enumerable, the fuzzy subset $-(s_1 \cap s_2) = -s_1 \cup -s_2$ is recursively enumerable. Thus $s_1 \cap s_2$ is decidable. Finally, the fuzzy subset $s_1 \cup s_2$ is decidable since it is the complement of the decidable fuzzy subset $-s_1 \cap -s_2$. □

Also, the notion of recursive enumerability for fuzzy subsets enables us to define a notion of computability for fuzzy functions.

Definition 3.6. Let S_1 and S_2 be two sets which admit a coding. Then we say that a fuzzy function $f : S_1 \sim\!> S_2$ is *computable* or *partial recursive* if f is a recursively enumerable fuzzy subset of $S_1 \times S_2$.

The proof of the following proposition is trivial:

Proposition 3.7. *Let s be a fuzzy subset of S. Then the following claims are equivalent:*

i) *s is recursively enumerable,*
ii) *s is the domain of a partial recursive function,*
iii) *s is the codomain of a partial recursive function.*

4. ENUMERABILITY BY DISCRETE TOPOLOGY

In defining the notion of recursive enumerability we can also interpret the convergence in the equality $s(x) = lim_{n\to\infty} h(x,n)$ with respect to the discrete topology in U. In this way we get a different notion of recursive enumerability (see Biacino and Gerla [1987]).

Definition 4.1. A fuzzy subset $s : S \to U$ is *d-recursively enumerable* if a recursive map $h : S \times N \to \ddot{U}$ exists, increasing with respect to the second variable, such that, for any $x \in S$,

$$s(x) = lim_{n\to\infty} h(x,n) \tag{4.1}$$

where the limit is taken with respect to the discrete topology.

By definition, (4.1) is equivalent to saying that

$$\forall x\, \exists m\, \forall n\, (n \geq m \Rightarrow h(x,n) = s(x)). \tag{4.2}$$

Consequently, the truth-values assumed by a *d*-recursively enumerable fuzzy subset are rational numbers. So, in this section we will consider only fuzzy subsets with values in $\ddot{U}$.

Proposition 4.2. *Every d-recursively enumerable fuzzy subset is recursively enumerable while a recursively enumerable fuzzy subset* $s : S \to \ddot{U}$ *exists which is not d-recursively enumerable.*

Proof. The first part of the proposition is trivial. To prove the second part, let $k : S \times N \to \ddot{U}$ be the function defined by

$$k(x,n) = \begin{cases} n/(2n+1) & \text{if } \phi_{c(x)}(x) \text{ does not converge in less that } n \text{ steps,} \\ r/(2r+1) & \text{if } \phi_{c(x)}(x) \text{ converges in } r \text{ steps and } r \leq n \end{cases}$$

where $c : S \to N$ is a coding of S. Then, since k is recursive and increasing with respect to n, the fuzzy subset

$$s(x) = lim_{n\to\infty} k(x,n) = \begin{cases} 1/2 & \text{if } \phi_{c(x)}(x) \text{ diverges,} \\ r/(2r+1) & \text{if } \phi_{c(x)}(x) \text{ converges in } r \text{ steps} \end{cases}$$

is recursively enumerable. Suppose that s is d-recursively enumerable, and let $h(x,n)$ be a recursive and increasing map satisfying (4.2). Then

$$\{x \in S : \phi_{c(x)}(x) \text{ divergent}\} = \{x \in S : s(x) = 1/2\}$$
$$= \{x \in S : n \text{ exists such that } h(x,n) = 1/2\}.$$

Hence, the set $\{x \in S : \phi_{c(x)}(x) \text{ divergent}\}$ is recursively enumerable and this contradicts the recursive unsolvability of the halting problem. □

Proposition 4.3. *For every fuzzy subset $s : S \to \ddot{U}$ the following are equivalent :*
(a) *s is d-recursively enumerable.*
(b) *The set $E(s) = \{(x,\lambda) \in S \times \ddot{U} : s(x) \geq \lambda\}$ is recursively enumerable.*
(c) *A recursive map $h : \ddot{U} \to N$ exists such that, for any $\lambda \in \ddot{U}$,*

$$C(s,\lambda) = W_{h(\lambda)}.$$

(d) *A recursive map $h : S \times N \to \ddot{U}$ exists such that, for every $x \in S$,*

$$s(x) = Max\{h(x,n) : n \in N\}. \tag{4.3}$$

Proof. (a) $\Rightarrow$ (b). Let $h(x,n)$ be a recursive map increasing with respect to n such that $s(x) = Max\{h(x,n) : n \in N\}$ and define the partial recursive function $g : S \times \ddot{U} \to N$ by setting

$$g(x,\lambda) = \begin{cases} 1 & \text{if } h(x,n) \geq \lambda \text{ for a suitable } n \in N, \\ \text{divergent} & \text{otherwise.} \end{cases}$$

Then, since

$$(x,\lambda) \in E(s) \Leftrightarrow s(x) \geq \lambda \Leftrightarrow Max\{h(x,n) : n \in N\} \geq \lambda$$
$$\Leftrightarrow \exists\, n \in N \text{ such that } h(x,n) \geq \lambda \Leftrightarrow (x,\lambda) \in Dom(g),$$

we have that $E(s) = Dom(g)$. This proves that $E(s)$ is recursively enumerable.
(b) $\Rightarrow$ (c). Let g be a partial recursive map whose domain is $E(s)$ and, by the s-m-n-theorem, let h be a recursive map such that $\phi_{h(\lambda)}(x) = g(x,\lambda)$. It is evident that

$$x \in C(s,\lambda) \Leftrightarrow (x,\lambda) \in E(s) \Leftrightarrow g \text{ converges in } (x,\lambda)$$
$$\Leftrightarrow \phi_{h(\lambda)} \text{ converges in } x \Leftrightarrow x \in W_{h(\lambda)}.$$

(c) $\Rightarrow$ (d). By hypothesis,

$$s(x) = Max\{\lambda \in \ddot{U} : x \in C(s,\lambda)\} = Max\{\lambda \in \ddot{U} : x \in W_{h(\lambda)}\}.$$

Set

$$g(x,j,\lambda) = \begin{cases} \lambda & \text{if } \phi_{h(\lambda)} \text{ converges in } x \text{ in fewer than } j \text{ steps,} \\ 0 & \text{otherwise.} \end{cases}$$

Then $g(x,j,\lambda)$ is recursive and $s(x) = Max\{g(x,j,\lambda) : j \in N, \lambda \in \ddot{U}\}$. Let $\pi : N \to N \times \ddot{U}$ be any recursive one-one map and set $h(x,n) = g(x,\pi(n))$. Then h is total recursive and $s(x) = Max\{h(x,n) : n \in N\}$.
(d) ⇒ (a). Assume that h is a recursive map such that $s(x) = Max\{h(x,n) : n \in N\}$ and set $k(x,n) = h(x,1) \vee ... \vee h(x,n)$. Then k is a recursive map increasing with respect to n such that $s(x) = Max\{k(x,n) : n \in N\}$. □

The characterization given by (b) shows that the notion of *d*-recursive enumerability fits well the definition of fuzzy point which is proposed in Kerre [1980] and in Pu Pao-Mine and Liu Ying-Ming [1980]. Instead, recall that the notion of recursive enumerability fits the definition of fuzzy point proposed in Wong [1974] well. Also, observe that Proposition 4.2 entails the existence of a fuzzy subset s such that $K(s) = \{(x,\lambda) \in S \times \ddot{U} : s(x) > \lambda\}$ is recursively enumerable and $E(s) = \{(x,\lambda) \in S \times \ddot{U} : s(x) \geq \lambda\}$ is not recursively enumerable. The characterization given by (c) shows that we can identify the *d*-recursively enumerable fuzzy subsets with the "effective" continuous chains of recursively enumerable subsets.

Definition 4.4. We say that a fuzzy subset s is *d-recursively co-enumerable* if its complement $-s$ is *d*-recursively enumerable. We say that s is *d-decidable* if s is both *d*-recursively enumerable and *d*-recursively co-enumerable.

The following theorem shows that the notion of *d*-decidability is related to finite-steps computation processes and not necessarily to infinite (effective) approximation processes.

Theorem 4.5. *A fuzzy subset s is d-decidable iff s is a (classically) computable function from S to Ü.*

Proof. Assume that s is *d*-decidable and let h and k be two recursive functions such that

$$s(x) = Max\{h(x,n) : n \in N\} = Min\{k(x,n) : n \in N\}$$

where h is increasing and k is decreasing with respect to n. Then, to calculate $s(x)$, we have

- to generate the sequence $h(x,1)$, $k(x,1)$, $h(x,2)$, $k(x,2)$, ... until $h(x,i) \neq k(x,i)$
- to stop if i is the first integer such that $h(x,i) = k(x,i)$
- to give as output the number $h(x,i) = k(x,i)$.

This proves that s is a computable function from S to $\ddot{U}$.

Conversely, if s is computable, then also its complement $-s$ is computable. Then both s and $-s$ are *d*-recursively enumerable. □

The next proposition shows that decidable fuzzy sets whose degrees of membership belong to $\ddot{U}$ are not necessarily *d*-decidable.

Proposition 4.6. *A decidable but not d-decidable fuzzy set* $s : S \to \ddot{U}$ *with values in* $\ddot{U}$ *exists.*

Proof. Let $k : S \times N \to \ddot{U}$ be the map defined by setting

$$h(x,n) = \begin{cases} n/(n+1) & \text{if } \phi_{c(x)}(x) \text{ does not converge in fewer than } n \text{ steps,} \\ m/(m+1) & \text{if } \phi_{c(x)}(x) \text{ converges in } m \leq n \text{ steps.} \end{cases}$$

Then h is a computable function with rational values increasing with respect to n. As a consequence, the fuzzy subset s defined by

$$s(x) = lim_{n\to\infty} h(x,n) = \begin{cases} 1 & \text{if } \phi_{c(x)}(x) \text{ diverges,} \\ m/(m+1) & \text{if } \phi_{c(x)}(x) \text{ converges in } m \text{ steps.} \end{cases}$$

is recursively enumerable. Furthermore, s is decidable, indeed s is the limit of the decreasing, recursive function

$$k(x,n) = \begin{cases} 1 & \text{if } \phi_{c(x)}(x) \text{ does not converges in } n \text{ steps,} \\ m/(m+1) & \text{if } \phi_{c(x)}(x) \text{ converges in } m \text{ steps where } m \leq n. \end{cases}$$

Since $s(x) \neq 1$ if $\phi_{c(x)}(x)$ converges and $s(x) = 1$ if $\phi_{c(x)}(x)$ diverges, the map s cannot be computable. □

5. KLEENE HIERARCHY

In this section we extend the classical Kleene hierarchy to fuzzy relations in a domain S. As in the classical case, we define, for every $n \in N_0$, the classes Σ_n, Π_n and Δ_n of fuzzy relations by induction on n through the projection and co-projection operators. We define the *projection* of a $(p+1)$-ary fuzzy relation $r' : S^{p+1} \to U$ as the p-ary fuzzy relation $r : S^p \to U$ such that

$$r(x_1,\ldots,x_p) = Sup\{r'(x_1,\ldots,x_p,x_{p+1}) : x_{p+1} \in S\}.$$

Dually, we define the *co-projection* of r' as the fuzzy relation r' such that

$$r(x_1,\ldots,x_p) = Inf\{r'(x_1,\ldots,x_p,x_{p+1}) : x_{p+1} \in S\}.$$

Then we set

- $\Sigma_0 = \Pi_0$ equal to the class of the fuzzy relations which are computable functions with values in $\ddot{U}$,
- Σ_{n+1} equal to the class of the projections of fuzzy relations in Π_n,
- Π_{n+1} equal to the class of the co-projections of fuzzy relations in Π_n,
- $\Delta_n = \Sigma_n \cap \Pi_n$.

We say that a fuzzy relation r *belongs to the Kleene hierarchy* if there is an integer n such that $r \in \Sigma_n$. The basic results of classical hierarchy theory are extended by the following proposition whose proof can be found in Biacino and Gerla [1989]:

Proposition 5.1. *Let* $n \geq 1$, *then*:

(a) *The projection of an element of* Σ_n *belongs to* Σ_n.

(b) *The co-projection of an element of* Π_n *belongs to* Π_n.

(c) $\Sigma_n \cup \Pi_n \subseteq \Delta_{n+1}$ *but* $\Sigma_n \cup \Pi_n \neq \Delta_{n+1}$.

(d) Σ_n *and* Π_n *are closed under finite unions and intersections.*

(e) $\Sigma_n = \{-r : r \in \Pi_n\}$ *and* $\Pi_n = \{-r : r \in \Sigma_n\}$.

(f) Δ_n *is closed with respect to the complement, the finite unions and the finite intersections.*

Assume that $S = N$. Then the next proposition characterizes the classes Σ_n and Π_n in terms of iterated limits. To this aim, given a function $f : N^q \to U$, and an integer $r \leq q$, we define the function $lim^r f$ from N^{q-r} to U by setting

$$lim^1 f(n_1,...,n_{q-1}) = lim_{n\to\infty} f(n_1,...,n_{q-1},n)$$
$$lim^r f(n_1,...,n_{q-r}) = lim^1(lim^{r-1} f(n_1,...,n_q)).$$

Theorem 5.2. *For every* $n \geq 1$ *a fuzzy relation* r *in* N *belongs to* Σ_n (Π_n) *iff, for every* $x \in N^p$,

$$r(x) = lim^n\, h(x,y_1,...,y_n) \tag{5.1}$$

where $h : N^p \times N^n \to \ddot{U}$ *is computable and* $h(x,y_1,...,y_n)$ *is increasing (decreasing) with respect to* y_i *in the case* i *is odd and decreasing (increasing) otherwise.*

Proof. We proceed by induction on n. If $r \in \Sigma_1$, then r is the projection of an element r' of Σ_0, i.e., $r(x) = Sup\{r'(x,n) : n \in N\}$. This means that (5.1) holds with $h(x,n) = Sup\{r'(x,j) : j \leq n\}$. Likewise, if $r \in \Pi_1$, then r is the co-projection of an element r' of Σ_0 and r is the limit of the sequence $k(x,n) = Inf\{r'(x,j) : j \leq n\}$. Suppose $r \in \Sigma_n$, with $n \neq 1$. Then r is the projection of a fuzzy relation r' in Π_{n-1}, that is, $r(x) = Sup\{r'(x,y_1) : y_1 \in N\}$. By induction hypothesis, there exists a computable map $h' : N^p \times N \times N^{n-1} \to \ddot{U}$ such that $r'(x,y_1) = lim^{n-1}h'(x,y_1,y_2,...,y_n)$. Further, for every $i \geq 2$, the map $h'(x,y_1,y_2,...,y_n)$ is decreasing with respect to y_i if i is even, increasing otherwise. Set

$$h(x,y_1,y_2,...,y_n) = Sup\{h'(x,j,y_2,...,y_n) : j \leq y_1\}.$$

Then $h(x,y_1,y_2,...,y_n)$ is computable, increasing with respect to y_i is i is odd and decreasing if y_i is even. We have

$$\begin{aligned} lim^n h(x,y_1,y_2,...,y_n) &= lim^1(lim^{n-1} h(x,y_1,y_2,...,y_n)) \\ &= lim^1(lim^{n-1}(Sup\{h'(x,j,y_2,...,y_n) : j \leq y_1\})) \\ &= lim^1(Sup\{lim^{n-1} h'(x,j,y_2,...,y_n) : j \leq y_1\})) \\ &= lim^1(Sup\{r'(x,j) : j \leq y_1\}) \\ &= Sup\{r'(x,y_1) : y_1 \in N\} = r(x) \end{aligned}$$

where we used the fact that if $(\lambda_n)_{n\in N}$ and $(\mu_n)_{n\in N}$ are two sequences of elements of U both decreasing or increasing, then

$$lim_{n\to\infty}(\lambda_n \vee \mu_n) = (lim_{n\to\infty} \lambda_n) \vee (lim_{n\to\infty} \mu_n),$$
$$lim_{n\to\infty}(\lambda_n \wedge \mu_n) = (lim_{n\to\infty} \lambda_n) \wedge (lim_{n\to\infty} \mu_n).$$

We proceed similarly in the case $r \in \Pi_n$. The converse part of the proposition is obvious. □

The proof of the following proposition is trivial:

Proposition 5.3. *Σ_1 is the class of recursively enumerable fuzzy relations, Π_1 is the class of recursively co-enumerable fuzzy relations and Δ_1 is the class of decidable fuzzy relations.*

In order to examine the class Δ_2 in the above-defined hierarchy, we call a fuzzy subset s *limit decidable* if a recursive function $h : S \times N \to \ddot{U}$ exists such that $s(x) = lim_{n\to\infty}h(x,n)$ for every $x \in S$. We say that s is *by-frequency decidable* if a recursive function $k : S \times N \to \{0,1\}$ exists such that, for every $x \in S$,

$$s(x) = lim_{n\to\infty} (k(x,1)+...+k(x,n))/n.$$

Finally, we say that s is *with limit-decidable cuts* if a computable function $c : S \times \ddot{U} \times N \to \{0,1\}$ exists such that

$$s(x) > \lambda \;\Rightarrow\; lim_{n\to\infty} c(x,\lambda,n) = 1 \quad ; \quad s(x) < \lambda \Rightarrow lim_{n\to\infty} c(x,\lambda,n) = 0.$$

The following proposition, whose proof we omit, gives some characterization of the class Δ_2 (see Biacino and Gerla [1989]).

Proposition 5.4. *The following are equivalent:*
- (a) $s \in \Delta_2$.
- (b) *s is limit decidable.*
- (c) *s is by-frequency decidable.*
- (d) *s is with limit decidable cuts.*

Observe that Propositions 5.4 and 3.4 show that Δ_1 and Δ_2 are related to two different levels of effectiveness with regard to the notion of limit: the effective convergence of an effective sequence and the classical convergence of an effective sequence.

We conclude this section by noting that the Kleene hierarchy is useful in classifying the fuzzy relations definable within a quantificational multivalued logic.

Theorem 5.5. *Let $\mathcal{L}$ be a multivalued logic whose valuation structure is U and consider a model of $\mathcal{L}$ whose domain S is codified in some way and in which the relation symbols are interpreted by elements of Σ_0, i.e., by computable functions. Then each formula $\alpha(x_1,...,x_n)$ is interpreted by a fuzzy relation belonging to a suitable level of the hierarchy. Such a level depends on the number of alternations of the quantifiers in a prefix normal form for $\alpha(x_1,...,x_n)$.*

Proof. It is enough to observe that in any multivalued logic the existential and universal quantifiers are interpreted by the projection and co-projection operators, respectively. □

As an example, the interpretation of an existential formula, i.e., a formula like $\exists x_i\alpha(x_1,...,x_i,...,x_n)$ with α without quantifiers, is a recursively enumerable fuzzy relation. The interpretation of a universal formula is a co-enumerable fuzzy relation, and so on.

6. GÖDEL NUMBERING AND CHURCH THESIS

In the following ψ_1, ψ_2, ... denotes a Gödel numbering of the partial recursive functions from $S \times N$ to $\ddot{U}$.

Definition 6.1. We say that a recursive function $h : N \to N$ is a *Gödel numbering* for the class of the recursively enumerable fuzzy subsets if, for every $i \in N$,

(a) $\psi_{h(i)} : S \times N \to \ddot{U}$ is total and increasing with respect to the second variable,

(b) if s is any recursively enumerable fuzzy subset then, for any $x \in S$,

$$s(x) = lim_{n\to\infty}\ \psi_{h(i)}(x,n).$$

A Gödel numbering enables us to assign a code number to any recursively enumerable fuzzy subset. Indeed, given an index i, we denote by s_i the fuzzy subset defined by setting, for every $x \in S$,

$$s_i(x) = lim_{n\to\infty} \psi_{h(i)}(x,n). \tag{6.1}$$

Theorem 6.2. *A Gödel numbering $h : N \to N$ of the recursively enumerable fuzzy subsets exists such that if $\psi_i : S \times N \to \ddot{U}$ is total and increasing with respect to the second variable, then, for any $x \in S$,*

$$lim_{n\to\infty}\ \psi_i(x,n) = lim_{n\to\infty}\ \psi_{h(i)}(x,n).$$

Proof. Let $\pi : N^2 \to N$ be a coding of N^2 and ψ the function defined by

$$\psi(x,r,t,i) = \begin{cases} \psi_i(x,r) & \text{if } \psi_i(x,r) \text{ converges in fewer than } t \text{ steps,} \\ 0 & \text{otherwise.} \end{cases}$$

Moreover, set $\Psi^*(x,n,i) = Sup\{\psi(x,r,t,i) : \pi(r,t) \le n\}$. Then Ψ^* is recursive and increasing with respect to n. By the *s-m-n*-theorem a recursive map h exists such that $\psi_{h(i)}(x,n) = \Psi^*(x,n,i)$. In order to prove that

$$lim_{n\to\infty}\ \psi_i(x,n) = lim_{n\to\infty}\ \psi_{h(i)}(x,n),$$

where ψ_i is any total function increasing with respect to n, we must prove that

$$Sup\{\psi_i(x,r) : r \in N\} = Sup\{\psi^*(x,n,i) : n \in N\}. \tag{6.2}$$

Now, for every $r \in N$, let a_r be the number of the steps in which $\psi_i(x,r)$ converges and set $n = \pi(r,a_r)$. Then $\Psi^*(x,n,i) \ge \psi_i(x,r)$ and

$$Sup\{\Psi^*(x,n,i) \ge Sup\{\psi_i(x,r) : r \in N\}. \tag{6.3}$$

On the other hand, for every $n \in N$, either $\Psi^*(x,n,i) = 0$ or j, $t \in N$ exist such that $\pi(j,t) \le n$, $\psi_i(x,j)$ converges in fewer than t steps, and $\Psi^*(x,n,i) = \psi_i(x,r)$ for a suitable $r \in N$. At any rate, for every $n \in N$ there exists $r \in N$ such that $\Psi^*(x,n,i) \le \psi_i(x,r)$ and hence,

$$Sup\{\Psi^*(x,n,i) : n \in N\} \le Sup\{\psi_i(x,r) : r \in N\}. \tag{6.4}$$

From (6.3) and (6.4) we get (6.2). Since (b) is evident, the proof is complete. □

We can try to define a *Gödel numbering for the d-recursively enumerable fuzzy subsets* as a recursive function $k : N \to N$ such that, for every $i \in N$,

(a) $\psi_{k(i)} : S \times N \to \ddot{U}$ is total, increasing with respect to the second variable, and convergent with respect to the discrete topology,

(b) for every recursively enumerable fuzzy subset $s : S \to U$, an index $i \in N$ exists such that:

$$s(x) = lim_{n\to\infty}\ \psi_{k(i)}(x,n)$$

for any $x \in S$. In such a case it should be possible to represent the class of all the *d*-recursively enumerable fuzzy subsets by the sequence s_1, s_2, ... defined by setting, for every $i \in N$ and $x \in S$,

$$s_i(x) = lim_{n\to\infty}\psi_{k(i)}(x,n).$$

Unfortunately, this is not the case as the following theorem proves:

Theorem 6.3. *No Gödel numbering exists for the d-recursively enumerable fuzzy subsets.*

Proof. Assume that such a numbering k exists. Let $f : \ddot{U} \to \ddot{U}$ be an order-preserving recursive map without fixed points. As an example, we can set $f(x) = x^2/2 + 1/4$. Define the fuzzy subset s by setting

$$s(x) = lim_{n\to\infty}f(\psi_{k(x)}(x,n))$$

for any $x \in S$. It is evident that s is *d*-recursively enumerable and therefore that an index i exists such that $s = s_i$. Then,

$$f(lim_{n\to\infty}(\psi_{k(i)}(i,n))) = lim_{n\to\infty}f(\psi_{k(i)}(i,n)) = s(i) = s_i(i) = lim_{n\to\infty}\psi_{k(i)}(i,n).$$

This contradicts the hypothesis that no fixed point for f exists. □

Church's Thesis for fuzzy set theory. The claim that the notion of a partial recursive function (equivalently, the notion of a recursively enumerable subset) provides a satisfactory counterpart to the informal notion of computability is known as *Church Thesis* (see, e.g., Rogers [1967], pag. 20). Then, the following question arises:

does our definition of recursive enumerability give the correct formal counterpart of the intuition and experience of fuzzy people about fuzzy computability ?

We can call *Extended Church Thesis* the positive answer to this question. By this thesis we admit that a fuzzy algorithm is an infinitary step-by-step approximation process and therefore that fuzzy computability is related to recursive analysis rather than to recursive arithmetic. As in the classical case, it is not possible to give a proof for such a thesis. Nevertheless, several considerations and evidences exist in its favour. As an example, our definition is a proper extension of the definition proposed by E. S. Santos by its class of fuzzy Turing machines. Such an extension is necessary since if s is recursively enumerable in accordance with the definition of Santos, then the codomain of s is finite. Indeed, the truth values that s can assume belongs to the finite lattice generated by the set of truth values occurring in the machine. This is unsatisfactory since, for example, in the fuzzy logic proposed for the heap paradox the fuzzy set of theorems takes as values all the numbers of the sequence 0.9^n. Also, our definition extends the definition proposed in Harkleroad [1984] which appear to be too restrictive. As an example, the class of fuzzy subset which are recursively enumerable in accordance with such a definition is not closed with respect to finite unions and intersections (see Biacino and Gerla [1987]). This is an unsatisfactory departure from the classical theory of recursive enumerability.

Finally, observe that, while Theorem 6.2 gives a reason in favour of Extended Church Thesis, Theorem 6.3 shows that we cannot substitute recursive enumerability with discrete recursive enumerability in this thesis. In fact, Theorem 6.3 entails that no universal language or universal machine can exist for the whole class of the *d*-recursively enumerable fuzzy subsets.

7. REDUCIBILITY AND UNIVERSAL MACHINES

Given two subsets A and B of S, we say that A is *m-reducible* to B if a recursive map $d : S \to S$ exists such that $A = d^{-1}(B)$, that is

$$x \in A \Leftrightarrow d(x) \in B$$

(see Rogers [1967]). If d is a one-one mapping, then we say that A is *one-one reducible* to B. To extend these notions to the fuzzy subsets of S, we define the *inverse image* of a fuzzy subset s' of S via a map d as the fuzzy subset s such that,

$$s(x) = s'(d(x))$$

for every $x \in S$. In this case we write $s = d^{-1}(s')$.

Definition 7.1. We say that a fuzzy subset s is *m-reducible* to a fuzzy subset s', in brief $s \leq_m s'$, if s is the inverse image of s' via a recursive map $d : S \to S$. In the case in which d is one-one, we say that s is *one-one reducible* to s' and we write $s \leq_1 s'$.

The following propositions summarize the main properties of m-reducibility. The same properties are satisfied by one-one reducibility.

Proposition 7.2. *The m-reducibility is a preorder relation and*

$$s \leq_m s' \Rightarrow -s \leq_m -s'.$$

Moreover, if $s \leq_m s'$ and m is recursively enumerable (d-recursively enumerable, decidable, d-decidable), then s is recursively enumerable (d-recursively enumerable, decidable, d-decidable, respectively).

Proposition 7.3. *Assume that $s \leq_m s'$. Then :*

(a) *For every $\lambda \in U$,* $C(s,\lambda) \leq_m C(s',\lambda)$.
(b) *For every $\lambda \in U$,* $O(s,\lambda) \leq_m O(s',\lambda)$.
(c) *s' is crisp $\Rightarrow$ s is crisp.*
(d) *$s' \in \Sigma_n$, (Π_n or Δ_n) $\Rightarrow$ $s \in \Sigma_n$ (Π_n or Δ_n, respectively).*

Proof. In order to prove (a) note that

$$x \in C(s,\lambda) \Leftrightarrow s(x) \geq \lambda \Leftrightarrow s'(d(x)) \geq \lambda \Leftrightarrow d(x) \in C(s',\lambda).$$

One demonstrates (b) in the same way. Property (c) is evident and (d) is a consequence of Proposition 5.2. □

In Rogers [1967] a subset K of S is called 1-1-*complete* if K is recursively enumerable and every recursively enumerable subset of S is one-one reducible to K. The existence of a complete subset is equivalent, in a sense, to the existence of a universal machine or to the existence of a universal programming language. The extension of such a notion to the fuzzy subsets is obvious.

Definition 7.4. We say that a fuzzy subset k of S is 1-1-*complete* if :
- k is recursively enumerable
- every recursively enumerable fuzzy subset is one-one reducible to k.

In other words, k is 1-1-complete if k is a greatest element of Σ_1 with respect to the relation $\leq_1$. The characteristic function c_K of a (classically) 1-1-complete subset K is not a 1-1-complete fuzzy subset. Indeed, it is obvious that only the crisp recursively enumerable fuzzy sets are one-one reducible to c_K. In the following we call *level set* of a fuzzy subset s the set $L(s,\lambda) = \{x \in S : s(x) = \lambda\}$.

Proposition 7.5. *Let k be an* 1-1-*complete fuzzy subset and* $\lambda \in \ddot{U}-\{0\}$, *then* :
(a) *The level set* $L(k,\lambda)$ *is a* 1-1-*complete set.*
(b) *The closed cut* $C(k,\lambda)$, $\lambda \neq 0$, *is a* 1-1-*complete set.*
(c) *The open cut* $O(k,\lambda)$, $\lambda \neq 1$, *is a* 1-1-*complete set.*

Proof. In order to prove (a), let W be any recursively enumerable subset of S and define s by setting $s(x) = \lambda$ if $x \in W$ and $s(x) = 0$ if $x \notin W$. Then s is recursively enumerable and therefore $s \leq_1 k$, i.e., a recursive one-one map d exists such that $s(x) = k(d(x))$. Then

$$x \in W \Leftrightarrow s(x) = \lambda \Leftrightarrow k(d(x)) = \lambda \Leftrightarrow d(x) \in L(k,\lambda).$$

In order to prove (b), observe that, since the characteristic function c_W of W is a recursively enumerable fuzzy set, we have $c_W \leq_1 k$ via a suitable map d. Hence,

$$x \in W \Leftrightarrow c_W(x) = 1 \Leftrightarrow k(d(x)) = 1 \Leftrightarrow d(x) \in C(k,\lambda).$$

One demonstrates (c) in the same way. □

In the following we denote by $s_1, s_2, \ldots$ the Gödel numbering of all the recursively enumerable fuzzy subsets of S obtained by Theorem 6.2. Moreover, we codify the elements of $S \times N$ by the elements of S; that is, we consider a recursive one-one map $\pi : S \times N \rightarrow S$ and two recursive maps $\pi' : S \rightarrow S$ and $\pi'' : S \rightarrow N$ in such a way that $\pi'(\pi(x,i)) = x$ and $\pi''(\pi(x,i)) = i$ for all $x \in S$.

Theorem 7.6. *Let* k_0 *be the fuzzy set defined by setting, for every* $x \in S$,

$$k_0(x) = s_{\pi''(x)}(\pi'(x)). \tag{7.1}$$

Then k_0 *is a* 1-1-*complete fuzzy subset of S.*

Proof. By definition, we have

$$k_0(x) = lim_{n\to\infty}\psi_{h(\pi''(x))}(\pi'(x),n)$$

and the function $\psi(x,n) = \psi_{h(\pi''(x))}(\pi'(x),n)$ is recursive and increasing with respect to n. It follows that $k_0 \in \Sigma_1$. Moreover, let $s_i \in \Sigma_1$ and set $d(x) = \pi(x,i)$. Then, since $d : S \to S$ is recursive and injective and, for all $x \in S$,

$$k_0(d(x)) = lim_{n\to\infty}\psi_{h(i)}(x,n) = s_i(x),$$

we have that $s_i \leq_1 k_0$. □

Theorem 7.7. *Let k be the fuzzy subset of S defined by setting, for every $x \in S$*

$$k(x) = s_{c(x)}(x), \tag{7.2}$$

then k is 1-1*-complete.*

Proof. It is evident that $k \in \Sigma_1$. In order to prove that k is 1-1-complete we prove that $k_0 \leq_1 k$. To simplify the proof we assume that $S = N$ and therefore that $c(x) = x$. Let $g : N^3 \to N$ be the recursive function defined by setting

$$g(x,y,n) = \psi_{h(\pi''(x))}(\pi'(x),n).$$

Note that such a function does not depend on the variable y. Also, by the *s-m-n*-theorem there is a recursive one-one map f such that

$$\psi_{f(x)}(y,n) = g(x,y,n). \tag{7.3}$$

Since $\psi_{f(x)}(y,n)$ is increasing with respect to n,

$$s_{f(x)}(y) = lim_{n\to\infty}\psi_{h(f(x))}(y,n) = lim_{n\to\infty}\psi_{f(x)}(y,n).$$

Since $s_{f(x)}$ is a constant map,

$$k_0(x) = s_{\pi''(x)}(\pi'(x)) = lim_{n\to\infty}\psi_{h(\pi''(x))}(\pi'(x),n) = lim_{n\to\infty}g(x,y,n)$$
$$= lim_{n\to\infty}\psi_{f(x)}(y,n) = s_{f(x)}(y) = s_{f(x)}(f(x)) = k(f(x)).$$

This proves that $k_0 \leq_1 k$ via f. □

Both Theorems 7.6 and 7.7 suggest that a universal machine and a universal programming language for the class of recursively enumerable fuzzy subsets can exist. It is an interesting open question to show concrete examples of such a machine or such a language.

8. EFFECTIVE ABSTRACT FUZZY LOGIC

It is natural to require that the deduction operator of a fuzzy logic satisfies some kind of "*computability*" property besides continuity. To this purpose, we start from the definition of enumeration operator given in Rogers [1967] for crisp operators. We expose such a definition in terms of sequents (see Sections 7, 8, 9 in Chapter 7). First, notice that, since $\mathcal{F}$ is codified, both the class of finite subsets $\mathcal{P}_f(\mathcal{F})$ of $\mathcal{F}$ and the class $SEQ_f = \mathcal{P}_f(\mathcal{F})\times\mathcal{F}$ of finite sequents can be codified. Consequently, the notion of recursively enumerable subset of SEQ_f is defined.

Definition 8.1. A consequence relation $\vdash$ is *recursively enumerable* if $\vdash$ is the compact extension of a recursively enumerable consequence relation $W \subseteq SEQ_f$.

An operator $H : \mathcal{P}(\mathbb{F}) \to \mathcal{P}(\mathbb{F})$ is an *enumeration operator* if H is the operator associated with a recursively enumerable consequence relation.

In other words, $H : \mathcal{P}(\mathbb{F}) \to \mathcal{P}(\mathbb{F})$ is an enumeration operator if a recursively enumerable subset W of SEQ_f exists such that, for any $X \in \mathcal{P}(\mathbb{F})$,

$$H(X) = \{x \in \mathbb{F} : X_f \text{ exists such that } (X_f,x) \in W \text{ and } X_f \subseteq X\}. \quad (8.1)$$

It is easy to verify that an enumeration operator H is compact and that

$$X \text{ recursively enumerable} \Rightarrow H(X) \text{ recursively enumerable.}$$

Proposition 8.2. *An operator $H : \mathcal{P}(\mathbb{F}) \to \mathcal{P}(\mathbb{F})$ is an enumeration operator iff H is compact and*

$$W_H = \{(X_f,x) \in SEQ_f : x \in H(X_f)\}$$

is a recursively enumerable relation.

Proof. Assume that H is an enumeration operator and let W be as in Definition 8.1. Then, for any $X_f \in \mathcal{P}_f(\mathbb{F})$,

$$H(X_f) = \{x \in \mathbb{F} : X'_f \text{ exists such that } (X'_f,x) \in W \text{ and } X'_f \subseteq X_f\}.$$

Let g be a computable function whose codomain is W. Then, since

$$(X_f,x) \in W_H \Leftrightarrow n \in N \text{ exists such that } g(n) = (X'_f,x) \in W \text{ and } X'_f \subseteq X_f,$$

W_H is recursively enumerable. The converse implication is self-evident. □

To extend Definition 8.1 to the fuzzy operators, we set $\underline{SEQ}_f = \mathcal{F}_f(\mathbb{F}) \times \mathbb{F}$ and we call *finite fuzzy sequents* the elements of $\underline{SEQ}_f$. Moreover, in accordance with the notations in Chapter 7, given any fuzzy set $w : \underline{SEQ}_f \to U$ of sequents and $(X,\alpha) \in \underline{SEQ}_f$, we write $w(X \vdash \alpha)$ to denote the value $w(X,\alpha)$. Since $\underline{SEQ}_f$ can be codified, the notion of recursively enumerable fuzzy subset of $\underline{SEQ}_f$ is well defined.

Definition 8.3. We say that a fuzzy operator $H : \mathcal{F}(\mathbb{F}) \to \mathcal{F}(\mathbb{F})$ is an *enumeration fuzzy operator* if a recursively enumerable fuzzy subset $w : \underline{SEQ}_f \to U$ exists such that

$$H(s)(x) = Sup\{w(s_f \vdash x) : s_f \ll s\}. \quad (8.2)$$

Obviously, H is an enumeration fuzzy operator if a recursive map $k : \underline{SEQ}_f \times N \to \ddot{U}$ exists such that, for any $x \in \mathbb{F}$ and $s \in \mathcal{F}(\mathbb{F})$,

$$H(s)(x) = Sup\{k(s_f, x, m) : m \in N \text{ and } s_f \ll s\}.$$

Theorem 8.4. *Any enumeration fuzzy operator $H : \mathcal{F}(\mathbb{F}) \to \mathcal{F}(\mathbb{F})$ is continuous. Moreover,*

$$s \text{ recursively enumerable} \Rightarrow H(s) \text{ recursively enumerable.}$$

Proof. Let $(s_i)_{i\in I}$ be a directed family of fuzzy subsets. In order to prove that $H(\cup_{i\in I} s_i) = \cup_{i\in I} H(s_i)$, recal that, if s_f is an element in $\mathcal{F}_f(S)$ such that $s_f \ll \cup_{i\in I} s_i$ then $i \in I$ exists such that $s_f \subseteq s_i$. Then,

$$H(\cup_{i\in I} s_i)(x) = Sup\{w(s_f \vdash x) : s_f \ll \cup_{i\in I} s_i\} = Sup\{w(s_f \vdash x) : s_f \ll s_i, i \in I\}$$

$= Sup_{i \in I} H(s_i)(x)$.

Assume that s is a recursively enumerable fuzzy subset and therefore that $s(x) = lim_{n \to \infty} h(x,n)$, where $h : \mathbb{F} \times N \to \ddot{U}$ is a recursive map increasing with respect to the second variable. Moreover, let $k : \underline{SEQ_f} \times N \to \ddot{U}$ a recursive map such that

$$w(s_f \vdash x) = Sup_{m \in N} k(s_f, x, m)$$

for any $x \in \mathbb{F}$ and $s_f \in \mathcal{F}_f(\mathbb{F})$. For any $n \in N$, let h_n be the fuzzy subset of $\mathbb{F}$ defined by setting $h_n(x) = h(x,n)$ for any $x \in \mathbb{F}$. Then, since H is continuous and $(h_n)_{n \in N}$ is a directed family, we have

$$H(s)(x) = H(\cup_{n \in N} h_n)(x) = Sup_{n \in N} H(h_n)(x).$$

By observing that $H(h_n)(x) = Sup\{w(s_f \vdash x) : s_f \ll h_n\}$,we get

$$H(s)(x) = Sup\{k(s_f, x, m) : s_f \ll h_n \; n, m \in N\}.$$

Define r by setting $r(x,i,n,m) = k(s_f, x, m)$ if i is the code number of s_f and $s_f \ll h_n$ and otherwise, $r(x,i,n,m) = 0$. Then r is a recursive map and

$$H(s)(x) = Sup_{i \in N} Sup_{n \in N} Sup_{m \in N} r(x,i,n,m).$$

This demonstrates that $H(s)$ is recursively enumerable. □

Theorem 8.5. *Let $H : \mathcal{F}(\mathbb{F}) \to \mathcal{F}(\mathbb{F})$ be a fuzzy operator and define the fuzzy relation $w_H : \underline{SEQ_f} \to U$ by setting, for any $x \in \mathbb{F}$ and $s_f \in \mathcal{F}_f(\mathbb{F})$,*

$$w_H(s_f \vdash x) = H(s_f)(x). \tag{8.3}$$

Then H is an enumeration operator iff H is continuous and w_H is a recursively enumerable fuzzy relation.

Proof. Let H be an enumeration operator and let $k : \underline{SEQ_f} \times N \to \ddot{U}$ be a recursive map such that

$$H(s)(x) = Sup\{k(s_f, x, m) : m \in N \text{ and } s_f \ll s\}$$

for any $x \in \mathbb{F}$ and $s \in \mathcal{F}(\mathbb{F})$. Then

$$w_H(s_f \vdash x) = H(s_f)(x) = Sup\{k(s'_f, x, m) : s'_f \ll s_f,\; m \in N\}.$$

Define the function h by setting $h(s'_f, x, s_f, m) = k(s'_f, x, m)$ if $s'_f \ll s_f$ and otherwise, $h(s'_f, x, s_f, m) = 0$. Then

$$w_H(s_f \vdash x) = Sup\{h(s'_f, x, s_f, m) : s'_f \in \mathcal{F}_f(\mathbb{F}) \text{ and } m \in N\},$$

and this proves that w_H is recursively enumerable.

Conversely, if w_H is recursively enumerable and H is continuous, then

$$H(s)(x) = Sup\{H(s_f)(x) : s_f \ll s\} = Sup\{w(s_f \vdash x) : s_f \ll s\}.$$

This proves that H is an enumeration operator. □

We reach the main definition in this section.

Definition 8.6. We call *effective abstract fuzzy logic* an abstract fuzzy logic $(F, \mathcal{D})$ such that $\mathcal{D}$ is an enumeration fuzzy operator.

9. FUZZY LOGIC = ENUMERATION FUZZY CLOSURE OPERATOR

It is natural to require some kind of "*effectiveness*" to a fuzzy deduction apparatus and therefore to assume that the fuzzy subset of logical axioms is recursively enumerable and that the fuzzy inference rules are "computable". Also, as showed in Biacino and Gerla [2000]a, in the case of infinite inference rules we have to require that the algorithms for these rules are given in a *uniform* way. So, we propose the following definition. We assume that:

- r'_1, r'_2, ... is an effective coding of all the partial recursive functions from a Cartesian product $\mathbb{F}^n$ of $\mathbb{F}$ to $\mathbb{F}$ (where n varies in $\mathbb{N}$),
- d_1, d_2, ... is an effective coding of all the partial recursive functions from a Cartesian product $\ddot{U}^n$ of $\ddot{U}$ to $\ddot{U}$ (where n varies in $\mathbb{N}$).

Moreover, we set

$$r''_i(x_1,...,x_n) = Sup\{d_i(\lambda_1,...,\lambda_n) : \lambda_1 < x_1,..., \lambda_n < x_n, \text{ and } (\lambda_1,...,\lambda_n) \in Dom(d_i)\}. \quad (9.1)$$

Then each r''_i is a total function satisfying the continuity condition. All the usual triangular norms can be obtained in such a way, i.e., extending by (9.1) a recursive operation defined on the rational numbers.

Definition 9.1. A fuzzy H-system $(a, \mathbb{R})$ is *effective* provided that :

(a) two recursive maps $h : \mathbb{N} \to \mathbb{N}$ and $k : \mathbb{N} \to \mathbb{N}$ exist such that $d_{k(i)} : \ddot{U}^n \to \ddot{U}$ is a total function satisfying the continuity condition and

$$\mathbb{R} = \{(r'_{h(i)}, r''_{k(i)}) : i \in \mathbb{N}\},$$

(b) the fuzzy set a of logical axioms is recursively enumerable.

Observe that each $r''_{k(i)}$ is an extension of $d_{k(i)}$ and therefore $r''_i(\lambda_1,...,\lambda_n)$ is a rational number whenever $\lambda_1,...,\lambda_n$ are rational numbers.

The constructive point of view imposes a more precise definition of a proof. Indeed, recall that the justification of a formula α_j in a proof $\alpha_1,...,\alpha_n$ is one of the following claims:

i) *α_j is a logical axiom*

ii) *α_j is a hypothesis*

iii) *α_j is obtained by the n-ary rule $r_{h(i)}$ applied to the formulas $\alpha_{s(1)},...\alpha_{s(n)}$*.

If we represent the cases i), ii) and iii) by *la*, *hy* and $(i, s(1),...,s(n))$, respectively, then a justification is an element in the set

$$\{la, hy\} \cup \{(i, s(1),...,s(n)) : r_{h(i)} \text{ is } n\text{-ary}\},$$

and therefore we can assign a code number to any justification. In accordance, a *proof* is a sequence $<\alpha_1,i_1>,...,<\alpha_n,i_n>$ of elements in $\mathbb{F} \times \mathbb{N}$ such that, for $j = 1,...,n$, if i_j is the code number of a justification like $(i,s(1),...,s(n))$, then $s(1)<j,..., s(n)<j$ and $\alpha_j = r'_{h(i)}(\alpha_{s(1)},...,\alpha_{s(n)})$. It is evident that a partial recursive function exists such that, given any sequence $<\alpha_1,i_1>,...,<\alpha_n,i_n>$, the function converges if such a sequence is a proof, it diverges otherwise. Then the set of proofs is recursively enumerable. If the domain of each rule $r'_{h(i)}$ is recursive and a is decidable, then the set of proofs is decidable.

The following theorem is the constructive counterpart of Theorem 2.6 in Chapter 4. It shows that the theory of effective abstract deduction systems coincides with the theory of effective fuzzy Hilbert systems, i.e.,

effective approximate reasoning = theory of enumeration fuzzy closure operators.

Theorem 9.2. *The deduction operator $\mathcal{D}$ of an effective fuzzy Hilbert system $\mathcal{S}$ is an enumeration fuzzy closure operator. Conversely, let $\mathcal{D}$ be an enumeration fuzzy closure operator. Then an effective fuzzy Hilbert system exists whose deduction operator coincides with $\mathcal{D}$.*

Proof. Let $\mathcal{D}$ be the deduction operator of an effective Hilbert system $(a, \mathbb{R})$ and define $w : \mathbb{F} \times \mathcal{F}_f(\mathbb{F}) \to U$ by setting $w(s_f \vdash x) = \mathcal{D}(s_f)(x)$ for any $x \in \mathbb{F}$ and $s_f \in \mathcal{F}_f(\mathbb{F})$. Then, because of the continuity of $\mathcal{D}$, we have only to prove that w is recursively enumerable. Indeed, let $a' : \mathbb{F} \times N \to \ddot{U}$ be a computable function such that, for every formula x,

$$a(x) = Sup\{a'(x, n) : n \in N\}, \qquad (9.2)$$

and, for every $n \in N$, define the fuzzy subset a_n by setting $a_n(x) = a'(x,n)$. Moreover, given a proof π, let $Val(\pi, s_f, n)$ be the valuation of π in the fuzzy H-system $(a_n, \mathbb{R})$ obtained by assuming a_n instead of a as a fuzzy set of logical axioms. Since the values of s_f and a_n are rational numbers, $Val(\pi, s_f, n)$ is a rational number. We will prove that

$$Val(\pi,s_f) = Sup\{Val(\pi, s_f, n) : n \in N\} \qquad (9.3)$$

by induction on the length $l(\pi)$ of $\pi = \alpha_1, \alpha_2, ..., \alpha_m$. In fact, in the case $l(\pi) = 1$, and, more generally, in the case that the last formula α in π is assumed either as a hypothesis or as a logical axiom, (9.3) is evident. Suppose that α is obtained by an inference rule r, namely that $\alpha = r'(\alpha_{s(1)},...,\alpha_{s(p)})$; then

$$\begin{aligned} Val(\pi,s_f) &= r''(Val(\pi_{s(1)},s_f),...,Val(\pi_{s(p)},s_f)) \\ &= r''(Sup\{Val(\pi_{s(1)}, s_f, n) : n \in N\},...,Sup\{Val(\pi_{s(p)}, s_f, n) : n \in N\}) \\ &= Sup\{r''(Val(\pi_{s(1)}, s_f, n_1),...,Val(\pi_{s(p)}, s_f, n_p) : n_1,...,n_p \in N\} \\ &= Sup\{r''(Val(\pi_{s(1)}, s_f, n),...,Val(\pi_{s(p)}, s_f, n) : n \in N\} \\ &= Sup\{Val(\pi, s_f, n) : n \in N\}, \end{aligned}$$

where we used the inductive hypothesis, the fact that r'' preserves the joins and the fact that the quantities $Val(\pi_{s(j)}, s_f, n)$ are increasing with respect to n. From (9.3) it follows that

$$\begin{aligned} \mathcal{D}(s_f)(\alpha) &= Sup\{Val(\pi,s_f) : \pi \text{ is a proof of } \alpha\} \\ &= Sup\{Val(\pi, s_f, n) : \pi \text{ is a proof of } \alpha \text{ and } n \in N\}. \end{aligned}$$

Let $\pi_1, \pi_2, ...$ an effective enumeration of all the proofs and define the function $h : \mathcal{F}_f(\mathbb{F}) \times \mathbb{F} \times N \times N \to \ddot{U}$ by setting $h(s_f, \alpha, i, n) = Val(\pi_i, s_f, n)$ if π_i is a proof of α and otherwise, $h(s_f,\alpha,i,n) = 0$. Then h is a computable map and

$$\mathcal{D}(s_f)(\alpha) = Sup\{h(s_f,\alpha,i,n) : i \in N \text{ and } n \in N\}.$$

This proves that w is recursively enumerable and therefore that $\mathcal{D}$ is an enumeration fuzzy operator.

Let $\mathcal{D}$ be an enumeration fuzzy closure operator and let h: $\mathcal{F}_f(\mathbb{F}) \times \mathbb{F} \times N \to \ddot{U}$ be a recursive map such that, for any $s_f \in \mathcal{F}_f(\mathbb{F})$ and $x \in \mathbb{F}$,

$$\mathcal{D}(s_f)(x) = Sup\{h(s_f,x,n) : n \in N\}.$$

To define a suitable Hilbert system, we associate with any $\alpha \in \mathbb{F}$, $m \in N$ and $s_f \in \mathcal{F}_f(\mathbb{F})$, $s_f \neq \varnothing$, the fuzzy rule (r',r'') defined as follows. Let $\alpha_1,...,\alpha_n$ be the formulas in $Supp(s_f)$, then we set

$$r'(x_1,...,x_n) = \begin{cases} \alpha & \text{if } x_1 = \alpha_1,..., x_n = \alpha_n, \\ \text{undefined} & \text{otherwise.} \end{cases}$$

Also, define $d : \ddot{U}^n \to \ddot{U}$ by setting

$$d(\lambda_1,...,\lambda_n) = \begin{cases} h(s_f,\alpha,m) & \text{if } \lambda_1 > s_f(\alpha_1), ..., \lambda_n > s_f(\alpha_n), \\ 0 & \text{otherwise.} \end{cases}$$

Such a map satisfies the continuity condition and therefore the map r'' defined by (9.1) is an extension of d. Also, both r' and d are partial recursive functions whose algorithms depend uniformly on α, s_f, m. Consequently, two recursive functions $h : \boldsymbol{N} \to \boldsymbol{N}$ and $k : \boldsymbol{N} \to \boldsymbol{N}$ exist such that $r'_{h(i)} = r'$ and $d_{k(i)} = d$ where i is the code number of (α, s_f, m). We indicate by $\mathcal{S}$ the effective Hilbert fuzzy system whose fuzzy set of logical axioms is $\mathcal{D}(\emptyset)$ and such that $\mathbb{R} = \{(r'_{h(i)},r''_{k(i)}) : i \in \boldsymbol{N}\}$. To prove that $\mathcal{D}$ is the deduction operator of $\mathcal{S}$, we prove that a fuzzy set of formulas τ is a theory of $\mathcal{S}$ iff τ is a fixed point of $\mathcal{D}$, i.e., by Theorem 6.3 in Chapter 2, $\tau \supseteq \mathcal{D}(s_f)$ for any $s_f \in \mathcal{F}_f(\mathbb{F})$ such that $s_f \ll \tau$. Indeed, let τ be a theory. Then in the case $s_f = \emptyset$ we have that $\tau \supseteq \mathcal{D}(s_f)$ by hypothesis. If $s_f \neq \emptyset$, let $\alpha_1,...,\alpha_n$ be the elements in $Supp(s_f)$, $m \in N$, $\alpha \in \mathbb{F}$ and let (r',r'') be the inference rule associated with α, s_f and m. Then,

$$\tau(\alpha) = \tau(r'(\alpha_1,...,\alpha_n)) \geq r''(\tau(\alpha_1),..., \tau(\alpha_n))$$
$$= Sup\{d(\lambda_1,...,\lambda_n) : \lambda_1<\tau(\alpha_1),...,\lambda_n<\tau(\alpha_n)\}.$$

Since $\lambda_1, ...,\lambda_2$ exist such that $\tau(\alpha_1) > \lambda_1 > s_f(\alpha_1)$, ..., $\tau(\alpha_n) > \lambda_n > s_f(\alpha_n)$, we have

$$\tau(\alpha) \geq d(\lambda_1,...,\lambda_n) = h(s_f,\alpha,m).$$

Consequently, $\tau(\alpha) \geq Sup\{h(s_f,\alpha,n) : n\in N\} = \mathcal{D}(s_f)(\alpha)$ and $\tau \supseteq \mathcal{D}(s_f)$.

Let τ be a fixed point of $\mathcal{D}$, then $\tau = \mathcal{D}(\tau) \supseteq \mathcal{D}(\emptyset)$. Moreover, let (r',r'') be any rule and assume that (r',r'') is defined by $s_f \neq \emptyset$, $m\in N$ and $\alpha \in \mathbb{F}$. We claim that

$$\tau(r'(\alpha_1,...,\alpha_n)) \geq d(\lambda_1,...,\lambda_n)$$

for any $\lambda_1< \tau(\alpha_1),..., \lambda_n< \tau(\alpha_n)$ and therefore that

$$\tau(r'(\alpha_1,..., \alpha_n)) \geq r''(\tau(\alpha_1),..., \tau(\alpha_n)),$$

where $\alpha_1,...,\alpha_n$ are the elements in $Supp(s_f)$. Indeed, if $d(\lambda_1,...,\lambda_n) \neq 0$ then $\lambda_1 > s_f(\alpha_1),...,\lambda_n > s_f(\alpha_n)$. Consequently,

$$\tau(r'(\alpha_1,..., \alpha_n)) = \tau(\alpha) = \mathcal{D}(\tau)(\alpha) \geq \mathcal{D}(s_f)(\alpha) \geq h(s_f,\alpha,m) = d(\lambda_1,...,\lambda_n).$$

This means that τ is a theory of $\mathcal{S}$. □

We call *axiomatizable* a fuzzy theory admitting a decidable fuzzy subset of axioms. The next theorem extends a basic feature of classical logic to fuzzy logic.

Theorem 9.3. *Consider an effective fuzzy Hilbert logic. Then any axiomatizable theory is recursively enumerable. If the logic is with negation, and* v *is a decidable fuzzy set of axioms, then* $\mathcal{D}(v)$ *is recursively enumerable and* $\mathcal{D}(v)^{\perp}$ *is recursively co-enumerable. Moreover, any axiomatizable and complete theory is decidable.*

Proof. Theorem 9.2 proves that any axiomatizable theory is recursively enumerable. Assume that the fuzzy logic under consideration is with negation and that v is decidable. Then, a recursive map $h : F \times N \to \ddot{U}$ exists such that h is increasing with respect to the second variable and $\mathcal{D}(v)(\alpha) = lim_{n\to\infty} h(\alpha,n)$ for any formula α. As a consequence,

$$\mathcal{D}(v)^{\perp}(\alpha) = 1-lim_{n\to\infty} h(\neg\alpha,n) = lim_{n\to\infty} -h(\neg\alpha, n) = lim_{n\to\infty} k(\alpha,n)$$

where we have set $k(\alpha,n) = 1-h(\neg\alpha,n)$. The map k is computable and increasing with respect to the second variable. Then, $\mathcal{D}(v)^{\perp}$ is recursively enumerable.

Assume that v is complete, then, since $\mathcal{D}(v) = \mathcal{D}(v)^{\perp}$ the fuzzy set $\mathcal{D}(v)$ is both recursively enumerable and recursively co-enumerable. □

It is easy to prove that in the logics with optimal proofs the axiomatizable theories are *d*-recursively enumerable. The following theorem is a consequence of Theorem 6.3.4 in Hájek [1998] (see also Scarpellini [1962]):

Theorem 9.4. An effective Hilbert system exists with an axiomatizable theory τ whose cut $C(\tau,1)$ is Π_2-complete and therefore not recursively enumerable.

Obviously, such a theorem does not contradict Theorem 9.3. It means only that, given any formula α, while we are able to produce an increasing sequence of rational numbers converging to $\tau(\alpha)$, we are not able to decide if $\tau(\alpha)$ is equal to 1 or not. This phenomenon is not a characteristic of fuzzy logic. It arises whenever a constructive approach for a theory involving real numbers is proposed. Indeed, it is not decidable if two recursive real numbers are equal or not. Then it is not surprising that we know an algorithm to compute the real number r and that, at the same time, we are not able to decide if r is equal to 1 or not.

Remark. The set of signed formulas of an axiomatizable theory τ is not necessarily recursively enumerable. Indeed, Proposition 4.3 says that this set is recursively enumerable iff τ is *d*-recursively enumerable.
This fact is an argument against the reduction of fuzzy logic to a (crisp) calculus of signed formulas.
Indeed, any logic is an effective process to generate information from a piece of available information. So, a reduction of a logic $\mathcal{L}_1$ to a logic $\mathcal{L}_2$ needs to take into account the effectiveness of the corresponding deduction apparatus. On the other hand, the notion of effectiveness in fuzzy logic is related to the continuous structure [0,1] and, therefore, to the idea of an (effective) infinite approximation process. Instead, in any crisp logic the notion of effectiveness is related to the discrete structure {0,1} and, therefore, to the idea of a finite-steps and terminating computation. Then, in our opinion, no reduction of fuzzy logic to crisp logic is possible (recall also the observation about the compactness at the end of Section 7 in Chapter 4).

We conclude this section by giving an example of a continuous truth-functional modal semantics which is axiomatizable by a Hilbert system but not axiomatizable by an effective Hilbert system (see Biacino and Gerla [2000]a).

Theorem 9.5. *A continuous truth-functional fuzzy semantics exists which is not effectively axiomatizable.*

Proof. Consider any continuous truth-functional fuzzy semantics $\mathcal{M}$ such that $\neg$ is interpreted by an injective computable map and add a new unary connective $\Diamond$ to the language of $\mathcal{M}$. We interpret $\Diamond$ by a unary function $f: U \rightarrow U$ by obtaining a new semantics $\mathcal{M}_\Diamond$ for a fuzzy modal logic in which, as usual, $\Diamond(\alpha)$ means *"α is possible"*. We define f as follows: Let W be a subset of N and assume that W is not recursively enumerable. Then f is the continuous function such that, for any $i \in N$

$f(x) = (2i+1)/(2i^2+2i)$ if $x = 1/(i+1)$ and $i \in W$,

$f(x) = x$ if either $x = 0$ or $x = 1$ or $x = 1/(i+1)$ and $i \notin W$,

f is linear in each interval $[1/(i+1), 1/i]$.

Since $(2i+1)/(2i^2+2i)$ is the average between $1/(i+1)$ and $1/i$, we have that

$$1/(i+1) < (2i+1)/(2i^2+2i) < 1/i.$$

Then, it is easy to verify that $f(1/(i+1)) < f(1/i)$ for any $i \in N$ and therefore, that f is an injective, order-preserving mapping. The function f is not computable. Since f is order-preserving, $f(0) = 0$, $f(1) = 1$ and $f(\lambda) \geq \lambda$ for every $\lambda \in U$, f represents a plausible interpretation of $\Diamond$ in a modal logic. Consider now the initial valuation $v : \mathcal{F} \rightarrow U$ defined by setting:

$$v(x) = \begin{cases} 1/(i+1) & \text{if } x = p_i\,, \\ \sim(1/(i+1)) & \text{if } x = \neg p_i\,, \\ 0 & \text{otherwise.} \end{cases}$$

By Proposition 2.1 in Chapter 8, such a valuation has only a model m_v, namely the truth-functional model such that $m_v(p_i) = 1/(i+1)$. Consequently, for any formula α,

$$Lc(v)(\alpha) = m_v(\alpha).$$

Since,

$$m_v(\Diamond(p_i)) = f(m_v(p_i)) = f(1/(i+1)),$$

we have

$$Lc(v)(\Diamond(p_i)) = \begin{cases} 1/(i+1) & \text{if } i \notin W, \\ (2i+1)/(2i^2+2i) & \text{if } i \in W. \end{cases}$$

Assume that $\mathcal{M}_\Diamond$ is effectively axiomatizable, i.e. that there is a suitable *H-system* whose deduction operator $\mathcal{D}$ coincides with the logical consequence operator Lc of $\mathcal{M}_\Diamond$. In such a case the decidability of v implies that the fuzzy subset $\mathcal{D}(v) = Lc(v)$ is recursively enumerable. Then, a recursive function $h : \mathcal{F} \times N \rightarrow \ddot{U}$ exists such that h is increasing with respect to the second variable and

$$Lc(v)(\alpha) = \mathcal{D}(v)(\alpha) = lim_{n\rightarrow\infty} h(\alpha,n)$$

for any formula α. In particular,

$$lim_{n\to\infty} h(\Diamond(p_i),n) = 1/(i+1) \qquad \text{if } i \notin W,$$

$$lim_{n\to\infty} h(\Diamond(p_i),n) = (2i+1)/(2i^2+2i) \qquad \text{if } i \in W.$$

This means that W is the set $\{i \in N : \exists\, n\, h(\Diamond(p_i),n) > 1/(i+1)\}$ and therefore, that W is the projection of the decidable relation

$$R(i,n) \equiv h(\Diamond(p_i),n) > 1/(i+1).$$

This contradicts the hypothesis that W is not recursively enumerable. □

10. CREATIVE FUZZY SETS AND GÖDEL THEOREMS

In Rogers [1976] a subset X of S is called *productive* if a partial recursive function $e : S \to S$ exists such that

$$W_i \subseteq X \Rightarrow e(i) \in X\text{-}W_i.$$

In other words, X is productive if we can prove in an effective way that X is different from any recursively enumerable subset of S. A subset X is *creative* if X is recursively enumerable and its complement is productive. The notions of productive subset and creative subset are very important in first order logic. Indeed, we can summarize two basic limitative theorems as follows.

(i) *The set V of true formulas of elementary arithmetic is a productive set. Consequently, V cannot be axiomatized.*

(ii) *The set T of theorems of Peano arithmetic is creative. Consequently, T is undecidable and incomplete.*

A first step to extend such results to fuzzy logic is to give a suitable definition of productive and creative fuzzy subset.

Definition 10.1. We define *productive* a fuzzy set s for which a partial recursive function $e : S \to S$ exists such that

$$s_i \subseteq s \Rightarrow s(e(i)) \neq s_i(e(i)).$$

We say that s is *creative* if s is recursively enumerable and $-s$ productive.

In other words, s is productive if we can prove in an effective and uniform way that, given any recursively enumerable fuzzy set s_i, s is different from s_i.

Lemma 10.2. *Let $f : U \to U$ be any computable upper-semicontinuous function such that*

- $f(x) \neq 1 - x$ for every $x \in U$,

- $f(x)$ is rational for x rational,

- $f: \ddot{U} \to \ddot{U}$ is computable.

Let k be the 1-1-complete fuzzy subset given in Theorem 7.7. Then the fuzzy subset $s = f \circ k$ is creative.

Proof. k is recursively enumerable and therefore there exists a computable map $h : S \times N \to \ddot{U}$ which is increasing with respect to n and such that $k(x) = Sup\{h(x,n) : n \in N\}$ for every $x \in S$. Then, since

$$s(x) = f(k(x)) = f(Sup\{h(x,n) : n \in N\}) = Sup\{f(h(x,n)) : n \in N\},$$

s is a recursively enumerable fuzzy subset. In order to prove that $-s$ is productive, assume that $s_i \subseteq -s$ and let $c : S \to N$ a coding and $t : N \to S$ its inverse. Then,

$$s_i(t(i)) \le -s(t(i)) = 1 - f(s_i(t(i))).$$

Since by hypothesis $s_i(t(i)) \ne 1 - f(s_i(t(i)))$, we have that $s_i(t(i)) \ne -s(t(i))$. This demonstrates that s is creative. □

Lemma 10.2 entails the existence of creative fuzzy sets.

Proposition 10.3. *If s, p and c are fuzzy subsets, then:*
(i) *p productive and $p \le_1 s \Rightarrow s$ productive,*
(ii) *c creative, s recursively enumerable and $c \le_1 s \Rightarrow s$ is creative.*

Proof. (i) Assume that $p(x) = s(f(x))$ for every $x \in S$, where $f : S \to S$ is total and recursive and let $e : S \to S$ be as in Definition 10.1. Then from $s_i \subseteq s$ it follows that $s_i(f(x)) \le s(f(x)) = p(x)$. By recalling that $s_i(f(x)) = lim_{n\to\infty}\psi_{h(i)}(f(x),n)$, we define the recursive function $\phi(i,x,n)$ by setting $\phi(i,x,n) = \psi_{h(i)}(f(x),n)$ and by the *s-m-n*-theorem we set $\psi_{g(i)}(x,n) = \phi(i,x,n)$ where $g : N \to N$ is a suitable recursive function. Since $\psi_{g(i)}(x,n)$ is recursive and increasing with respect to n,

$$s_{g(i)}(x) = lim_{n\to\infty}\psi_{h(g(i))}(x,n) = s_i(f(x))$$

and this means that $s_i \circ f = s_{g(i)}$. Then $s_i \subseteq s$ implies $s_{g(i)} \subseteq p$, i.e., $e(g(i))$ is convergent and $p(e(g(i))) > s_{g(i)}(e(g(i))$. Therefore $f(e(g(i)))$ is convergent and $s(f(e(g(i)))) > s_i(f(e(g(i))))$. Thus, s is productive via the function $f \circ e \circ g$.

(ii). Observe that from $c \le_1 s$ it follows that $-c \le_1 -s$. Since $-c$ is productive, by (i), $-s$ is productive. Thus, s *is creative.* □

In recursion theory one proves that a subset X is creative iff it is 1-1-complete. The next proposition shows that we cannot extend such a result to fuzzy subsets.

Proposition 10.4. *Every* 1-1-*complete fuzzy set is creative while a creative fuzzy subset exists which is not* 1-1-*complete.*

Proof. Let k be a 1-1-complete fuzzy subset and c a creative fuzzy subset. Then, since $c \le_1 k$, by Proposition 10.3 k is creative. Let f be defined by setting $f(x) = 1$ if $x \ne 0$ and $f(x) = 0$ if $x = 0$ and let $c = f \circ k$ be the creative fuzzy subset defined in Lemma 10.2. Then, c coincides with the characteristic function of the open cut $O(k,0)$. Since c is crisp, a non-crisp fuzzy subset cannot be one-one reducible to c. This shows that c cannot be 1-1-complete (in the whole class of fuzzy sets). □

Consider a fuzzy logic such that F is the set of formulas of a first order language with *"numerals"*, i.e., with a name $\underline{n}$ for every integer n. Given a model m, we say that a fuzzy subset s of N *is represented in* m by the formula α if, for every $n \in N$, $s(n) = m(\alpha(\underline{n}))$. Likewise, given a fuzzy theory τ, we say that a fuzzy subset s of N *is represented in* τ *by* α if, for every $n \in N$, $s(n) = \tau(\alpha(\underline{n}))$.

Theorem 10.5. *Consider a fuzzy Hilbert logic with a model m able to represent a productive fuzzy subset. Then m is productive and no axiomatization for it exists. Namely, given any recursively enumerable fuzzy system of axioms v for m, a formula* α *exists such that* $\mathcal{D}(v)(\alpha) < m(\alpha)$.

Proof. Assume that a productive fuzzy set c is represented in m by the formula α. Then c is one-one reducible to m via the function associating every $n \in N$ with the formula $\alpha(\underline{n})$. By Proposition 10.3 this proves that m is productive. □

Theorem 10.6. *Let* τ *be an axiomatizable fuzzy theory of an effective fuzzy H-system with a negation and assume that* τ *is able to represent a creative fuzzy subset. Then* τ *is creative and, hence, undecidable and incomplete.*

Proof. From Proposition 10.3 it follows that τ is creative and, hence, undecidable. From Proposition 9.3 it follows that τ is incomplete. □

11. SHARPENED AND SHADED VERSIONS: LIMITATIVE THEOREMS

Given two elements λ_1 and λ_2 in U, we set $\lambda_2 \preceq \lambda_1$ provided that

$$\lambda_1 > 1/2 \Rightarrow \lambda_2 \geq \lambda_1 \quad \text{and} \quad \lambda_1 < 1/2 \Rightarrow \lambda_2 \leq \lambda_1.$$

In particular, this means that $\lambda \preceq 0.5$ for any $\lambda \in U$. Moreover, if s and s' are two fuzzy subsets of a set S, then we set $s \preceq s'$ if $s(x) \preceq s'(x)$ for every $x \in S$. In such a case we say that s is a *sharpened version* of s' or that s' is a *shaded version* of s (see A. De Luca and S. Termini [1972]). The fuzzy set $s^{0.5}$ constantly equal to 0.5 is the greatest element of U with respect to the sharpness relation, while the crisp subsets are the minimal elements.

Let's examine the following question:

given an undecidable fuzzy subset k of S, is it always possible to modify k so as to get a decidable (or a recursively enumerable) sharpened version of k ?

Now, if k is crisp, since there is no proper sharpened version of k, the answer is negative in a trivial way. Then, we are only interested in the case in which k is not crisp. In fact, in this case we can try to modify k in order to obtain a decidable sharpened version. The more favorable case is to have infinite x such that $k(x) = 1/2$. We say that a fuzzy subset s of a set S is *infinitely undetermined* if $L(s,1/2) = \{x \in S : s(x) = 1/2\}$ is infinite. Then, we can reformulate the above question as:

is it always possible to modify an infinitely undetermined fuzzy subset k to get a decidable (or a recursively enumerable) sharpened version of k ?

Proposition 11.1 and Theorem 11.2 give a negative answer:

Proposition 11.1. *Let k be the* 1-1-*complete fuzzy subset defined in Proposition 7.7. Then, k is a recursively enumerable infinitely undetermined fuzzy subset such that no sharpened version of k is decidable.*

Proof. At first, observe that k is infinitely undetermined. Indeed, let $s^{1/2}$ be the fuzzy set constantly equal to 1/2. Then, since $s^{1/2} \leq_1 k$, a recursive one-one map h

exists such that $k(h(x)) = 1/2$ for every $x \in S$. Let s be a sharpened version of k and W any recursively enumerable set which is not decidable. Since k is 1-1-complete, a recursive map d exists such that $c_W(x) = k(d(x))$ for every $x \in S$. We now have that $s(y) = k(y)$ everywhere $k(y) \in \{0,1\}$ and this entails that $c_W(x) = s(d(x))$. Then W is one-one reducible to s. Thus, since W is not decidable, we can conclude that s is not decidable. □

Theorem 11.2. *An infinitely undetermined fuzzy set* $g : S \rightarrow U$ *exists such that no sharpened version of g belongs to the Kleene hierarchy.*

Proof. Since S is codified in N, an ordering is defined in S. Let $E_1, E_2, \ldots$ be an enumeration of the infinite subsets of S belonging to the Kleene hierarchy and define the sequence $(x_1,y_1,z_1), (x_2,y_2,z_2), \ldots$ by setting

(x_1,y_1,z_1) equal to the first three elements in E_1

...

$(x_{i+1},y_{i+1},z_{i+1})$ equal to the first three elements in E_{i+1} greater than z_i .

Then, $x_1 < y_1 < z_1 < x_2 < y_2 < z_2 < \ldots$ and x_i, y_i and z_i are elements of E_i. Define the fuzzy set g by

$$g(x) = \begin{cases} 1 & \text{if } x = x_i \text{ for a suitable } i \in N, \\ 0 & \text{if } x = z_i \text{ for a suitable } i \in N, \\ 1/2 & \text{otherwise.} \end{cases}$$

Since the sequence $y_1, y_2, \ldots$ is infinite, g is infinitely undetermined. Let s be a sharpened version of g. Then, because $L(s,0) \supseteq L(g,0) \supseteq \{z_i \in S : i \in N\}$, the set $L(s,0)$ is infinite. If s belongs to the Kleene hierarchy, then $L(s,0)$ belongs to this hierarchy, too. As a consequence, $L(s,0) = E_i$ for a suitable index i. Thus, as $x_i \in E_i$, it is $s(x_i) = 0$ while $g(x_i) = 1$. This is a contradiction. □

We now want to examine another question:

is it always possible to shade an undecidable fuzzy set in order to get decidability ?

Now, since any fuzzy set s may be shaded in the totally undetermined fuzzy set and this set is decidable, we have to reformulate the question as:

is it always possible to shade an undecidable fuzzy set in order to get decidability without falling into an infinitely undetermined fuzzy subset ?

The answer is also negative in this case.

Proposition 11.3. *Let W be any recursively enumerable undecidable subset of S. Then every decidable shaded version of W is infinitely undetermined.*

Proof. If s is a decidable shaded version of W, then,

$$s(x) > 1/2 \Rightarrow x \in W \text{ and } s(x) < 1/2 \Rightarrow x \notin W.$$

Let h and h' be two computable functions such that

$$lim_{n\rightarrow\infty} h(x,n) = s(x) = lim_{n\rightarrow\infty} h'(x,n)$$

with h increasing and h' decreasing with respect to the second variable. Then:

$s(x) > 1/2$ iff an integer n exists such that $h(x,n) > 1/2$,
$s(x) < 1/2$ iff an integer n exists such that $h'(x,n) < 1/2$.

Suppose that $L(s,1/2)$ is finite. Then the following is an algorithm to compute the characteristic function c_W of W:

- if $x \in L(s,1/2)$, then set $c_W(x) = 1$ if $x \in W \cap L(s,1/2)$ and $c_W(x) = 0$ if $x \notin W \cap L(s,1/2)$.
- if $x \notin L(s,1/2)$, then compute the two sequences

$$h(x,1), h(x,2), \ldots \text{ and } h'(x,1), h'(x,2), \ldots$$

and set

$$c_W(x) = \begin{cases} 1 \text{ if } n \text{ exists such that } h(x,n) > 1/2, \\ 0 \text{ if } n \text{ exists such that } h'(x,n) < 1/2. \end{cases}$$

Since W is not decidable, we can conclude that $L(s, 1/2)$ is not finite. □

We say that s is *almost-everywhere undetermined* if $\{x \in S : s(x) \neq 1/2\}$ is finite. An almost-everywhere undetermined fuzzy subset whose values are computable real numbers is decidable. Trivially, any fuzzy subset s admits a decidable almost-everywhere undetermined shaded version. We say that s' is a *proper shaded version* of s if s' is a non almost-everywhere undetermined shaded version of s.

Theorem 11.4. *A set T exists such that no proper shaded version of T belongs to the Kleene hierarchy.*

Proof. Let $E_1, E_2, \ldots$ be an enumeration of all the infinite subsets of S belonging to the Kleene hierarchy and define the sequence $(x_1,y_1), (x_2,y_2), \ldots$ by setting

- x_1, y_1 equal to the first two elements in E_1
- x_{i+1}, y_{i+1} equal to the first two elements in E_{i+1} greater than y_i.

Then, $x_1 < y_1 < x_2 < y_2 < \ldots$ and, for every $i \in N$, x_i and y_i are elements of E_i. We denote the set of elements of the sequence $x_1, x_2, \ldots$ by T. Assume that s is a fuzzy subset in the Kleene hierarchy such that $T \preceq s$. Then, both $\{x \in S : s(x)<1/2\}$ and $\{x \in S : s(x)>1/2\}$ belong to the Kleene hierarchy and

$$x \in T \Rightarrow s(x) \geq 1/2 \text{ and } x \notin T \Rightarrow s(x) \leq 1/2.$$

To prove that s is almost-everywhere undetermined, we must prove that both the sets $\{x \in S : s(x)>1/2\}$ and $\{x \in S : s(x)<1/2\}$ are finite. Now, assume that $\{x \in S : s(x)>1/2\}$ is infinite. Then $i \in N$ exists such that $E_i = \{x \in S : s(x)>1/2\}$. In particular, since $y_i \in E_i$, $s(y_i)>1/2$. This contradicts the fact that, since $y_i \notin T$, $s(y_i) \leq 1/2$. In a similar way one proves that $\{x \in S : s(x)<1/2\}$ is finite. □

REFERENCES

Aguzzoli S. [1998], The Complexity of McNaughton Functions of One Variable, *Advances in Applied Mathematics*, **21**, 58-77.

Aguzzoli S., Ciabattoni A. [2000], Finiteness in infinite-valued Łukasiewicz logic, *Journal of Logic Language and Information*, **9**, 5-29.

Babuška R. [1998], *Fuzzy Modeling for Control*, Kluwer Academic Publishers, Dordrecht.

Bacchus F. [1990], *Representing and Reasoning With Probabilistic Knowledge*, MIT Press.

Biacino L. [1993], Generated envelopes, *J. Math. Anal. Appl.*, **172**, 179-190.

Biacino L., Gerla G. [1984], Closure systems and L-subalgebras, *Inform. Sci.*, **32**, 181-195.

Biacino L., Gerla G. [1987], Recursively enumerable L-sets, *Zeitschr. f. math. Logik und Grundlagen d. Math.*, **33**, 107-113.

Biacino L., Gerla G. [1988], Decidability and recursive enumerability for fuzzy subsets, in: B. Bouchon, L. Saitta, R.R. Yager (eds), *Uncertainty and Intelligent Systems*, Lectures Notes in Computer Science (Springer-Verlag) Berlin, 55-62.

Biacino L., Gerla G. [1989], Decidability, recursive enumerability and Kleene hierarchy for L-subsets, *Zeitschr. f. math. Logik und Grundlagen d. Math.*, **35**, 49-62.

Biacino L., Gerla G. [1990], Rational valuations, *"Third International Conference IPMU"*, Paris, 62-64.

Biacino L., Gerla G. [1992]a, Fuzzy subsets: a constructive approach, *Fuzzy Sets and Systems*, **45**, 161-168.

Biacino L., Gerla G. [1992]b, Generated necessities and possibilities, *Internat. J. of Intell. Systems*, **7**, 445-454.

Biacino L., Gerla G., [1993], Closure operators for fuzzy subsets, *EUFIT '93, First-European Congress on Fuzzy and Intelligent Technologies*, Aachen, 1441-1447.

Biacino L., Gerla G., [1996], An extension principle for closure operators, *J. of Math. Anal. Appl.*, **198**, 1-24.

Biacino L., Gerla G., [1998], Logics with approximate premises, *International J. of Intell. Syst.*, **13**, 1-10.

Biacino L., Gerla G., [1999], Closure Operators in Fuzzy Set Theory, in *Fuzzy Sets in Approximate Reasoning and Information Systems*, Edited by J. C. Bezdek, D. Dubois, H. Prade, Kluwer Academic Publishers, Dordrecht, 243-278.

Biacino L., Gerla G., [2000]a, Axiomatizability, continuity and computability for multivalued logics, unpublished paper.

Biacino L., Gerla G., [2000]b, Boolean logic and capacities, unpublished paper.

Biacino L., Gerla G., Ying M. [2000], Approximate reasoning based on similarity, *Math. Log. Quart.*, **46**, 77-86.

Birkhoff G. [1948], *Lattice Theory*, Amer. Math. Soc., New York.

Bison P., Chemello G., Sossai C., Trainito G., [1995], An Algebraic Logic for Autonomous Systems, *Intelligent Autonomous Systems*, IAS-4, Karlsruhe, 182-189.

Boldrin L., Sossai C. [1995], An algebraic semantics for possibilistic logic. Proceedings of the *International Conference on Uncertainty in Artificial Intelligence (UAI 95)*, Montreal, Canada.

Brown D. J., Suszko R. [1973], Abstract Logics, *Dissertationes Mathematicae*, **102**, 9-42.

Castro J.L. [1994], Fuzzy logics as families of bivaluated logics, *Fuzzy Sets and Systems*, **64**, 321-332.

Castro J.L., Trillas E. [1993], The management of the inconsistency in expert systems, *Fuzzy Sets and Systems*, **58**, 51-57.

Castro J.L., Trillas E., Cubillo S. [1994], On consequences in approximate reasoning, *J. of Applied Non-classical Logics*, **4**, 91-103.

Chang C.C., Keisler H.J. [1966], *Continuous Model Theory*, Princeton, Princeton University Press.

Chang C.C., Keisler H.J. [1973], *Model Theory*, North-Holland, Amsterdam.

Chakraborty M.K. [1988], Use of fuzzy set theory in introducing graded consequence in multiple valued logic, in *Fuzzy Logic in Knowledge-Based Systems, Decision and Control*, M.M. Gupta and T.Yamakawa (eds), North-Holland, Amsterdam. 247-257.

Chakraborty M.K. [1994], Graded consequence: further studies, *J. of Applied non-classical Logic*, **5**, 227-237.

Cignoli R., D'Ottaviano I., Mundici D. [1995], *Algebras of many-valued sentential calculi of Łukasiewicz* (Portuguese), Collection CLE, Center of Logic, Epistemology and History of Science vol. 12, State University of Campinas, SP, Brazil.

Cignoli R., D'Ottaviano I., Mundici D. [2000], Algebraic Foundations of Many-valued Reasoning, Kluwer Academic Publishers, Dordrecht.

Clares B., Delgado M., [1987], Introduction to the concept of recursiveness for fuzzy functions, *Fuzzy Sets and Systems*, **21**, 301-310.

Cleave J.P. [1974], The notion of logical consequence in the logic of inexact predicates, *Zeitschr. f. math. Logik und Grundlagen d. Math.*, **20**, 307-324.

Cohn P. M. [1965], *Universal Algebra*, Harper and Row, New York.

Conrad F., [1980], Fuzzy topological concepts, *J. Math. Anal. Appl.*, **74**, 432-440.

De Luca A., Termini S. [1972], A definition of nonprobabilistic entropy in the setting of fuzzy set theory, *Information and Control*, **20**, 301-312.

Dempster A.P. [1967], Upper and lower probabilities induced by a multivalued mapping, *Ann. Math. Statist.*, **38**, 325-339.

De Finetti B. [1964], Foresight: Its Logical Laws, Its Subjective Sources, in H. E. Kyburg, Jr. and H.E. Smokler, Eds, *Studies in Subjective Probability*, Wiley, New York.

Di Nola A. , Gerla G. [1986], Fuzzy models of first order languages, *Zeitschr. f. math. Logik und Grundlagen d. Math.*, **32**, 331-340.

Di Nola A., Gerla G. [1987], Lattice valued algebras, *Stochastica*, **11**, 137-150.

Dubois D., Esteva F., Garcia P., Godo L., Prade H. [1996], *A logical approach to interpolation based on similarity relations*, Report de Recerca IIIA, Campus de la Universitat Autonoma de Barcelona.

Dubois D., Lang J., Prade H. [1986], Automated reasoning using possibilistic logic: semantics, belief revision and variable certainty weights. *Proc. 5th Workshop on Uncertainty in Artificial Intelligence*, Windsor, Ont., 81-87.

Dubois D., Lang J., Prade H. [1991], Fuzzy sets in approximate reasoning, part 2: logical approaches, *Fuzzy Sets and Systems*, **40**, 203-244.

Dubois D., Prade H. [1985], A review of fuzzy set aggregation connectives, *Inform. Sci.*, **36**, 85-121.

Dubois D., Lang J., Prade H. [1994], Possibilistic logic, In: D.Gabbay, C.Hogger and J.Robinson (eds.), *Handbook of Logic in Artificial Intelligence and Logic Programming*, vol. 3. Clarendon Press.

Dubois D., Prade H. [1987], Necessity measures and the resolution principle, *IEEE Trans. Systems Man and Cybernetics*, **17**, 474-478.

Dubois D., Prade H. [1988], *Possibility Theory*, Plenum Press, New York and London.

Dubois D., Prade H. [1990]a, Resolution principles in possibilistic logic, *Inter. J. of Approximate Reasoning*, **4**, 1-21.

Dubois D., Prade H. [1990]b, Reasoning with inconsistent information in a possibilistic setting, *Proc. of the Europ. Conf. on A.I. (ECAI-90)*, Stockholm, 259-261.

Dubois D., Prade H. [1992], Combination of fuzzy information in the framework of possibility theory, In: Abidi and Gonzales (eds.), *Data Fusion in Robotics and Machine Intelligence*, Academic Press, 481-505.

Dubois D., Prade H. [1994], A survey of belief revision and updating rules in various uncertainty models, *Internat. J. of Intell. Systems*, **9**, 61-100.

Dubois D., Prade H. [1996], What are fuzzy rules and how to use them, *Fuzzy Sets and Systems*, **84**, 169-185.

Ferrante F., Gerla G. [1998], Grasping Infinity by Finite Sets, *Math. Log. Quart.*, **44**, 383-393.

Gerla G. [1982], Sharpness relation and decidable fuzzy sets, *IEEE Trans. on Automatic Control*, AC-27, **5**, 1113.

Gerla G. [1985], Pavelka's fuzzy logic and free L-subsemigroups, *Zeitschr. f. math. Logik und Grundlagen d. Math.*, **31**, 123-129.

Gerla G. [1986], The category of the fuzzy models and Lowenheim-Skolem theorem, In *Mathematics of Fuzzy Systems*, Di Nola and Ventre (eds), Verlag TUV Rheinland, Germany, 121-141.

Gerla G. [1987]a, Decidability, partial decidability and sharpness relation for L-subsets, *Studia Logica*, **46**, 227-238.

Gerla G. [1987]b, Code theory and fuzzy subsemigroups, *J. Math. Anal.Appl.*, **128**, 362-369.

Gerla G. [1989], Turing L-machines and recursive computability for L-maps, *Studia Logica*, **48**, 179-192.

Gerla G. [1992], Fuzzy grammars and recursively enumerable fuzzy languages, *Inform. Sci.*, **60**, 137-144.

Gerla G. [1994]a, Comparing fuzzy and crisp deduction systems, *Fuzzy Sets and Systems*, **67**, 317-328.

Gerla G. [1994]b, Inferences in probability logic, *Artificial Intelligence*, **70**, 33-52.

Gerla G. [1994]c, Fuzzy refutations for probability and multivalued logics, *International J. of Approximate Reasoning*, **4**, 257-279.

Gerla G. [1994]d, An extension principle for fuzzy logics, *Mathematical Logic Quarterly*, **40**, 357-80.

Gerla G. [1996], Graded consequence relations and closure operators, *J. of Applied non-classical Logic*, **6**, 369-379.

Gerla G., [1997], Probability-Like Functionals and Fuzzy Logic, *J.. Math. Anal.Appl.*, **216**, 438-465.

Gerla G., [1999], Closure operators, fuzzy logic and constraints, in *Fuzzy Sests, Logics and Reasoning About Knowledge*, (D. Dubois, H. Prade E.P. Klement, Eds), Kluwer Academic Publishers, Dordrecht, 101-120.

Gerla G., [2000], Fuzzy Control as a Fuzzy Deduction System, *Fuzzy Sets and Systems*, to appear.

Gerla G., Scarpati L. [1998], Extension principles for fuzzy set theory, *J. of Inform. Sci.*, **106**, 49-69.

Gerla G., Tortora R. [1990], Fuzzy natural deduction, *Zeitschr. f. math. Logik und Grundlagen d. Math.*, **36**, 67-77.

Goguen J.A. [1968/69], The logic of inexact concepts, *Synthese*, **19**, 325-373.

Gottwald S. [1993], *Fuzzy Sets and Fuzzy Logic*, Wiesbaden: Vieweg.

Grätzer G. [1968], *Universal Algebra*, D. Van Nostrand Company, Toronto.

Hähnle R. [1993], *Automated Deduction in Multiple-valued Logics*, Clarendon Press, Oxford.

Halpern J.Y. [1991], An analysis of first-order logics of probability, *Artificial Intelligence*, **46**, 311-350.

Harkleroad L. [1984], Fuzzy recursion, RET's and isols, Z. Math. Logik Grundlag. Math., **30**, 425-436.

Hájek P. [1993], On logics of approximate reasoning, *Neural Network Word* 6/1993, 733-744.

Hájek P. [1995], Fuzzy logic and arithmetical hierarchy, *Fuzzy Sets and Systems*, **3**, 359-363.

Hájek P. [1997], Fuzzy logic and arithmetical hierarchy II. *Studia Logica*, **58**, 129-141.

Hájek P. [1998], *Metamathematics of Fuzzy Logic*, Kluwer Academic Publishers, Dordrecht.

Hájek P. [1999]a, On the metamathematics of fuzzy logic, in *Discovering World with Fuzzy Logic: Perspectives and Approaches to Formalization of Human-Consistent Logical Systems*, (V. Novák, I. Perfilieva, editors), Springer-Verlag, Heidelberg.

Hájek P. [1999]b, Ten questions and one problem on fuzzy logic, *Annals of Pure and Applied Logic*, **96**, 157-165.

Höhle U. [1983], Fuzzy filters: a generalization of credibility measures, in *Fuzzy Information, Knowledge Representation and Decision Analysis*, E.Sanchez and M.M.Gupta (eds), Peragmon Press, Elmsford, NY.

Höhle U. [1991], Foundations of fuzzy sets, *Fuzzy Sets and Systems*, **40**, 257-296.

Höhle U. [1996], On the Fundamentals of Fuzzy Set Theory, *J. Math. Anal. Appl.*, **201**, 786-826.

Horn A., Tarski A. [1948], Measures in Boolean algebras, *Trans. Am. Math. Soc.*, **64**, 467-497.

Kerre E.E. [1980], Fuzzy Sierpinski space and its generalizations, *J.Math. Anal. Appl.*, **74**, 318-324.

Klir G., Yuan B. [1995], *Fuzzy sets and fuzzy logi; theory and applications*, Prentice Hall.

Kosko B. [1993], *Fuzzy Thinking: The New Science of Fuzzy Logic*, Hyperion, New York.

Ivanov L.L. [1986], *Algebraic recursion theory*, Ellis Horwood limited Publishers, Chichester.

Lallement G. [1979], *Semigroups and Combinatorial Applications*, J.Wiley and Sons, New York 1979.

Lloyd J.W. [1987], *Foundations of Logic Programming*, Springer-Verlag, Berlin.

Lu J. J., Murray N. V., Rosenthal E. [1998], Signed Formulas and Fuzzy Operator Logics, *J. Automat. Reason.* **21**, 39-67.

Łukasiewicz J. [1970], *Selected works*, Studies in Logic and Foundations of Mathematics, North-Holland Publ.Comp., Amsterdam and PWN, Warszawa.

Mamdani, E.H., Gaines, B.R. eds. [1981], *Fuzzy Reasoning and its Applications*, Academic Press, London.

McNaughton R. [1951], A theorem about infinite-valued sentential logic, *J. of Symbolic Logic*, **16**, 1-13.

Mendelson E. [1964], *Introduction to Mathematical Logic*, D. Van Nostrand Company, Princeton.

Menger K. [1951], Ensembles flous et fonctions aleatoires, C.R. Acad. Sci., **232**, 2001-2003.

Mundici D. [1987], Satisfiability in many-valued sentential logic is NP-complete, *Theoretical Computer Science*, **52,** 145-153.

Mundici D. [1992], Normal forms in infinite-valued logic: the case of one variable, *Lecture Notes in Computer Science*, **629**, 272-277.

Mundici D., [1994], A constructive proof of McNaughton's theorem in infinite-valued logic, *J. of Symbolic Logic*, **59**, 596-602.

Murali V. [1991], Lattice of fuzzy subalgebras and closure systems in I^X, *Fuzzy Sets and Systems*, **41**, 101-111.

Murali V. [1993], L-fuzzy Subalgebras in *Advances in Fuzzy Theory and Technology*, vol. 1, Ed. P.P. Wand, Bookwrights Press, Durham, North Carolina, 77-90.

Murray N. V., Rosenthal E. [1991], Resolution and path dissolution in multiple-valued logics, *Proc. of the 6th International Synposium on Methodologies for Intelligent Systems,* Springer vol. 542, 570-579.

Negoita C.V., Ralescu D.A. [1975]a, Representation theorems for fuzzy concepts, *Kybernetes*, **4**, 169-174.

Negoita C.V., Ralescu D.A. [1975]b, *Applications of Fuzzy Sets to Systems Analysis*, Birkhauser Verlag, Basel.

Nilsson N.J. [1986], Probabilistic logic, *Artificial Intelligence*, **28**, 71-97.

Novák V. [1987], First-Order fuzzy logic, *Studia Logica*, **46**, 87-109.

Novák V., [1990], On the syntactico-semantical completeness of first-order fuzzy logic I, *Kybernetika*, **26**, 47-66.

Novák V., [1990], On the syntactico-semantical completeness of first-order fuzzy logic II, Kybernetika, **26**, 134-152.

Novák V., Perfilieva I, Mockor J. [1999], *Mathematical Principles of Fuzzy Logic*, Kluwer Academic Publishers, Dordrecht.

Owen G. [1982] *Game Theory*, Academic Press, New York.

Panti G., [1995] A geometric proof of the completeness of the Łukasiewicz calculus, *J. of Symbolic Logic*, **60**, 563-578.

Papamarcou A., Fine T.L. [1986], A note on undominated lower probabilities, *The Annals of Probability*, **14**, 710-723.

Pavelka J. [1979]a, On fuzzy logic I: Many-valued rules of inference, *Zeitschr. f. math. Logik und Grundlagen d. Math.*, **25**, 45-52.

Pavelka J. [1979]b, On fuzzy logic II: Enriched residuated lattices and semantics of propositional calculi, *Zeitschr. f. math. Logik und Grundlagen d. Math.*, **25**, 119-134.

Pavelka J. [1979]c, On fuzzy logic III: Semantical completeness of some many-valued propositional calculi, *Zeitschr. f. math. Logik und Grundlagen d. Math.*, **25**, 447-464.

Pawlak Z. [1982], Rough sets, *International J.of Information and Computer Science*, **11**, 341-356.

Pour-El M. B., Richards I. [1989], *Computability in Analysis and Physics*, Springer-Verlag.

Pu Pao-Ming, Liu Ying-Ming [1980], Fuzzy topology I, Neighborhood structure of a fuzzy point and Moore-Smith convergence, *J. Math. Anal. Appl.*, **76**, 571-599.

Ramik J. [1983], Extension principle and fuzzy-mathematical programming, *Kybernetika*, **19**, 513-25.

Rasiowa H. [1974], *An algebraic Approach to Non-classical Logics*, PWN and North-Holland.

Rogers H. [1967], *Theory of Recursive Functions and Effective Computability*, McGraw-Hill, New York.

Rosenfeld A. [1971], Fuzzy groups, *J. Math. Anal. Appl.*, **35**, 512-517.

Ruspini E. [1977], A theory of fuzzy clustering. *Proc. IEEE Conf. on Decision and Control*, New Orleans, 1378-1383.

Ruspini E. [1991], On the semantics of fuzzy logic, *Int. J. of Approximate Reasoning*, **5**, 45-88.

Santos E. S. [1968], Maximin, Minimax, and Composite Sequential Machines, *J. of Math. Anal. Appl.*, **24**, 246-259.

Santos E. S. [1969], Maximin Sequential Chains, *J. of Math. Anal. Appl.*, **26**, 28-38.

Santos E. S. [1969], Maximin sequential-like machines and chains, *Math. System Theory*, **3**, 300-309.

Santos E. S. [1970], Fuzzy algorithms, *Inform. and Control*, **17**, 326-339.

Santos E. S. [1972], Max-Product Machines, *J. of Math. Anal. Appl.*, **37**, 677-686.

Santos E. S. [1976], Fuzzy and probabilistic programs, *Inform. Sci.*, **10**, 331-335.

Santos E. S., Wee W. G. [1968], General formulation of sequential machines, *Information and Control*, **12**, 5-10.

Scarpellini B. [1962], Die Nichaxiomatisierbarkeit des unendlichwertigen Prädikatenkalküls von Łukasiewicz, *J. of Symbolic Logic*, **27**, 159-170.

Schrijver A. [1986], *Theory of linear and integer programming*, John Wiley & Sons, New York.

Schweizer B., Sklar A. [1983], *Probabilistic Metric Spaces*, North –Holland, New York.

Shafer G. [1976], *A Mathematical Theory of Evidence*, Princeton University Press, Princeton.

Sikorski R. [1960], *Boolean Algebras*, Springer-Verlag Berlin.

Stone M. H. [1937], Topological representation of distributive lattice and Brouwerian logics, *Cas. Mat. Fys.*, **67**, 1-25.

Suppes P., Zanotti M. [1989], Conditions on upper and lower probabilities to imply probabilities, *Erkenntnis*, **31**, 323-345.

Tarski A. [1956], *Logic, semantics and metamathematics*, Clarendon Press, Oxford.

Trillas E., Alsina C. [1993], Logic: Going farther from Tarski?, *Fuzzy Sets and Systems*, **53**, 1-13.

Trillas E., Valverde L. [1987], On inference in Fuzzy logic, *Proc. 2nd. IFSA Congress, Tokyo*, 294-297.

Turunen E. [1995], Well-Defined Fuzzy Sentential Logic, *Mathematical Logic Quarterly*, **41**, 236-248.

Turunen E. [1999], *Mathematics Behind Fuzzy Logic*, Physica-Verlag.

Valverde L. [1985], The structure of F-indistinguishability operators, *Fuzzy Sets and Systems*, **17**, 313-328.

Vickers J.M. [1976], *Belief and Probability*, D.Reidel Publishing Com., Dordrecht Holland.

Walley P., Fine T.L. [1979], Varietes of modal (classificatory) and comparative probability, *Synthese*, **41**, 321-374.

Ward M., [1942], The closure operators of a lattice, *Annals of Mathematics*, **43**, 191-196.

Weichselberger K. , Pohlmann S. [1990], *A methodology for Uncertainty in Knowledge-Based Systems*, Springer-Verlag, Berlin.

Weston T. [1987], Approximate truth, *J. of Philosophical Logic*, **16**, 203-227.

Wong, C.K. [1974], Fuzzy points and local properties of fuzzy topology, *J. Math. Anal. Appl.*, **46**, 316-328.

Wójcicki R. [1970], Some remarks on the consequence operation in sentential logics, *Fund. Math.*, **68**, 269-279.

Wójcicki R. [1973], Matrix approach in methodology of sentential calculi, *Studia Logica*, **32**, 7-37.

Wójcicki R. [1988], *Theory of Logical Calculi: Basic Theory of Consequence Operations*, Kluwer Academic Publishers, Dordrecht.

Ying M.S. [1994], A logic for approximate reasoning, *J. of Symbolic Logic*, **59**, 830-837

Zadeh L.A. [1965], Fuzzy Sets, *Information and Control*, **12**, 338-353.

Zadeh L.A. [1968], Fuzzy algorithms, *Information and Control*, **5**, 94-102.

Zadeh L.A. [1971], Similarity relations and fuzzy orderings, *Information Sci.* **3**, 159-176.

Zadeh, L.A. [1975]a, Fuzzy logic and approximate reasoning, *Synthese*, **30**, 407-428.

Zadeh, L.A. [1975]b The concept of a linguistic variable and its application to approximate reasoning I, II, III, *Inf. Sci.* **8**, 199-275, 301-357; **9**, 43-80.

Zadeh, L.A. [1997], Toward a theory of fuzzy information granulation and its centrality in human reasoning and fuzzy logic, *Fuzzy Sets and Systems*, **90**, 111-127.

INDEX

LIST OF SYMBOLS

TRENDS IN LOGIC

1. G. Schurz: *The Is-Ought Problem.* An Investigation in Philosophical Logic. 1997
ISBN 0-7923-4410-3
2. E. Ejerhed and S. Lindström (eds.): *Logic, Action and Cognition.* Essays in Philosophical Logic. 1997 ISBN 0-7923-4560-6
3. H. Wansing: *Displaying Modal Logic.* 1998 ISBN 0-7923-5205-X
4. P. Hájek: *Metamathematics of Fuzzy Logic.* 1998 ISBN 0-7923-5238-6
5. H.J. Ohlbach and U. Reyle (eds.): *Logic, Language and Reasoning.* Essays in Honour of Dov Gabbay. 1999 ISBN 0-7923-5687-X
6. K. Došen: *Cut Elimination in Categories.* 2000 ISBN 0-7923-5720-5
7. R.L.O. Cignoli, I.M.L. D'Ottaviano and D. Mundici (eds.): *Algebraic Foundations of many-valued Reasoning.* 2000 ISBN 0-7923-6009-5
8. E.P. Klement, R. Mesiar and E. Pap: *Triangular Norms.* 2000
ISBN 0-7923-6416-3
9. V.F. Hendricks: *The Convergence of Scientific Knowledge.* A View From the Limit. 2001 ISBN 0-7923-6929-7
10. J. Czelakowski: *Protoalgebraic Logics.* 2001 ISBN 0-7923-6940-8
11. G. Gerla: *Fuzzy Logic.* Mathematical Tools for Approximate Reasoning. 2001
ISBN 0-7923-6941-6

KLUWER ACADEMIC PUBLISHERS – DORDRECHT / BOSTON / LONDON

MIX
Papier aus verantwortungsvollen Quellen
Paper from responsible sources
FSC® C105338
FSC
www.fsc.org

If you have any concerns about our products,
you can contact us on
ProductSafety@springernature.com

In case Publisher is established outside the EU,
the EU authorized representative is:
Springer Nature Customer Service Center GmbH
Europaplatz 3, 69115 Heidelberg, Germany

Printed by Libri Plureos GmbH
in Hamburg, Germany